The World of Hannah Heaton

Hannah Cook Heaton

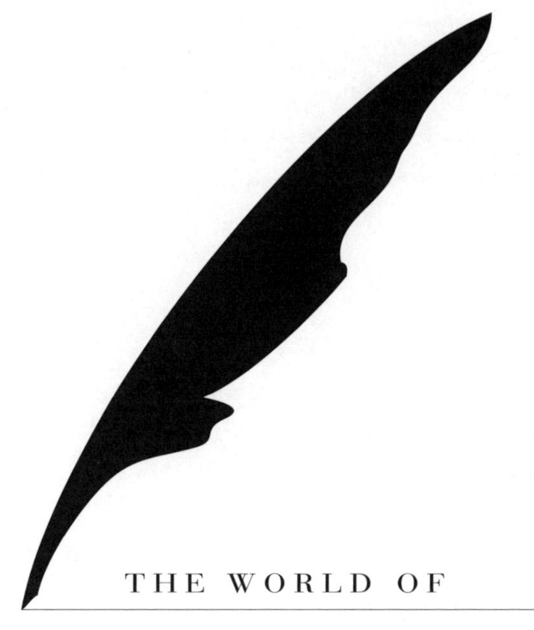

THE WORLD OF

Hannah Heaton

The Diary of an Eighteenth-Century New England Farm Woman

Edited by Barbara E. Lacey

NORTHERN ILLINOIS UNIVERSITY PRESS

DeKalb

Published by the Northern Illinois University Press, DeKalb, Illinois 60115

Manufactured in the United States using acid-free paper

Design by Julia Fauci

Library of Congress Cataloging-in-Publication Data

Heaton, Hannah, 1721–1793

The World of Hannah Heaton : the diary of an

eighteenth-century farm woman/edited by

Barbara E. Lacey.

p. cm.

Includes bibliographical references and index.

ISBN 0-87580-312-1 (acid-free paper)

1. Heaton, Hannah, 1721–1793—Diaries. 2. Christian

biography—Connecticut—North Haven—Diaries. 3. Dissenters,

Religious—Connecticut—North Haven—Diaries. I. Lacey, Barbara E., 1937–.

II. Title.

BR1720.H38 A3 2003

974.6/702/092 B 21

2002044876

Contents

* Bracketed numbers refer to Heaton's pagination

Foreword

As I reflect upon the history and content of the diary of Hannah Heaton, I find great pleasure and satisfaction in the fact that it is now published. There is a lesson here for us—whether or not we are historians and genealogists.

What begins as an individual's story soon becomes the story of a family. Heaton recorded her spiritual experiences for her own sake, but her family came into possession of the document and accepted the trust that such possession bestows. Through the years, the descendants of Mrs. Heaton kept the diary, and it grew to be a part of their story. The last family member to preserve the diary and to accept that accompanying responsibility was Mrs. Winifred N. Lincoln.

By the time of Mrs. Lincoln's death, the diary had become more than just a family matter. Anyone reading this wonderful account of Hannah Cook Heaton's walk with God will recognize that it is a document for all. It is a matter of some gratitude to me that, just prior to her death, Mrs. Lincoln had come to the same conclusion and had instructed her daughter to present the diary to the Whitney Library of the New Haven Colony Historical Society. By this action, all who have an interest in Heaton's story benefit greatly.

We learn thereby that our personal stories, our family stories, and our community stories are never just our own. Time has a way of identifying the broader constituencies for such diaries, letters, and other seemingly personal accounts. That constituency is nothing less than all of us.

Now published, Heaton's diary can be acknowledged for what it truly is—a part of our common story—in the most inclusive sense.

As the one charged with keeping watch over the diary, I appreciate the work of Barbara Lacey and the commitment of Northern Illinois University Press. We can learn from this labor and commitment not only more facts about the Great Awakening or women's history or North Haven in the eighteenth century. We can learn more about our connections—to the past and to each other. Not a bad legacy for Hannah Heaton.

James W. Campbell
Librarian and Curator of Manuscripts
The Whitney Library
New Haven Colony Historical Society

Acknowledgments

Hannah Heaton's diary came to my attention in 1981 as I prepared a dissertation on women and religion in eighteenth-century Connecticut under the direction of George A. Billias at Clark University. I stumbled upon a microfilm version of the diary at the Connecticut Historical Society in Hartford and realized I had found a rare and remarkable document. Ronald Formisano, also of Clark, thought I could send a brief piece on the diary to the "Notes and Documents" section of the *William and Mary Quarterly*. In response to my submission, *Quarterly* editor Michael McGiffert suggested that I submit for consideration a full-length article based on the diary.

In the composition of that article, I used both the original diary, then owned by Mrs. Winifred Lincoln, and a typescript of the diary, prepared by Juliana Losaw for the North Haven Historical Society, which was made available to me by Gloria Furnival, Curator of the Society. I also engaged in new research in the probate records at the Connecticut State Library. The article, "The World of Hannah Heaton: The Autobiography of an Eighteenth-Century Connecticut Farm Woman," appeared in the April 1988 issue of the *Quarterly*.

The full text of the diary has not been published previously, perhaps because for many years women's writings were undervalued, and a potential audience seemed limited. However, Northern Illinois University Press was venturesome, and with the editorial leadership of Kevin Butterfield, aided by Kelly Parker, and assisted by readers Cornelia Hughes Dayton and Michael McGiffert, the project moved forward. Now Hannah Heaton can be heard in her own voice.

I am thankful to James W. Campbell, Librarian and Curator of Manuscripts, New Haven Colony Historical Society, where the manuscript diary now is located, for providing a photocopy of the manuscript and for permission to publish it. I am indebted to the resourceful staff, especially Kathleen Kelley, of the Saint Joseph College Library and to Mark Paradis of the college's Information Technology Department, as well as to librarians and collections at the Connecticut Historical Society, North Haven Historical Society, and the Connecticut State Library.

My family has encouraged my endeavors, and I am grateful to my husband, Jim, my best wordsmith; my daughter, Beth, for technical help; and my parents, Arthur and Irene Ellson.

Considerable effort has been made to ensure the accuracy of the transcription of the diary; however, I accept responsibility for any errors that may have crept in.

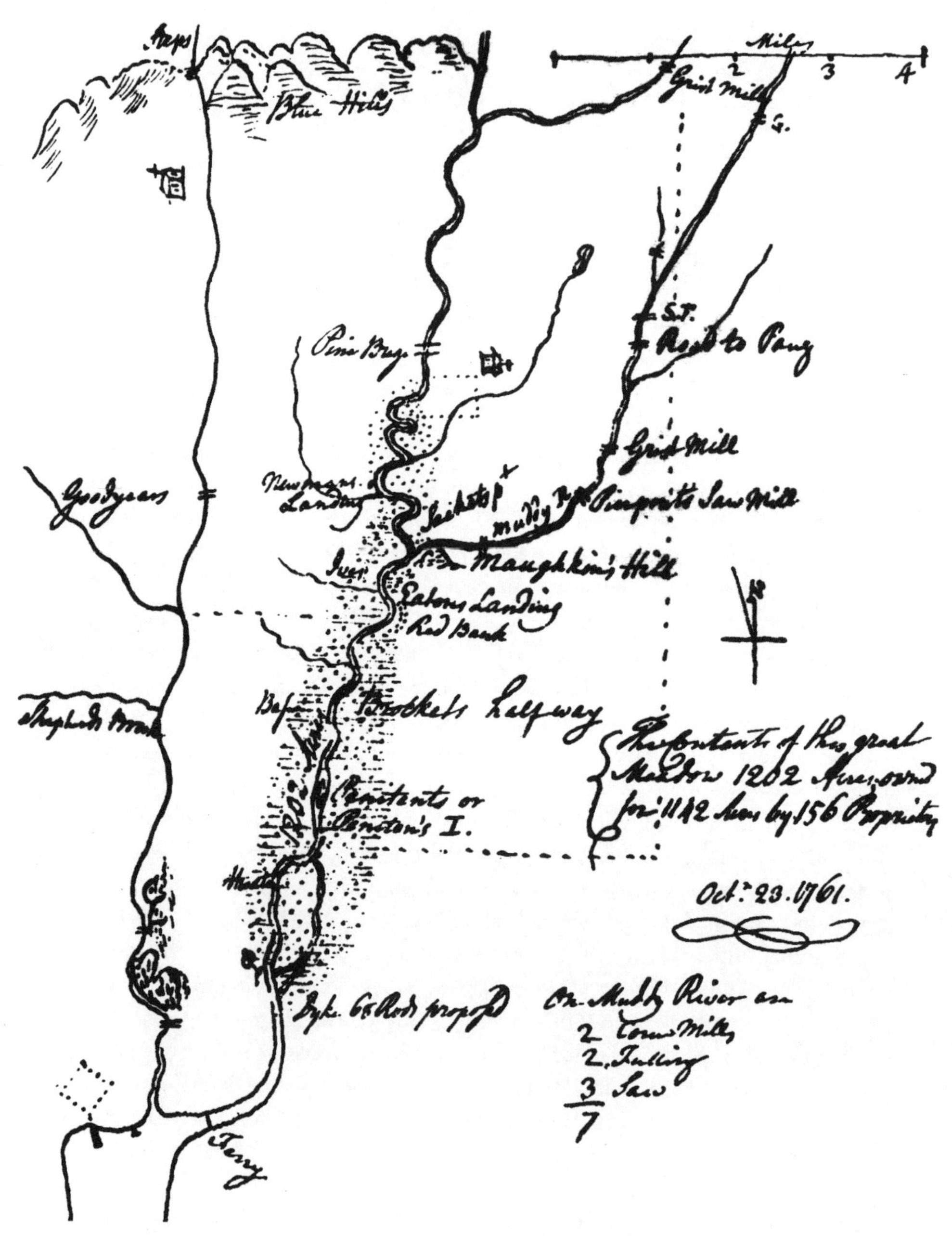

Map of North Haven drawn by Ezra Stiles, who reports there were about 175 families in the parish in 1760, increasing to 208 families by 1768. *Extracts from the Itineraries . . . of Ezra Stiles,* ed. Franklin Bowditch Dexter (New Haven, Conn: Yale University Press, 1916), p. 154. Courtesy of the American Antiquarian Society.

Introduction

"Come and hear all ye that fear god and i will declare what he hath done for my soul" (Psalms 66:16). With this call for an audience Hannah Heaton (1721–1794), a farm wife of North Haven, Connecticut, began her diary, recording events and reflections from the time of the Great Awakening to the early years of the Republic, in a work that has remained unpublished for two centuries. At first glance, the diary of Hannah Heaton does not make much of an impression, concerned as it is with an eighteenth-century Connecticut woman whose life revolved prosaically around family, farm, and faith. Yet a careful reading of the approximately four-hundred-page account that Heaton kept over a forty-year period reveals a compelling personal world of reflections, emotions, and visions, as well as arresting insight into the times in which she lived.

This extraordinary account is neither an autobiography, which looks back and gives order to a substantial portion of life, or a diary, but a complex document with characteristics of both. The classic autobiography can be defined as an author's attempt to stand apart from life in order to judge. The diary by contrast demands daily composition, an emersion in the text that parallels the immersion in life.[1] Unlike the prototypical diary, Heaton's work is not always a daily record. The early undated pages contain remembrances of childhood on Long Island and a visit to Connecticut during the Great Awakening, when she experienced religious conversion. Sometime in the 1750s, after she married and settled in North Haven, she began occasional dated entries summarizing spiritual experiences and worldly events that had occurred over a period of a year or more. During the momentous years of the American Revolution and the early Republic, she made entries every few months. The last date recorded is August 1793, entered shortly before her death at age seventy-three. Because she wrote about events that occurred days, months, or years in the past, the diary has many of the selective, ordered, and purposeful qualities of an autobiography. In addition to the term "autobiography," I use the terms "diary" and "journal" in describing Heaton's writing because she herself used them, in addition to calling it "a spiritual history of my experiences" [p. 219]. Because of its length, its span of time, themes of both spiritual and family life, and the rarity of such a document, the diary merits attention.

Little information about Hannah Heaton other than that found in the autobiography has survived. Born in the Meacox area of Southampton, Long Island, in 1721, she was the oldest of the ten children of Jonathan Cook, a "chirergeon," and Temperance Rogers Cook. On October 12, 1743, she was married to Theophilus Heaton, Jr. (1719–1791), of New Haven by Isaac Stiles, minister of the North Haven Congregational church.[2] Hannah, Theophilus, and their two sons, Jonathan and Calvin, spent their lives on farms in North Haven, an area of "cultivated and pleasant fields, orchards and gardens."[3] In this rural setting, Heaton experienced a remarkably complex inner life, which she recorded in her diary.

Heaton acknowledged that she wrote for several reasons. For one, it was her "duty from gods word to keep a journal or spiritual history" of her experience. She wanted to follow worthy biblical models, offering Moses as an example because he wrote about the exodus of the children of Israel; she also noted that "david paul & john & many others writ the dealings of god by his spirit & providences with them." Also, she benefited from reviewing her experiences: "sometimes in my darknes when i have read things in my diary that i had forgot it has seemd strengthning & quickning to me." Finally, she wanted to leave her account to her children, who she hoped might learn from it and remember her:

> My dear children
> I leave you here a little book for you to look upon that you may see your mothers face when she is dead and gone. [p. 219]

In addition, because she occasionally invoked "christian readers," she may have hoped that her work would be published, or at least shared with others, and she offered herself to her imagined readers as a model worthy of imitation. Writing an autobiography such as Heaton's involves complex literary and psychological processes that require selection of audience, arrangement of detail, and the finding of a voice. The particular way in which the Heaton diary addresses these literary concerns shows the writer at work actively constructing a "self" to inhabit a space within socially prescribed boundaries.[4]

The "territories of the self" underwent rapid and radical transition in the early eighteenth century. Previously, according to Mechal Sobel, neither men nor women thought of themselves as autonomous individuals but as part of an interdependent social whole. By the last third of the eighteenth century, however, as the Atlantic world was engaged in a vast project in which new territories were being mapped and colonized, the project extended a split along gender lines between the ideal of a separate, autonomous, objective, male self and a relational, connected, and empathic female self. Self-narratives were the records of the great changes occurring in the self and were agents of change in and of themselves. Creating narratives of their lives gave men and women coherence and purpose and provided structure to the self. Furthermore, according to Charles Taylor, the multifaceted nature of the modern self owes much to the driving forces behind the Protestant Reformation—individualism, self-responsible independence, and total commitment—which were no longer only the duty of an elite, but were demanded of all Christians. The hallowing of life took place not only at the periphery, but was a change that could penetrate the full extent of mundane life.[5]

The composition of Heaton's autobiography was as important to her as any of the experiences she set down. It was an act of introspection and an examination of the workings of God's grace within her, which helped her to rise above a world of sin into union with God. It was also a way for her to describe her struggle to endure and overcome adversity, and in the process she established her strength and rightness, both in her own eyes and for future readers. At times she sought relief from desperate loneliness in writing and may have described what she suffered so that others could identify with her and not be alone in their suffering. Essentially, her account shows how intensely she lived in a private world, often in outright opposition to society, in a condition that sometimes can be described as heroic and self-reliant, and sometimes as lonely and alienated.

Near the beginning of the diary, Heaton gives an account of her conversion experience at the time of the Great Awakening. She also describes her life of daily meditation, reading, and prayer and her nightly dreams. Heaton sought to join with individuals who thought as she did and to be part of a "pure" church, and in this quest she recorded her views of the standing ministry and the Separates. Heaton's outward experiences shed light on her inner life and unintentionally reveal the social conditions of her time. The diary portrays tempestuous relationships with her husband, children, and neighbors. She also gives accounts of events of the American Revolution as experienced by her sons or as reported by neighbors, newspapers, and sermons. Finally, she recorded revivals that occurred near the end of her life, during the early years of the republic; they seemed to her to announce the millennium. Throughout the diary, however, much of her daily life was filtered out because it got in the way of the path to salvation. When the business of the world was admitted into the narrative, or allowed to impinge upon it, the incidents were construed as benevolent guidance from God. Heaton, it will be observed, turned away from experiences in the outer world in order to write about them in her diary, where she examined them. They were understood as benevolent guidance from God, or as omens and trials that were part of God's plan.

In the opening pages Heaton observed that her early years were carefree and happy. She was close to her parents, especially her father, who was a deeply pious man. At age ten she began to be concerned for her soul, and in 1741 at age twenty she crossed Long Island Sound to New Haven to hear the inspirational preaching of revivalists George Whitefield and Gilbert Tennent. At the same time, in Southampton, the Awakening produced a Separate church that flourished in the 1750s under the ministry of Elisha Paine. Heaton's father and other members of the Cook family were active in this church.[6]

The religious revival had a profound impact on Heaton and many of her contemporaries throughout the colonies. From Whitefield's tour until 1743, the period when the revival was at its peak, thousands were converted. The Awakening reached many people because the colonists placed great importance on the experience of conversion. In Calvinist theology, which influenced much of the population, the crucial event was the bestowal of divine grace. It assured admission to God's presence in the hereafter and prepared a person for a full life on earth. However, not everyone agreed that God was working in the revival. Efforts on the part of the established ministry to subdue the new religious enthusiasm soon provoked a revolt. Out of the complex of issues that led to open rupture, almost a hundred Separate churches emerged, marking a permanent shattering of the Congregational establishment in New England.[7]

After hearing the eminent Whitefield and Tennent, Heaton, under her father's guidance, prayed and occasionally experienced "encouragement." One day, after attending a meeting of Southampton Separates and feeling complete despair about salvation, she suddenly grew calm and quiet, and words came to her: "seek and you shall find come to me all you that are weary and heaviladen and i will give you rest . . . Me thot i see jesus with the eyes of my soul." She saw "a lovely god man with his arms open ready to receive me his face was full of smiles he lookt white and ruddy and was just such a saviour as my soul wanted." Her heart went out with love and thankfulness, because she had found, she said, both a new soul and a new body [p. 6]. By recording her subsequent spiritual reflections, she hoped to keep alive the sense of assurance she had experienced, although she repeatedly lapsed into uncertainty and doubt.

The Calvinist framework of conversion that undergirds the Heaton diary was learned from reading the works of John Bunyan and others who impose a pattern on the course of devotion and self examination. After conviction of sin, and other accepted stages of conversion, the believer's assurance of faith might be marked by a number of experiences leading to acceptance of Christ's righteousness. However, since illumination might have an emotional impact but not a biblical basis, Heaton records it with a degree of doubt and hesitation.[8] Such swings in mood between elation and depression about religion are quite evident throughout the Heaton diary and in the work of others such as Nathan Cole and Sarah Osborn.

Heaton's description of her vision is also related to the Calvinist reconstruction of the Christ image found in some sermons of the time. Jonathan Edwards, for example, in *Religious Affections,* criticizes those of the awakened who took comfort in visions of Christ's "smiling" face, his outstretched arms, and his "pleasant voice." Such phrases are similar to those used by Heaton to describe the appearance of Christ. Some Separates, according to historian Alan Heimert, even told of seeing Christ on the cross, a spear piercing his side, with blood streaming out to them. The prevalence of such baroque visions was a tribute to the kind of preaching in which Cotton Mather and Edward Taylor, among others, had indulged earlier in the century, versions of which could still be heard in the Connecticut countryside as well.[9]

Like other evangelical writers, Heaton described her spiritual coldness, doubts, and sometimes her despair. For example, on one occasion she felt "sensible withdrawings of the lord from me. He did not leave me at once but by degrees by little & little" [p. 24]. To alleviate her sense of sin and guilt, she would pray or turn to her Bible, but often without relief. At one point she recorded, "O my god i have been crying 13 years to thee and hant got the victory yet over one corruption" [p. 60]. Although her father had reminded her that she was no worse than any other sinner because "we all pertook equally alike" in Adam's sin [p. 84], she remained unconvinced. While her actual misconduct as a child seems to have been relatively minor—she recalled stealing watermelons, frolicking with friends in a deserted house, and being disobedient to her parents—these memories produced much regret in later years [p. 385]. Thwarted in her desire for inner peace, sometimes she had "athistical thots" [p. 73] and wondered if God existed at all.

During some of these periods of gloom, Heaton experienced encounters with Satan. As early as age thirteen, she said, she lay awake afraid of the devil [p. 7]. When grown, she recorded times when the devil would try to keep her from prayer: "leave of leave of says satan you have prayed anuf your worldly conserns suffers while you pray so much pray no longer" [p. 62]. The devil would send "darting" thoughts to her, as when she felt particularly melancholy, she thought she felt the devil had twitched her clothes and had once whispered, "hang your self" [p. 4]. Even at the age of sixty she admitted that, although she was "ashamed to say it," many nights she was afraid to go to bed "for fear of the devil that he would distress me" [p. 326]. These exhausting encounters with Satan, which end in her victory, indicate important aspects of her thought. Deity and devil were vividly alive in her mind, and she was free to choose between them. She wanted to side with God, but the devil could make a formidable case. In these experiences, Satan was a disembodied voice, sending thoughts into her mind that engaged her in a self-contained dialectical argument. The conversations were ordinary and practical but were stated with compelling logic and posed a profound threat to her mental stability by challenging her religious beliefs. Puritans of the time understood such experiences as attacks on the soul by means of assaults on the body; because women's bodies were weaker, the devil could reach women's souls more easily. Not only was the body the means of gaining control of the soul, it was the very expression of the devil's attack.[10]

Heaton's inner life also included a sense of reassurance and faith. Stanzas of joyous poetry can be found throughout the diary, attesting to her confidence and trust in God. For example, in the early pages she writes that she "must stop & offer a song of praise and admiration to that god that has done great things for me." Indeed the act of praying could be "so delightsome" that she regretted having to stop [p. 248]. At age sixty-five she could look back and see how often God had helped her when she was in "great straits," and she knew "that if i should be past labour and could not help my self," there was no need to fear because "god would still be the same to take care of me" [p. 419]. She saw herself as set apart by the Lord as his special possession, sanctified as she walked with him in faith and obedience. Such passages of hope and faith occur less frequently than those of doubt, but this does not necessarily mean that Heaton spent most of her life in misery. Unpleasant occurrences and negative thoughts prompted her to turn to the diary and write in order to gain control over her feelings and overcome her sense of isolation. On the other hand, a sense of peace and oneness with God did not require remedial action and could be cherished without recourse to her diary.

Other aspects of Heaton's inner life include her interpretations of dreams, natural occurrences, and unusual incidents in the community. Fearful dreams of a snake, a burning house, a dead dry tree, and a great mad bull were reminders of ever-present dangers that indicated the need for God's preserving care. She also made note of fires, comets, eclipses, and thunderstorms, as well as deaths from drowning, suicide, and public execution; each event was viewed as a call to New Englanders to repent. Heaton, like Susanna Anthony and other pious women who were beginning to acknowledge their own emotions and their selves, interpreted dreams and remarkable occurrences as portents or indicators of tasks to bring about change in her life or in her self.

American Indians appear in the diary, sometimes as "the dear indiens" who testified to their religious experiences in her Long Island church [p. 230], and sometimes as "cruel barbarous enemies" who made "mischief," as in 1754, when Indians appeared in the "upper towns," killing some citizens and carrying off others. In the latter instance, Heaton viewed the raid as a warning from God concerning the "backslidings" of his people who needed to repent if all New England were not to be destroyed [p. 84]. Elsewhere, Heaton expresses pity for Indians, including one jailed for murder, because, she said, except for God's mercy she might find herself in the same benighted condition. The Indian, then, is characterized in her writing both as a potential convert and as a graceless savage.

Throughout the diary, Heaton emphasizes her devotion to meditation and prayer. Some of her periods of prayer were preceded by fasting.[11] Other times she would take her Bible and diary, and sometimes a bottle of beer and some food, to the barn, the cow house, the woods, or the swamp in order to pour out her heart to God for hours at a time without interruption. She also had spiritual reflections while she went about her work, both in the house and on the farm: one time while making her bed, "how refresht" she felt remembering a "lovely dream" about God and an angel [p. 125]. Another time, while burdened with a "sence of expected trouble," she went to fetch water and heard words that strengthened her flagging spirits: "let us indure hardness as good soldiers" [p. 332]. Even the prosaic life of the barnyard held lessons for her: "I lookt out at the door & see a company of chickens scratching and picking in the mud. How they prattled it took my mind to see how content they was with what god had provided for them. I see that all of gods creatues answer the end of creation better than poor sinfull man" [p. 386].

While these images of farm life give a homely character to her religious writing and set her apart from contemporaries, such as Esther Edwards Burr, who were beginning to cultivate a genteel style, Heaton was capable of fairly sophisticated rhetoric as well. While often it is not clear whether a passage is largely of her own composition or drawn from her reading, the choice of words reflects her literary taste. She is particularly effective when using repetitive phrases and parallel constructions, probably having learned these forms from reading the Old Testament. In the 1770s she prayed, "Lord i thank thee that i am out of hell i thank thee that i am here where i may read thy blessed word. I thank thee that i am here where i may pray to thee groan to thee weep to thee kneel on the ground to thee ring my hands to thee sigh to thee. O let me feel and see thy salvation in my own sanctifycation and in thy turning my family to thee" [p. 199]. She also made effective use of balanced contrast, as in this passage written in the 1760s: "O methinks what a pyty it is that god that made the world should have so little of its service & satan that ondid mankind should be so willingly and expensively addored" [p. 158]. The simple imagery and short, strong phrasing, as well as much variation in vocabulary, are striking. Although her writing is stylistically interesting, it lacks the conventions of formal prose, and to appreciate the sophistication and extent of her literacy one must also examine her reading.

Heaton was a voracious reader of religious books; as she said, "i read constantly and find it teaching" [p. 253]. Her experience typifies the dramatic increase in female literacy, including reading and writing, that was taking place over the eighteenth century both in female academies and in basic schooling, even though the rate of female literacy was only half of that for men. By the time of the American Revolution an unprecedented number of books were published, including self-help books, textbooks, histories, children's books, novels, and religious works, testifying to a new social class that wanted to increase its literary skills and turn those skills to its own advantage.[12] Heaton probably acquired the rudiments of learning at home rather than at school, since she makes no reference to a school in her diary and was helped principally by a father who read aloud to his family. While it was not unusual for girls to be given less training in writing than boys, both sexes were expected to be able to read, principally in order to study the Bible on their own.[13] By keeping her Bible nearby while she worked, Heaton frequently enjoyed a "feast of reading" [p. 39] and committed much of what she read to memory. Scripture provided a vast reservoir of themes, phrases, meanings, and habits of thinking to Heaton and her contemporaries, who understood themselves to be God's people with direct access to the word of God as their most precious privilege. Heaton "lived" the Bible; she believed it foreshadowed what she experienced and provided insight into the present.[14]

In addition to the Bible, Heaton was familiar with many works of doctrinal and practical religion. At the time of her husband's death in 1791, probate records show that the family owned "1 old Watts Psalm Book / 1 old Faith Anatomized / 1 old Bible / Spiritual Logick / 11 small books." Supplementing this brief list is information in the diary indicating that books often were lent to her by neighbors [p. 57]. There is no record of her purchasing a book other than a Bible. She was familiar with works by David Brainerd, Thomas Brookes, John Bunyan, John Flavel, Thomas Harrison, James Janeway, Samuel Mather, "the Second Spira," Thomas Shepard, Solomon Stoddard, Isaac Watts, and Michael Wigglesworth. References to her reading increased over time, and by the period of the American Revolution she began reading newspapers as well, but as a rule throughout her life she read intensively traditional devotional books and contemporary works with devotional themes.[15] In these works she encountered narratives that filled out the meaning of being one of the Lord's own, moving from captivity to deliverance, from sin to redemption, and from weakness to triumph.

Heaton read from religious authors upon awakening and before retiring, and meditated on them throughout the day. They often prompted her to write her own spiritual reflections and personal prayers. Such a practice could lead, according to Cecile Jagodzinski, to the development of the private self, a self characterized by a sense of personal autonomy, a recognition of the differences between one's public and private roles, and a desire to conceal or keep secret the workings of the inner person.[16] Heaton called upon her spiritual exercises of reading, writing, and private meditation to see her through her daily duties and life crises and hoped they would help her along the path to salvation. In creating this pious self, however, she withdrew from the company of others, experienced frequent moods of melancholy, and came into conflict with ministers and civil authority because she refused to attend meetings of the established church.

After her marriage and settlement in North Haven, Heaton went to the regular Congregational church with her husband until she fell out with its minister, Isaac Stiles. She believed he was an unconverted minister and therefore a "blind guide." In this respect her opinion was one in the avalanche of accusations made against the established clergy following Whitefield's visit to New England during the Great Awakening. She also may have objected to Stiles's practice of admitting members into the church without their recounting a conversion experience, as well as to his extension of baptism and communion to the children and grandchildren of those who had been converted.[17]

While Theophilus Heaton continued to attend Stiles's church, Hannah began to associate with a Separate congregation that sprang up in the Muddy River area of North Haven, where it met at the home of Benjamin Beach until Beach, who served the group as minister, moved the meeting place to Wallingford in 1762.[18] Because Hannah was no longer attending the regular church meeting, she was visited by neighbors and officials who tried to convince her to rejoin their fold. In 1758 she was charged formally with breaking the Sabbath. Why Heaton was prosecuted on this charge is not clear. Later in the diary she gives examples of the more usual charge leveled at Separates—fining and imprisonment for not paying their rates toward the established minister's salary [p. 215].[19] Although records of the trial, other than those in the diary, have not survived, clearly it was one of the major events in Heaton's lifelong quest to worship according to her own light.

In the midst of a thunderstorm, Heaton arrived at court feeling "poor in body" and with "ague in my head." According to her diary, she was called by the justice to speak but could say only a few words before he bade her hold her tongue. She was not allowed to say anything more on her own behalf. Two women spoke in her defense, telling the court of her bodily infirmities, "but they was not regarded." Of two women summoned as witnesses against her, "one womans concience smote her," and she said little. The other said merely "that she did not remember that she had seen me to meeting lately." When the session ended Hannah pronounced her own judgment: she "told the justice there was a day a coming when justice would be done . . . there is a dreadful day acoming upon them that have no christ." She recorded that "the justice told me i talked sasse" [pp. 120–22].

The proceedings against Heaton frightened her husband, who paid the fine (twelve shillings) against her will [p. 123]. Later, Justice Davenport tried to bring the matter of Heaton and other Separates to the attention of the General Assembly in Hartford, but the assembly refused to hear his complaint [p. 136]. New charges of not attending meeting were brought against her, but they were dropped [p. 142]. Unrepentant, Heaton soon had the satisfaction of recording the judgment of God on those who had caused her to suffer. The grand juryman fell at a house-raising into a cellar of stones; a child of one of the prosecutor's witnesses became seriously ill; the other witness had a son pressed into marriage with a young pregnant woman; both the judge and his wife died shortly after the trial. Heaton recorded the sufferings and deaths of the hostile witnesses and members of their families and felt fully vindicated by such evidence of God's wrath toward her enemies [p. 344].

Isaac Stiles also did not live long after Heaton's trial, and in 1760 a new minister, Benjamin Trumbull, was installed. Though Trumbull would say in 1773 that in North Haven "we have the greatest religious, as well as civil privileges" and "no man is persecuted for his religion,"[20] Heaton did not find him much of an improvement over his predecessor. She thought he was too lenient toward those living "a corrupt life" and believed he was critical of dissenters [pp. 143, 148]. Heaton rejected the leadership of established ministers such as Stiles and Trumbull, who by her lights were "not converted" [p. 33]. Also, she had no patience with their ministerial style; their sermons, delivered from notes, seemed to her dry, repetitive, and lifeless. She favored the New Light and Separate ministers, who spoke extemporaneously by divine illumination and interwove biblical texts with practical applications in a forceful and animated method of address. Whenever possible, Heaton attended meetings where Separate ministers preached. On occasion, she heard three sermons in one day. Among the well-known preachers whom she heard were Jedediah Mills, Ebenezer Frothingham, and Eleazer Wheelock. Each time it was Hannah's "practice always . . . to pray for my ministers before i hear them" [p. 303].

While keenly aware of the dangers in being a Separate, after many years of occasionally attending Separate meetings, Heaton decided to apply for membership in the Wallingford congregation. It seemed to her that she was called to put her "neck under christs yoke," yet she wondered whether "it was fear made me go forward to join more than it was love to god" [p. 315]. When she asked her husband's opinion of this step, "he then seemd pleasant but his answer was—i shall not tell you. I begged again that he would tell me but his answer was the same again and in as much as he did not forbid me i went forward" [p. 316]. In July 1778 she made a brief relation of her religious experiences to the Separates, answered some questions, and was received as one of them. It was a happy moment: "When i was coming home i felt sweet peace in my soul." But her sense of peace was not complete, for she added, "ah alas at my first adventure to the lords table i felt low in mind" [p. 316]. The underside to the spiritual fulfillment that Heaton experienced through closeness with God was the ever-present, tormenting fear of separation from him.[21] Hannah's ambivalence and anxiety at this time also may be due to a constellation of troubling family matters: her sons' participation in the Revolutionary War, the death in childbirth of her daughter-in-law, and a limited sense of companionship with her husband. In joining with the Separates, she may have hoped to overcome feelings of being alone in the face of overwhelming family problems.

Hannah's struggle for independence of spirit led to adversarial encounters with nearly everyone, including her husband, children, and neighbors, as well as ministers. Her account of her marriage, for example, resembles the English confessions of "injured females," women who have been wronged.[22] Though not as formulaic as these sensational memoirs, Heaton's diary does describe a precipitate marriage with an unconverted man and a stormy life with him that continually tested her faith. She offered her life as a warning to others to show what evils needed to be avoided, particularly when choosing a spouse.

Heaton blamed herself for the tempestuous marriage. She had been warned by the preacher James Davenport not to marry an unconverted man but did so while in a "dark frame of mind" [p. 20]. She believed her subsequent afflictions were a punishment for disobeying God's word. Heaton loved and respected her husband and would not have exchanged him for "the best man on earth but alas the yoak is unequal" [p. 22]. She referred to his "userped athority" [p. 145] on one occasion, and called him "cruel faroah" [p. 166] on another, when he kept her from attending religious meetings. Furthermore, he threw her diary into the mud, hid her spectacles so she could not spend time in pious reading, and expressed suspicions about her fidelity when she spent long hours away from home with the Separates. They had battles that lasted for days; still, she prayed incessantly for him.

Despite adversarial encounters, Heaton held to the idea that she and her husband should be "best friends" [p. 103]. Her struggle for spiritual autonomy and companionate marriage can be viewed as a personal contention with the hierarchical ordering of the family in colonial America in which men were heads of household, charged with preserving social order, while married women were defined as dependents, or *feme covert,* before the law. In some ways, Heaton reflects the rise of the modern personality, which gave a new value to intimate personal relations. In modern culture, the companionate marriage and the demand for privacy rose together and were seen as a crucial part of what makes life worthy and significant.[23]

Theophilus Heaton remained an Old Light and a casual observer of the Sabbath to the end. He was alternately irritated by and accommodating to his wife's religious thinking but always maintained his own independence. Even on his death bed, as Heaton observed, "he seemd loth to talk about his soul. Once i asked him if he was afraid to die he said yes. I asked him if he had a heart to beg for mercy and i thot he said not much." She reflected, "I many times told him the necessity of having a christ or there can be no salvation but he is gone. I must leave him the judge of all the earth will do right" [p. 422]. She had spent a lifetime asserting her spiritual independence yet also hoped for close companionship with her husband—goals whose irreconcilability was painful to her.

Heaton acknowledged readily that her husband was a "good provider" [p. 166], and local records indicate he accumulated a sizable amount of wealth. She, too, was a good provider: milking cows, making butter and cheese, and tending chickens were some of her responsibilities. The couple's success is particularly impressive because they had begun with nothing. In the first year of their marriage, 1743, when they lived with Theophilus's family, Hannah "did reprove" his family for "wicked practices . . . the family was soon set against me and we was turnd out and no portion given us as the rest of the children had" [p. 20]. Somehow Theophilus obtained a farm in North Haven and through careful husbandry was able to purchase additional land and livestock. Perhaps his father relented in his decision; in his will he gave twenty acres "in addition to what I had already given."[24] Hannah observed later in her diary that "god in his providence has caused our inheritance to increase" [p. 208].

In 1755, twelve years into their marriage, when Theophilus Heaton was thirty-six, the tax list for North Haven showed his property valued at £52, very close to the average worth (£56) of the town's heads of household. At his death in 1791 at age seventy-two, his total wealth was valued at £628, of which £533 was in real estate. He left no will and there is no mention of his wife in the records of his estate, although they contain an agreement signed by his sons to observe a division of their father's property based on the court inventory. The inventory lists clothing, bedding and furniture, cows and sheep, as well as the homestead and considerable land. There is mention of "the land and right in a sabbath day house by the marketplace" and "the right in a fishing place" at Sackett Point and

Duck Cove, as well as "the house where Jonathan dwells" and "the house where Calvin dwells."[25]

Little else besides farming and family seem to have occupied Theophilus Heaton's attention. Voted a collector of town rates at a town meeting held in New Haven, December 16, 1766, he gave notice on April 13, 1767, of his refusal to serve.[26] We learn from the diary that in addition to his work, he enjoyed canoeing, fishing, and an occasional drink with a friend. Overall, he worked his farm, supported his family, and was moderate in all things, including piety. He can be seen as an example of the successful yeoman farmer, the type of Yankee who was creating "the land of steady habits" in rural Connecticut.[27]

Heaton addressed her autobiography to her two surviving children in the hope of providing an example that would encourage them to live good and godly lives. She also wanted them to know how they had hindered her own spiritual progress. Her children from childbirth to maturity were a great trial to her, causing fear, pain, and anxiety. Only late in life did she suggest they were a source of comfort as well.

While she was pregnant, she felt "exceedingly weeakly & almost overborn with vapours." It was "satans time to try to tempt me to sin against god by holding up gastly death before me" and saying, "you deserve to undergo more than ever any woman did and to die in your deficult hour when it is come" [pp. 71–72]. Her first child, Jonathan (1744–1799), was born a year after her marriage: "I was in travel from the evening extreem bad till next day at noon. All hope almost was gone from all of my being delivered" [p. 20]. Her second child, Hannah, died at age one and a half. About eight months after this child died, a son was born, but he too died suddenly when about three weeks old. At first she "mourned bitterly but god soon stilld me by sending the ague in my head and teeth. I was in extreem misery." She pleaded with God, telling him that if he would give her ease and strength, she would mourn no more for her children; soon after, the pain subsided [p. 26]. Her fourth and last birth came after seven hours of labor, during which time she was in "great distres," but at last a boy, Calvin (1755–1820), was born, and "in a moment" her soul "winged away to glory" [p. 112].

Heaton's accounts of childbirth, mingling bodily description with spiritual interpretation, are in accord with Puritan belief. The body, viewed as a site of both sin and salvation, and as a center of both faith and unbelief, could gauge the level of spirituality attained by the individual. Entry into faith was manifested in the body; spiritual identity was ritualized in the body; and falling away from faith, and sin, also transpired in the body.[28]

Heaton does not discuss the early upbringing of her children. Only when Jonathan was twenty years old does she describe her experiences with him, criticizing him because "he lives wholly without prayr and but little regard if any to the sabbath and but sildom reads & goes often to frolicks disobedient to parents" [p. 154]. Such disobedience she believed was her lot partly because of her own negligent behavior toward her parents when she was young. Calvin was as irreligious as his older brother, and often her sons and husband united in opposition to her, making her effort to live the godly life extremely difficult. She believed her family was a "house of belial for want of family government" and concludes, "wo wo to me i am daily reaping the fruit of my marrying contrary to the mind and will of god take warning o christians" [p. 205].

In 1777 Jonathan married Isabel Hitchcock, and Hannah hoped for the best: "o lord make them like isaac and rebekah like zacharias & elizabeth o make them help meets to each other in the way of heaven" [p. 302]. But Isabel died within a year, probably of puerperal fever, after giving birth to a son. Although her daughter-in-law died in an exemplary manner, Hannah was plagued with misgivings, thinking she had not sent for the doctor soon enough. Jonathan had the same thoughts, grew concerned about his own soul, and began attending Separate meetings.

Jonathan remarried in 1781, to Rhoda Hall, who was thoughtful and attentive to Hannah. In 1783 this couple lost a son who lived only four hours, and in 1785 Rhoda miscarried. They had two daughters who lived to maturity, one born in 1788 and the other in 1790. Jonathan died in 1799, at the age of fifty-five. Since he left no will, the court divided his estate equally among his widow and two daughters. His total worth, including clothing, bedding, farm equipment, livestock, house, and land, was £975, of which £736 was in real estate. Accounts of debts to his estate included "cash paid and time spent in going after the Negro wench that run away."[29] Hannah made only brief entries about Jonathan after his second marriage, but she was no doubt proud of the way he turned out: he became a Separate, opening his home to evangelists and facing threats of imprisonment for not paying his taxes to support the established church [p. 361].

In contrast, Calvin caused Hannah much grief. Yet she inscribed her autobiography to him and favored him in her will.[30] He was special to her for some reason, perhaps because he was her youngest—eleven years younger than Jonathan. While living at home, Calvin loved to go to parties, or "frolicks," and wanted to attend one held in a tavern on Christmas

Eve. When his mother refused to give him money so he could go, his father gave him some [p. 281]. When Calvin was about twenty-two, Theophilus bought and gave him forty acres of "good land at waterbury," but Calvin sold it and squandered the profits. Nevertheless, Hannah said, "ah he is my dear child still" [p. 406].

In 1762 Calvin married Esther Humiston. They eventually had four sons, three of whom survived childhood, but Calvin's family gave little comfort to Hannah and her husband. Calvin was eager to acquire another farm from his father, and he chose devious means to accomplish this end. In 1788 he brought a town officer to the house and attached his father's estate for a thousand pounds, called by Hannah a "book debt," for work that he had done after he had come of age. "O this awfull sight for a child to tach the land and body of his aged father," she grieved. Fortunately Theophilus got the help of Col. Street Hall, who labored to bring father and son together again. After lengthy discussion, Calvin acknowledged that he was wrong and begged to be forgiven [pp. 408–10].[31] Soon after, Calvin and his wife were given the use of a large house that was built on the site of the old house, which had been torn down and its timbers used to build a small house for Hannah and Theophilus. There they lived in cramped quarters for four years, waiting for Calvin to complete construction of two rooms for them in the new building. Finally, they accepted the repeated invitations of Jonathan and Rhoda to come live with them. Once again, Hannah forgave the prodigal, saying, "now when calvin see the team was come to carry away our goods i believe his concience smote him for he asked his father not to go but it was now twoo late ah poor calvin" [p. 405].

One way or another, Calvin came into a considerable fortune. In 1806, judging by tax records, he was the wealthiest man in North Haven. Some of his wealth derived from trade with the West Indies. He died in 1820 with assets of $20,624 and debts of $13,297. The probate inventory lists household goods, livestock, blacksmithing equipment, boards and timber, a distillery, dry goods in quantities for sale, as well as meadows and lots.[32] In sum, Calvin was a Yankee entrepreneur who seized opportunities and took risks to increase his wealth. Hannah hoped to tell him in her diary about her disappointment with his spiritual state. She wanted to stir his sense of guilt and prompt him to search his heart for evidence of God's grace. But while he was capable of regret and remorse, generally he seemed untroubled by the kind of Puritan conscience that his mother had in such great measure.

Beyond the circle of Heaton's family were female acquaintances with whom her associations were sometimes strained or contentious, because they centered on opinions about religion.[33] One relationship ended in a quarrel when a longtime confidante disclosed a derogatory remark Heaton had made concerning a minister who, she said, read over a "great pack of notes" to his people and called it preaching [p. 65]. Another friendship was undermined when a woman came to Heaton's home and spoke at great length of her spiritual distress until Heaton herself grew deeply disturbed [p. 168]. Because of unpleasant encounters such as these, Heaton often preferred to be alone. She declined to visit the neighbors [p. 74] and would not go to the door when someone knocked [p. 89]. Nevertheless, there were occasions of joyous companionship with women: once two young women visited her and they all "talked of religion & sung himns & read wattses vision" [p. 127]. However, female friendships did not loom large through much of Heaton's life because she lacked the leisure time that city women had to cultivate such relationships or the free time that farm women had when there was more than one woman in a household. More important, she chose her closest associates principally from among her kin rather than among her neighbors.

Heaton gave advice and preached repentance to those around her who would listen. One outstanding example of her tendency to sermonize is shown in her encounter with an Indian. On May 21, 1772, she visited "new haven prison to see the poor indien under the sentence of death for murdering a white man. He stood at the windo in irons with twoo books in his hands. I talked some to him but he seemd stupid & unconserned about his soul" [pp. 210–11]. Four months later, on September 2, 1772, she reports that "poor moses paul was hanged (the indien i mentioned before). I went to new haven and heard mr samson occom (an indien minister) preach from romans 6–23—the wages of sin is death but the gift of god is eternal life through jesus christ our lord. . . . People said the poor crimenal trembled—then i went to the gallows but when i see him stand upon his coffin with his hands lifted up praying just agoing to be turned off i turnd about and went away" [p. 214].[34] This episode shows Heaton participating in, even intruding in, the lives around her, as when she lectured the jailed man. It also shows her sensitivity in choosing not to see him executed. It indicates how she could visit others, as if making a pastoral call, and engage in a sort of public discourse, much as a minister would.

The sermons of ministers in the vicinity of North Haven were of crucial importance to Heaton's spiritual life and to her understanding of events in the larger world. Her sermon notes early in the diary are concerned only with Scripture. She might note the chapter and verse of the passage under discussion and indicate the ideas the preacher developed. She sometimes recorded the effect of the preaching on her or on other listeners. For example, of Ebenezer Frothingham, who preached at her house from Matthew 5:3–12, she wrote, "I believe he was the first ordained saparate minister that ever i heard it was in the year 1753. Afterwards i had a sweet sence of the joys of heaven" [p. 46]. In contrast, in 1754, she "heard mr troop of southold but i felt & see nothing but death" [p. 86].

Halfway through the diary, the sermons she recorded began to carry political as well as spiritual messages, concerned at first with the relationship of the "christians" (meaning the Separates) and those who opposed them (the colonial government). In 1759, for example, when the Baptist Morse of New London spoke, "he shewed how god loved his children how tender he was of them and he charged the opposing world to let them alone" [p. 139], and in the same year Daniel Miner of Lyme came to preach and "to see how his brethren fared in the war" [p. 140]. In 1774 she recorded a jeremiad by Benjamin Beach, who saw imminent war with the British as a judgment from God for persecution of Separates and their ministers. Heaton agreed with him up to this point, but when he said further that "he believed our nation would be destroyed without a reformation," she objected, writing, "without gods spirit what is all preaching[?]" [p. 251]. War with England, and earlier with France, caused many people to link religion with politics. As a result, religious culture was one element that helped lead to confrontation and revolution.[35]

As relations between England and the colonies deteriorated in the 1770s, Heaton's sermon notes reported more commentary on political events of the day and were interwoven in the diary with newspaper reports about English affairs, statements concerning disorders in nearby towns, accounts of her sons who had joined the army, and her own prayers for "this threatned north amarica" [p. 255]. The English, she decided, were clearly in the wrong: "o has king george forfited his coronation oath o does lord north govern king and perliment and must he govern amarica twoo by papists laws" [p. 263].

In July 1776, Jonathan went to New York to join the army that had moved into the area, but he soon came down with camp fever and had to return home. On January 14, 1777, Calvin joined the army in New York, but he too returned home sick before long. After recovering, Calvin resumed active duty, and Hannah then recorded a dream about a favorable future: "Now i dreamed i was talking to jonathan and calvin. I thot i told them to pray to god that they might live threw this deluge of war for when it was over i believed there would be greater glory seen in new england than ever was before. And while i was speaking i thot my soul was raised with a vew of the glory and so i awaked" [p. 300].[36]

In July 1779, as the British approached North Haven, Heaton wrote:

> There come news that the brittish troops was landed at new haven on purpose to burn the town but a good god did not suffer it. Only they burnt a few stores abused wemen plundered houses carried off what they pleasd at the same time some hundreds landed at east haven. O what a sorrowfull soul i felt to hear the cannon and guns roaring. I had some sense of the justice of god but i kept pleading for mercy not knowing how quick they would be in my house. But glory to god that said hither to shall thou come & no further. O i hope i hope that he that once asked a draft of water of the woman of samaria is about coming to set up his glorious kingdom in america. Come lord jesus come quickly. [p. 320][37]

Finally, Heaton recorded the conclusive victory at Yorktown in October 1781, after she heard that "coronal walles and his army is taken," but she was quick to add: "o what shooting frolicking and heathenish rejoysing is there in our towns instead of giving glory to god for this smiling providence" [p. 332].

Heaton had turned to ministers to understand the vicissitudes of the war because she believed the Revolution, like every human event, was part of a divine plan. She heard the sermons of Separate ministers who interpreted the Revolution as retribution for New England's sins; at least one of them, Daniel Miner, viewed the war as heralding a new age [p. 299].[38] Significantly, she also gave her own views on imperial policy and military engagements, at length and in a forthright manner, oblivious to the question of whether it was proper for a woman to be concerned about such matters. Heaton does not address the question of "woman's sphere,"[39] either before, during, or after the war, although in a passage about the biblical Rebekah and Isaac as bride and bridegroom, she wrote that "she was in her place and he in his" [p. 21]. Nor does her diary help resolve the question of the long-term impact of the American Revolution on women's lives, though it sheds some light on what life was like as the war reached Connecticut and demonstrates Heaton's strong patriotic and religious support of the Revolution.

In the last stage of her life, Heaton experienced declining health. The diary is a record of her illnesses throughout her adult life, and the entries provide insight into the felt relationship between mind and body among the Puritans. In the 1750s, while in her thirties, she noted that she could not sleep or eat while pregnant [p. 72]; felt "low" or in a "dark frame of mind" [p. 95]; was weak in body and mind [p. 102]; and was terrified with fear of death [p. 109]. In the 1760s, in her forties, she continued to feel weak [p. 232], while in the 1770s, in her fifties, because of her physical and mental infirmities, she prayed to god to "convert me again" [p. 259]. In addition, she records that her body was "poor weak and low my faith gone" [p. 285]. In the 1780s, in her sixties, the entries for poor health increase: she was taken sick with a fever [p. 342], experienced pain in her back and weakness in her joints [p. 348], and described herself as an object of pity [p. 360]. But she understood that she was being weaned by the Lord from "frames and feelings" so she would learn "to trust and live upon his naked word" [pp. 369, 396]. Such writing about pain integrates that condition into the construction of the self, and the inability to dominate pain tends to undermine the concept of individualism, self-reliance, and self-determination.[40] Consequently, in the 1790s, in her seventies, she turned from her bodily distress and described her mother's death, reported to her by relatives and chosen by her as a model for her own transition: "She lived in wonderful nearness to god i believe was a witness for him and at her last appeared not to have the least fear of death but wanted to have it hasten. Her last words was she lifted up her hands and said come lord jesus come quickly and fell asleep as quiet as a lamb. Now when i heard of it my heart cryed glory to god for his mercy and kindness to my dear mother" [p. 431]. Heaton's mother, residing on Long Island, lived to age 90, well beyond the life expectancy of her time.[41]

In eighteenth-century New England, old age was regarded as a time of preparation for the afterlife. It was accompanied by physical infirmities, changing family arrangements, and the death of family members and friends. The drama of revelation and redemption was played out in illness, old age, and death. While God's grace was related to good health, as a consequence of original and subsequent sin, the body endured ill health as punishment. Suffering could range from a lack of vivacity to severe physical afflictions. While one could endure all sorts of corporal miseries, the agony of the soul had a more lasting effect on a person's life.[42]

During the last stages of her life, Heaton recorded deaths as they occurred, both in her family and in North Haven. She spent more time in pious reading, morning and night. She noted when religious revivals occurred: on Long Island in 1784; in Newark, New Jersey, in 1785; in Wallingford and

Meriden in 1786; in "one place down the country" where four hundred people were converted in 1787; and in Westfield (part of Middletown) in 1790.[43] She hoped for such a revival in North Haven, which she thought could be the site of the beginning of the Second Coming: "it seems to me sometimes as if it [North Haven] would be the chief seat of christs kingdom" [pp. 37, 362, 424]. Heaton's expanded vision encompassed the day when "Amarica will be emanuels land" [p. 480] and the gospel would spread "from sea to sea and from the river to the ends of the earth"[44] [p. 351, 361, 408]. In contrast, she also experienced doubt about the millennium on earth, ending her diary[45] in 1793 with this stanza: "Here the law of sin and grace will jar / both dwelling in one room / the saints expect perpetual war / till they are sent for home" [p. 436]. Such a change in view may be related to Ruth Bloch's contrast in visionary hopes between the 1770s and the 1780s. In the first period, American Protestants highlighted the global dimensions of the millennium, even when they did not particularly stress the redemptive role of the American nation, because they envisioned the gospel itself spreading throughout the world. In the second period, Americans experienced growing doubts about the meaning and destiny of the new republican nation and looked to the revivals for a long-range change in moral effect. By 1800, no longer identified with a specific denomination, millennialism had become a general and diffuse cultural orientation.[46]

The mood of the diary changes over time, from emphasizing spiritual encounters to living a life of prayer and occasionally experiencing the "downpour" of God's spirit. References to the devil grow fewer in number, and the writing is smoother, easier to read, and more confident. The spiritual autobiography of Hannah Heaton illumines both the inner and outer lives of an eighteenth-century Connecticut woman, but it is also notable for what it does not tell. Intimate details of her personal life, as well as specific dates, names, and places, are often omitted, and little is said about such subjects as child rearing, education, or woman's work. Concerns such as these were included only if they related in some way to her spiritual life. By excluding so much, Heaton could create in her writing a free and solitary self in a direct and personal relationship with God. However, the structuring of the self found at the center of this genre of literature, according to Roger Payne, is ambivalent. On the one hand, the conversion narrative tradition required converts to speak a language of self-negation, while on the other, the more that the autobiographer struggled to speak the expected language, the more self-focused and even self-creative the narrative became. Slowly, the ancient Christian tradition of the embodied soul was giving way to a modern concept of the self as a thinking, experiencing, and more autonomous entity.[47]

Ironically, organized religion, while reinforcing traditional values, contributed to the very process by which those values slowly eroded. Although the initial intent was to discourage frivolity, a woman who was encouraged to be self-disciplined, sober, and reflective could also become a woman who thought for herself and questioned prevailing assumptions.[48] Empowered by her religious beliefs derived from her reading, Hannah Heaton challenged, to the best of her abilities, the hierarchies of power in the household, the church, and the state. Yet she also struggled with the devil, experienced long periods of spiritual drought, and suffered profoundly from loneliness. Her life, Heaton wrote, was "a warfare" [p. 231], where the forces of sin and grace contended. But she is not to be pitied. Her efforts to understand her suffering, her rigorous adherence to ethical standards, and her costly disregard for respectable conformity mark her as an extraordinary woman.

Note to the Reader

Early in the process of preparing this manuscript diary for publication, it became evident that modernizing Hannah Heaton's prose would give a false impression of the original and do substantial disservice to the text. I have therefore stayed as close to the manuscript as feasible. The major change was to create sentences by providing a capital letter at the beginning and a period at the end. The sentences remain faithful to the cadences of Heaton's prose.

The few random capital letters in the early pages have been deleted, and capitalization will be found only at the beginning of sentences. Some orthographic changes were made. The eighteenth-century use of "u" for "v" and "i" for "j" have been changed to the modern practice. Only when referring to God has "the" been changed to "thee." Otherwise, Heaton's variant spelling has been maintained. A small number of hyphens have been removed for consistency. Dashes have been added to set off biblical citations, which are referenced only when they first occur. Following Heaton's practice, a new line is used to indicate a new date or topic, and there is no paragraphing. Heaton's pagination has been maintained, as well as her use of both the word "and" and the ampersand. Brief editorial notes are enclosed in brackets; empty brackets denote omissions in the manuscript; bracketed page numbers refer to the manuscript diary. It is hoped that the reader, after a few pages, will become accustomed to Heaton's stylistic peculiarities and discover in them considerable charm.

The World of Hannah Heaton

1721–1740

[*p. 1*]

I was born i believe in the year 1721 at meacox belonging to south hampton on long island. My fathers name was jonathan cook a chirergeon. He married temperance rogers by whom he had ten children. Six remains now alive whose names are hannah elizabeth phebe daniel jonathan calvin. My father was a religious man. It was his custom to read before prayr and he would tell us children that if he read anything that we did not understand to ask him what it ment and he would stop and tell us. One saturday night father was reading a passage of a godly child how constant he was in prayr. One night he forgot it and got up and kneled down in his shirt and asked god to forgive him the sin of forgetfulnes. He would tell his brothers & sisters if they eat without asking a blessing the food might just lev them. He foretold the day of his death which come to pass. He was glad when it was come.[2] Now i was the oldest child my father had. I believe i was then about ten years old but o this story toock fast hold of my heart. I felt guilty because i had lived without secret prayer. I sat in the corner and cryed hertily but hid it as well as i could but resolved to begin that night to pray in secret which i did. But o how fraid was i of god. He appeard to me to be an angry terrible being. I was [illeg.] for some time but told no body and when i committed sin i was distressed till i had prayed. Then i was easier. At last i prest myself upon thy promise love them that love me & those that seek me shall find me.[3] And when i read about christs death and sufferings & use to pyty him to think how much he underwent for poor sinners i thot i loved him and i seekt him for i prayed to him and so shall certainly find him. Thus i a poor dog cetcth at childrens bread [illeg.]. Not that all the promises were in christ and when i could cry [illeg.] i prayed i was mitily pleased with it and i thot [illeg.] able righteousness. But o a good god did not let me stop for i [illeg.] extremly terrifyed with frightful dreams of danger death and destruction coming and make such ado in my sleep that my father would cetch me up out of bed in his arms and carry me out into the air not knowing what the matter was. Neither could i tell the foundation cause. Now about that time there was talk of war with the papist nations.[4] Father said if they over-come us we must expect fire and faggots except we turn papists. Now i felt great gilt and hearing this it struck me thot i to burn for christ. I dare not. If i turn papist and deny him i must go to hell so i was in dreadful distres in my mind. O i thot to be sure of an intrest in christ was worth a thousand worlds but how to obtain it i knew not for i dont remember that i ever heard a word in them days about conversion work, I knew not that there be such a thing and now my hope would rise and fall. The next [illeg.] or neglected duty about this time i was terrified of the indiens coming over the lake. Death was

[*p. 2*]

terrible thing to me o how fraid was i of going to hell. Now about this time i being i believe about thirteen years old i have lain awake all night for fear of the devil.[5] Near this time them lights was seen often in the north.[6] Sometimes it seemed almost as if there was living shapes of men in the air in arms moving after each other. One night about half the orrisen loockt red like blood. I was in dreadful distres of soul. I was afraid christ was a coming to judgment. My flesh burnt as if i was in a scorching fever. Some said it was a sign of a storm and it was imprest upon my mind that it was. But it was a storm of gods wrath a coming on the wicked world and it seemd as if i stood naked as it were to it. Now i could not live quietly upon my old promise. Sometimes i used to try to comfort myself with that in matthew 24-6—when ye shall hear of wars and rumors of wars see that ye be not troubled for the end is not yet. And if i could keep the day of judgment a great way of i felt easier. I spent much time in reading and use to try to examine and serch my heart & the more i serched the more concerned i was. Sometimes i was afraid i was a hypocrite & did not seek god aright and i had no right to that promise. It would come in my mind not to serch my heart it made me feel so ugly and put it of while you are old or while you are maried That will be a more convenient season. Now you will be dispised among your mates and felt hurred to go to frollicking. But there death would come into my mind so that i could hardly conceal my distres. And when i come home and was got to bed i would lie and cry and think [illeg.] would go no more. And i use to talk to my sister elisabeth that use to [illeg.] with me.

Sometimes i hoped nay thot i was better than she for i [illeg.] and she was not once. I nerrowly escaped of being drowned with several more. Also i had several hard fits of sicknes once in the long fever lay almost two months once so near gone that they were about to close [my] eyes. I think i have heard my father say he watched with me eleven nights going so slept a days because he was loth to trust me to anybody else. Now me thinks i must write glory glory to god that saved me a great sinner. I trust to make a great display of the riches of his free grace. O how did he follow me with his spirit where ever i went when i knew it not nor had a thot that god was at work with me. Once i had a doleful blasphemous thot about christ cast into my mind. I was terrifyed with it and feard it was the unpardonable sin for i concluded that was committed in thot. Sometimes it would dart into my mind with such power it would make me almost cry out. Sometimes i use to get it out of my mind by saying the lords prayer or by thinking of some good thoughts as i calld them. After a while i told a friend of mine of it privately and i was delivered at once. Now i was hurried to pray let me be a doing what i would i must leave it that moment and go and pray or it would not [illeg.]. I was greatly worried. It was not my wants that sent me [illeg.] but a strange hurry and when i had made out a prayer [illeg.].

[*p.* 3]

o poor me i did not then see the snare but i believe the devil cares not how much a person prays if he does but live out nay will hurry them to pray.

1741–1749

[*p. 3, cont.*]

Now after a while i went over to new haven in the fall just before that great work of god began which was in the year 1741. There i heard mr tennant and mr whitefeild[1] preach which awakened me much. Mr white-feild laid down the marks of an unconverted person. O strange it was such preaching as i never heard before. Dont you said he when you are at the house of god long service should be over that your minds may be about your worldly conserns and pleasures. Is it not a wearines to you said he if one days serving god is so wearisom to you. How could you endure to be in heaven with him forever where nothing but praises are. He said if you was carried to heaven in this condition the first prayer you would make would be that yould might go into hell for that would be more agreeable to your natures. O thot i i have found it a wearines to me many a time over and over again. Then i began to think my nature must be changed but how to attain it i knew not. When i was coming from meet-ing to my quarters which was about 6 miles my company began to worry me to sing. I put them of till i feard they would be offended. At last i sang some verses about a contented mi[]. I thot that was better than to sing a song but o they little thot how i felt. It was hard work for me to sing i felt in such distress in my mind but i went to frollicks all winter and stifeld the conviction i had of its being a soul ruining sin I was much for fine cloaths and fashons. In the spring in may i went to middletown to keep election.[2] One of the days while i was there i was at a tavern in a frolick. Then there come in a young man from long island belonging to the soci-ety that i did and told me how the work of god was carried on there and of several of my mates that was converted. My sister elisabeth also sent a letter. I trembled when i read it. She said her soul magnifyed the lord her spirit rejoysed in god her saviour. Her sighs was turned into songs the

frolicks = soul ruining sin ?

comforter is come.[3] I had a strong impression upon my mind to go home which i did in a few days. As soon as i got into my fathers house young people come in and began to talk. Sister elisabeth began to cry over me because i had no intrest in christ. That i wonderd at but the next morning father examined me and i was forst to tell

[*p. 4*]

my experiences as wel as i could. He told me when i had done what a dreadful condition i was in. It took hold of my heart. I kept going to the meetings and was more and more concerned. And o what crying out there was among the people what shall i do to be saved. Now it began to be whispered in my ear it is too late too late you had better hang your self. And when i see a convenient place o how it would strike me. I was afraid to go alone to pray for fear i should see the devil. Once when i was on the ground away alone at prayer trying to give up all to christ in great distres of soul i thot i felt the devil twitch my cloaths. I jumpt up and run in fixed with terror and o how did i look at the winders in the night to see if christ was not coming to judgment. O how i did invi toads or any creature that had no souls to perish eternally. Many a time i kneeled down to pray and my mouth was as it were stopt and i did vent out my anguish with tears and groans and a few broken speches. Now it cut me to think how i had spent my precious time in vanity and sin against god. My not regarding the sabbath no more was bitter to me now. I thot sometime i could be willing to burn in the flames of fire if i could be delivered from the anger of god or appease his wrath that was out against me. Now my heart and soul and all nature was set against nay loathed the way of salvation by christ. And it seemed to me if i should give up all to christ he would send me directly to hell. Sometimes my heart would quarrel with god thus why he knows i cant convert myself why then dont he convert me. Now i thot if i knew of any place on earth where i could hide from god o how would i run to it. But them words was terrible to me. In amos 9 chap read to the 5th verce—tho they dig into hell tho they climb up to heaven tho they hide in the top of carmel tho they be hid in the bottom of the sea &. O how it cut me to think i could not get away from god but appear before him i must and i lived in daily expectation of it. Now sometimes it would be cast into my mind thus you need not be so conserned you are not so great a sinner as some are some have murdered and done dreadful things but you pray and go to meeting and god will not have a heart to send you to hell. This i thot was the devil trying to beat me of. True i had no sence of the justice of god all this while nor could i think what conversion was unles it was this & i fancied it was. I thot a person must be in a sort of trance and be carried to heaven and see wonders there and then be brought back again—but now them words was terror to me

[*p.* 5]

where it says god is angry with the wicked every day and the day cometh that shall burn as an oven and all the proud & them that do wickedly shall be stubble and the day cometh that shall burn them up saith the lord of hosts that it shall leave them neither root nor branch.[4] Some years back i use to pray for many things that i was afraid god would hear and answer them but now i cryed for mercy mercy mercy lord o save me from thy wrath o save me from hell. This my soul wanted. I did not want to go to heaven. I thot i should be tired of singing praises nay i felt a hatred against it and it seemd impossible to me that christ was willing to save me that i could not believe. I was such a loathsome sinner and he such a holy god sometimes i thot i was willing but he was not. I could hear of others finding mercy but o how it would strike me for i feard greatly that while others was taken i should be left. Now the promises in the schriptures was terror to me for i thot they belonged to the children of god. I had no part in them and i felt such an enmity against the way of salvation by christ. I could see no way to escape damnation. Now i began to feel like one lost in the woods. I knew not what to do nor what course to take for my heart began to grow hard. Now i could not cry and pray as before when i thot of hell. It did not terrify me as before it use to. Me thot i envied the very devils for they believed and trembled but i did not. Nothing now semd to help me. I grew worse and worse. I thot it must be a gone case with me and i thot so the more because father never spoke one word to me about my soul in particular as i remember after he first examined me till after i had found comfort which was about three weeaks after. It being in the year 1741 june 20 i was then i suppose in my twentieth year. It was the lord's day. I went to our separate meeting in the school house. They i think read a book of joseph allins[5] but i felt so stupid and hardned and shut up that i could not hear nor keep my mind upon anything. I thot if i could have purchased a world by it i could not shed a tear. Now i feard i was hardend & seald down to damnation with a witnes (jarico[6] was straitly shut up when the walls fell).

[p. 6]

for i had lost all my consern and felt a heart of stone. Meeting being done i got away to go home. I thot i would not go to the night meeting which was to be at thomas sanfords for it would do me no good. I remember in the lot as i went i see strawberries and these thots past through my mind. I may as wel go to picking strawberries now as not its no matter what i do its a gone case with me. I fear i have committed the unpardonable sin and now herdned but as i was going home i considered at last. I turned and went to meeting. Soon after meeting began the power of god come down. Many were crying out the other side of the room what shall i do to be saved. I was immediately moved to pres through the multitude and went to them. A great melting of soul come up on me. I wept bitterly and pleaded hard for mercy mercy. Now i was brought to vew the justice of god due to me for my sin. It made me tremble my knees smote together then i thot of belshezer when he see the hand writing against him.[7] It seemd to me i was a sinking down into hell.[8] I thot the flor i stood on gave way and i was just a going but then i began to resign and as i resind my distres began to go of till i was perfectly easy quiet and calm. I could say lord it is just if i sink in to hell. I felt for a few moments like a creature dead. I was nothing i could do nothing nor i desired nothing. I had not so much as one desire for mercy left me but presently i heard one in the room say seek and you shall find come to me all you that are weary and heavy laden and i will give you rest.[9] I began to feel a thirsting after christ and began to beg for mercy free mercy for jesus sake. Me thot i see jesus with the eyes of my soul stand up in heaven. A lovely god man with his arms open ready to receive me his face was full of smiles he lookt white and ruddy and was just such a saviour as my soul wanted every way suitable for me. O how it melted my heart to think he had been willing all this while to save me but i was not willing which i never believed before now. I cryed from the very heart to think what a tender herted savior i had been refusing how often i turned a deaf ear to his gracious calls and invitations. All that had kept me from him was my will. Jesus appeared altogether lovely to me now.

[p. 7]

My heart went out with love and thankfulness and admiration. I cryed why me lord and leave so many. O what a fulnes was their in christ for others if they would come and give up their all to him. I went about the room and invited people to come to him.

June 20 1741. About nine oclock in the evening in the twentyeth year of my age. I got a way to go home from meeting. It was about a mile but o me thot the moon and stars seemd as if they praisd god with me. It seemd as if i had a new soul & body both. I felt a love to gods children. I thot that night that jesus was a precious jesus. It being late our famile went to bed but i sat up and walked about the chamber. It seemd as if i could not sleep while the heavens was fild with praises and singing. That night i was brought into the lords prayr. Before i was afraid to say it but now it seemd sweet to call god father. Yea my heart could say every word in it. Ah what sweet peace i felt while my mind was swallowed up in that scripture matthew 5 from 9 to the 14. And now methinks i must stop & offer a song of praise and admiration to that god that has done great things for me.[10]

1 The soul that doth a jesus seek; his cries & groans he heard;
hagers son cried under a shrub; and water did appear
2 He hears the ravens when they cry; & will he not hear me
for whom he shed his own hearts blood: & love beyond degree
3 Does god the little sparrows feed: my soul heel feast much more
because for me he suffered all in his purple gore
4 May i now with hannah of old: her soul was in distres
god gave her what she did desire: and what she did request
5 Abide my jesus now with me: and let me feel thy love;
my soul then like a minadab:[11] with swift delight shall move

[*p. 8*]

6 My soul now longs to be with thee: o let me see my christ
my soul shall then be free from sin: when i am in paradise
7 O when shall i eat of those grapes: that grow on ashcols hill
8 O come my jesus take me up that i may have my fill
9 My jesus he is cald a rock: on which my soul does rest;
hees cald the blest emanuel: which my soul does feast
10 Hosanna to the sacred three: with praises singing loud
to him who sits upon the throne: beyond the sterry clouds
11 O land me on the eternal shores: my jesus to behold
those crowned heads that dazel bright: beyond the shining gold
12 O there they are quite free from sin: thave got the victory
ore death and hell ore sorrow sin; triumphing gloriously
13 O the sweet musick that is heard: in heavens courts above

O there they sing most gloriously: of jesus dying love
14 Blest saints do there see eye to eye: in glorious liberty
there i shall be from sinning free: to all eternity
15 Sorrow for sin here almost makes: my heart to rend asunder
there i shall have nothing to do: but praise & love & wonder
16 O there they sing to christ their king: while i sit mourning here
there they rejoyce with heart & voice: while i am prest with fear
17 O glorious jesus how i long: to get fast hold of thee
to twine my heart & never part: threw a vast eternity

[*p.* 9]

18 O love amazing love indeed: o soul aluring love
o love is heaven fild with love: o lovely place indeed
19 O how i long to wing away: to that supernal throng
who swim in seas of boundles joys eternity along
20 O what a blessed happy place: are saints in glory crownd
o there they sing to christ their king: with one eternal sound
21 The wicked there from troubling cease: the weary are at rest
o there they swim in seas of love: and lean on jesus breast
22 Oppression there cant never come: to trouble or molest
there i shall see my jesus dear: and be forever blest
23 There cherubims and ceraphims: stretch forth their charming wings
o there they talk of nothing else: but lovely glorious things
24 In thee i trust & come i must: i long to be with thee
and there to sing to christ my king: to all eternitee
25 behold behold our jesus comes; with beauty in his eyes
to bless his saints here on this earth: and sinners to surprise
26 I long to have an angels tongue: that i might sing aloud
to him who shines upon a throne: above the starry cloud
27 When i lay open to gods wrath he said peace and be still
the blessed dove comes from above: the ollive in her bill
28 Wonder of wonders lovely lord: to set thy love on me
i know the only reason was: thy love alone was free
29 Halaluiah halaluiah: to christ who come to dye
for sinful me to set me free: from sin and misery

Hannah Heaton

[*p. 10*]

Now when i first found comfort i felt as if all my sins was gone. It seemd as if i had not one left and who could have made me believe but that i should feel so always. But o in a short time i began to feel a wicked heart and it was cast into my mind that if i went to heaven i should certainly commit sin and be sent down to hell like the fallen angels. I was in sore distres not being acquainted with those schriptures that proves the perceverance of the saints. Wel i knew not what to do. At last i told my father what distressing thots i had. His answer was: child he whom christ loves he loves to the end. Here the lord helpt me i was distrest with it no more. Soon after i had a sweet time of refreshing come from the presence of the lord. About this time i heard mr pasons[12] of lime preach from them words—i am the way the truth and the life. I was fild with the love of christ. Then i heard young mr jud[13] preach my soul travelled for sinners. In the meantime we believed there was one converted. Then i heard mr youngs[14] of southold preach. I believe the house was fild with the power of god. My heart rejoyced in god my saviour but not long after i got into the dark so that i hated prayer. I felt enmity in my heart against the dealings of god. Now i began to think i was not converted. These thots was cast into my mind that if god would keep in the dark not to pray to him but o free mercy brought me near again to god. I had many sweet revivals but after a while my soul was dejected again and i was talking with some christians about my stupidity. I said i wanted to see hell and see if it would not move me to praise the lord and that night i had a vew in my sleep of hell. Me thot i was by the side of a great mountain and there was a hollow or cave in the mountain. Me thot i see it full of burning flames like a gloing oven. Me thot i see a man in it which i very well knew and the devil in the shape of a great snake all on a

[*p. 11*]

flame with his sting out ran violently at the man and seemd to aim at the mans mouth. I knew he was a wicked prophane man so i awoke with a great sence of the dreadful state of the wicked and wrastled with god in prayer and felt distrest for them that their souls might be converted. O the condecention of the great god towards dust & ashes. Them words was imprest upon my mind behold i come quickly even so come lord jesus.[15] Now I heard mr deavenport[16] of southold preach 11 sermons. I felt the power of gods spirit almost all the while. I believe also the power of god was visible in the assemblyes sinners crying out for mercy for their souls some saints a praising god some exorting and praying over poor sinners. Sometimes mr deavenport would cry out with a great voice above the multitude and say come away come away to the lord jesus. One night after meeting mr deavenport come to me and warned me not to marry an unconverted man and told me them schriptures that speakes against it.[17] Oh poor me here i rebelled as i shall shew by and by (now i began to see more and more that the world flesh and devil was against me) I remember one night i sat by the fire my mother and sister was with me. These words come with power to me—i will shew thee how great things thou shalt suffer for my name sake.[18] I spoke and told them of it that i must suffer for christ. I had sweet vews of the love of god. My heart burnt with love to christ. I felt resind to suffer for jesus. Ah what a glory and lovelynes i see in it. If he stands by to suffer for him me thot was the sweetest of all. The most contrary to nature of all duties so brings more honour to the lord than other graces. They that suffer with him shall rein with him but after a while i got something dejected in my mind. O my cruel sins that pull me from my god. I went mourning till one day i was at meeting and there was some christian indians wonderfully fild. It affected my heart to think the set time to favour the poor heathen was come. O how wofully have they been neglected by us that had the byble but o what have we been better than christian heathen. It made me weep till my bodily strength was weakned. O to vew the condecention of god.

[*p.* 12]

About this time i had many sweet manifestations of the love of god in christ which i thot i should never forget. Now after a while i felt sin rage i felt a wicked proud heart which made me loath myself and weep bitterly. Methinks i was now like the disciples that did not know jesus when he walked on the water to them but thot it was a spirit and cryed out. My soul thirsted for the sweet love of god. I went to a night meting the power of god come down. A great number was struck sinners crying out saints praising and praying. I hant skill nor time to write what a meeting that was. My soul seemd on the mount. I vewd the promist land. Them works was sweet to me—the winter is past the rain is over and gone the flowers appear the time of the singing of birds is come the voice of the turtle is heard in the land.[19] Also this was a good word to me—they that trust in the lord shall be as mount zion which cannot be moved.[20] Glory to my god bless his name o my soul o let me live to thy glory. One day i saw some indians who was under concern my heart was moved for them. Those words come with power to me—behold the fig tree and all the trees when they shoot forth ye know that summer is nigh.[21] Me thot i see the signs of the lovely day of judgment approaching the heathen coming into the kingdom i prayed hard for christles souls. O why me why me lord and leave so many. O vile me and lovely jesus. This lasted not a great while before i was troubled again with sin. It seemd as if i had nothing nor was nothing but sin from the crown of my head to the soles of my feet no clean part in me. I felt a body of death. I would say to myself o my sins that naild jesus to the cursed tree and o my sins that peirce him still my heart was ready to burst with grief. O how many times did i plead and weep before the lord. This grieveing and mourning for sin and for the want of a new pardon seald by the comforter lasted a considerable time till one night i had given to me a spirit of prayr for the comfortable presence

[*p. 13*]

of god. I went to bed. I had vews that night of christs coming to judgment a glorious day a dreadful day. Next day i felt a calm serene mind. Glory to god in the highest for peace on earth and good will towards men.

When ive the helping hand of god: my work goes sweetly on
i travel towards canaans land: where my dear lord is gone
But when he hides his lovely face: to eguept i would go
i think on onians and garlick: that in them parts do grow.

Now i had a temptation to take delight in the things of the world but i found in it no food for the soul. On the sabbath i went to our saparate meeting that was set up when the work of god began in 1741. There the lord shined into my soul. These words was good to me—the lord is my shepperd i shall not want.[22] I had much comfort. I had joy and sorrow mingled. I was grieved for souls out of christ. O if they did but know the lovelynes there is in jesus. Now i use to know in these days the spot where i turned away from god and what it was for and now my deceitful treacherous heart was drawn away again. O my sins that causes the comforter so often to withdraw. But it is by his spirit he shews me my wicked heart and nature or i should never see it. O his faithfulness—i will never leave you nor forsake you.[23] Now i felt sorrow for sin none can tell but they that feel. I went to a night meeting. The children of god was fild and sinners struck with consern. It seemd as if i see a fountain open but i could not get to it. My soul fainted because the lord hid his face from me. It made me cry has thou forgot to be gracious. When i prayed it seemd as if god had no regard to me. Next day i went to meeting to hear mr mills[24] of ripton preach. The whole house was fild with the glory of god. I believe now my soul traveld for sinners but o i was twoo unwatchful. I began in a little time to feel an uncommon stupid mind and was troubled with temptations from satan. Then i began to be afraid i was not a converted. But them words come to me—he whoom he loves he chasteneth.[25] But o i felt enmity in my heart. I had wicked thots. And now it pleased god to visit me with ilnes of body. But o i felt a quarrelling mind. These words was prest upon my mind—satan hath desired to have thee that he may sift thee as wheat. But by and by that come more powerful but i have prayed for thee that thy faith fail not (i have prayed for thee)[26] toock fast

[*p. 14*]

hold of my heart. O now i admired the riches of free grace. I injoyed great comfort in the lord. And while i feasted that about the queen of sheba come to me: the one half was not told me[27] she admired an earthly king but i admired the king of heaven and earth. O how good it was to praise the lord. I lived in this frame for some time (but alas the vanities of the world began to creep in by little and little and as they come in the spirit of god as to the opperation of a comforter withdrew gradually). Now i felt sin strive for victory. I felt a quarreling heart because the lord did not give me strength to withstand the temptations of the devil. Now i hated to pray. It seemd as if god did not hear me. I was distrest with wicked thots not fit to mention. I wondred i was suffered to be out of hell. I was tryed about a falling from grace for i feard i had lost mine. I thot how god repented him of the evil he thot to bring on ahab david and ninevah[28] and it was cast into my mind thus who knows but god repents him of the good he thot to bring on me because i was so wicked. My heart was ready to break with greif. I went alone and poured out my distrest mind to god. That in luke 22 come to me and being in an agony he prayed more earnestly and john 1-48 when thou wast under the fig tre i saw thee it was sweet to me my heart burnt with love to jesus. The tempter fled away. O why me lord and leave so many. I felt a longing for the coming of christ. I could say come lord jesus come quickly. I had lovely vews of the heavenly world in the 5 of revelations. I heard mr meed[29] preach. His text was—without the law sin was dead.[30] I felt greived for sinners a few days after i hear mr meed was coming to preach again. Them words come to

[*p. 15*]

me—where twoo or three are met together in my name and fear there will i be in the midst of them[31] to hear and to bless them and he is faithful that hath promissed and will do it. So i went to meeting. He preacht concerning the prodigal son.[32] I see he was faithful who had promised: for the power of god come down in a wonderful manner. It seemd as if i was on the mount and vewd the promist land. Many of gods children was fild and sinners crying out what shall we do to be saved. My heart was much drawn out for unconverted ministers also. Them words seemd to take hold of my heart—when he beheld the city he wept over it.[33] I talked to sinners and told them of the willingnes i saw in christ to save them but here satan soon come upon me and told me all this is nothing to you you call upon others but you hant closed with christ yourself. Soon after one of my cousens dyed it struck a dread upon me. Death was a terror to me because i feard god would leave me or hide his face and that would be a hell to my soul. I thot how jesus cried when dying my god my god why hast thou forsaken me[34] when his father hid his face. And me thot the devil told me he would plauge [plague] me then as much as he could. But o how did all this send me to god for assurance assurance of his love and favour and if i was deceived that i might know it all lookt dark and dolesome. Now mr meed come to preach again. I went with a heavi heart to meeting ready almost to break. He preacht conserning the ten virgins[35] but i began to feel a hard heart. I was so shut up i could not pray nor mourn nor do nothing. When sermon was ended the opposers made great opposition against the preacher. Then i began to pray my bonds were losed the holy ghost come down in a wonderful manner and fild the house. My fears and temtations al vanisht. The word was sweet to me. I had sensible communion with god a long time after. When i was asleep the folks said they heard me cry out and tell of the love of jesus and pray for sinners. Ah i slept but my heart waked. It is the voice of my beloved that knocketh. I could say with david my tongue doth speak of thy righteousness; and of thy praise all the day long[36] for the salvation of soul. O why me why me and leave others why me & leave fallen angels who never refused an offerd saviour as i have done. O he loves because he will he will love glory to his name. Once i had a vew in these days of the damned suffering the pain of los they have lost friends lost al worldly comforts lost hope lost

[p. 16]

all means lost saints and angels and heaven lost god & christ that is infinitely more than all losses. One smileing word of his is worth more than a thousand worlds. O sinners think of this you will never have your wills one time in hell only to sin & blespheme to eternal ages. Now the next lords day i had a vew of the condecention of christ. Me thot if i had a thousand tongues i should employ them all in praising god. Them words in the 24 psalm come to me—lift up your heads o ye gates and be ye lifted up ye everlasting doors and the king of glory shall come in. But here the devil tried to destroy my peace. He told me he could fill my soul and bring places of schripture to me for he did to jesus. But glory to god i soon got help in a way of prayr to heaven. Some time after i met with outward trouble. I felt a quarrelling mind. I could not submit. I toock the byble cast my eyes first on this word—murmer not as some of them murmered and was destroyed of the destroyer.[37] But o it stopt my mind at once. Stand astonished o heavens & wonder o earth was ever love like this that all my sins will not cause the lord to cease to love me. O make me love the more sanctify me and make me holy for thy honour and name sake amen &. Now it was a trying time with me. Again my heart was shut up o my unwatchfulnes. I loathed myself. I hated my sins but cant get rid of them. I wondred he suffered me to breath in his air. The byble was a seald book to me. Now satan and my wicked heart told me to take comfort in the things of the world. Ah what a proud stubborn heart did i feel it made me cry. O that my head was water and my eyes a fountain. O tears that i may weep day and night for my sins that do so easily beset me. I longed for the time to be delivered from all sin. I did not scruple my intrest in christ but sin was a bitter thing to me for it was against a holy god. I went crying o who shall deliver me from this body of death o when shall i have no will but thine.

[p. 17]

As the hart pants after the water brooks so pants my soul after thee o god.[38] I thirsted after christ. It made me hope he was near. One day i went to the burying place and prayed god to make it take effect on my heart as it had done sometimes but presently my mind was taken with the vanities of the world and i almost forgot i was among the dead. O stupid sinner o who has such a heart as i. Art thou never weary. So i sat down and cried and got me up and went home. But i felt a strugling and striving in my soul. I begged lord search me and try me o let me come forth as gold tryed seven times. Now after i had waded through a long night i went to meeting on the sabbath heavi and sad but just as meeting was done i began to feel a thirsting after christ and presently that word come—he comes leaping up on the mountains skipping upon the hills.[39] It toock hold of my heart with power and sweetnes glory glory to god that the mountains and hils of sin dont keep him from me. I had a sweet vew of christs coming in them words—he comes in the clouds and every eye shall see him & those that peirced him shall mourn & the nations shall wail because of him & put off thy shoes from thy feet for the place where thou standest is holy ground. I stood as it were on pisga and vewed the goodly land labanon.[40] Ah it was a good sabbath to many as weell as to me. I come home with sweet pease in my soul. Glory to god o free grace o condecending love. A few days after i had a sence of the justise of god. I was astonished my heart seemd melted into thankfulness for the great things he had done for my soul. My will seemd sweetly resind to the will of god. I could say anything lord great trials or small thy will be done on earth as it is in heaven. Now i longed for the convertion of ministers especially twoo. I hope the lord has mercy in store for that. In john 14-24 was sweet if ye ask anything in my name i will do it come lord jesus come quickly. My soul longs to see thee face to face but what am i a poor sinful worm. Wonder of heavens stand as tonight o earth at the condecencion of a god o what love is this. I can say now with david psalms 69 let the heavens and earth praise him the sea and everything that moveth therein.

[*p. 18*]

The next sabbath i was low and at night i went away alone and pourd out my complaint to god. But ah he heard me then when i could hardly believe he heard at all. My soul was fild with the love of christ so that my bodily strength was weakned. I was toock and laid on the bed. Places of schripture was lovely to me. I had a vew of the judgment day. I cried lord fold time together and come quickly. It grieved me to the heart to see the rong that sinners did to the dear jesus how they dispised him. I warnd my poor sister and brothers. I told them of the torments in hell they must have if they held on refusing christ and it seemd to take effect upon them glory to god. O why me and leave others o who can utter the mighty acts of the lord or shew forth all his praise. O let me live to thy glory—but alas poor sinful me i let down my watch and my enemies come in like a flood upon me. I felt so disserted and felt such a weit of sin lying on me my very life was a burden to me and to die i was afraid. I thot my heart was worse than the devil. My sins seemd like mountains. I felt a stubborn will and i could not pray but this did not last long. Through rich mercy god sent his word and heald me. It was these (he was in all points tempted like as we are yet without sin) because i live ye shall live also.[41] O what a sutable saviour now had i a saviour that had been tempted and knew how to deliver me a saviour that had dyed for sin to deliver his from all sin a saviour that lives and says i shall live.[42] I had sweet discoveries of the love of christ to my soul. Me thot like mary i could even wash the feet of jesus with tears and wipe them with the hairs of my head.[43] I longed for his coming. I cryed why is thy chariot wheels so long a coming. Make no tarrying o my god geather in thy elect swiftly fold time together and come. O when shall i see thee face to face and know as i am known. O when shall i be delivered from this body of sin and death. O when shall the droos be purged out. O when shall the tin be taken away. O how it greives me to keep

[*p. 19*]

sinning against so good so kind so lovely a god. But this is his way with me. When i see myself undone lost and fit for hell and destruction o then he comes to me with his lovely pardons with healing in his wings. Glory glory glory to his sacred name. It is free all free grace from the foundation to the top stone. I write things as short as i can for want of time. Now i think it was about 1743 one evening i was at prayr wrastling earnestly for myself. Them words came to me—if this be done to the green tree what shall done to the dry.[44] I had a vew of the sufferings of christ and the next morning the news came that my ant abygal rogers was dead. She was a very godly woman she lived in great nearness to god. She got up in the morning and drest herself and was going acroos the room and fell down on the bed and dyed. She loockt pleasant as if she was asleep. They kept her about seven days some of the time in a warmd bed till she began to alter and then she was buried after a sermon was preacht by mr meed from these words—the time is coming and now is when the dead which are in their greaves shall hear the voice of the son of god and those that hear shall live.[45] Lovely young mr meed dyed soon after. Alas if god takes away all his green trees dry trees will burn furiously. O sinners think of this this was a great tryal to me. It made me search and pray and mourn for sin. I believe god had forgiven me but i could not forgive myself. Now i thot of mourning peter that could not make haste to the sepulcher it seems as if he cryed as he went along. Now dying was a terror to me. The devil told me if i prayed he would come and terrify me but i was constant in prayr and found him a lyer. I had some comfort by reading the 51 of isaiah but still my mind was tryed a long time. But when comforting time was come i had such a sence of christs love i could say my lord and my god all lookt sweet and clear i longed to be made holy. I see it was good for me to be afflicted. O if a smile from thee can so heal and comfort a wounded distrest soul what is heaven full of smiles that never will be done no sin is there. After a while i felt my sins strugling again for life pride enmity unwillingnes to submit to gods will and selfishnes. I thot nobody was afflicted with such

[p. 20]

a heart as i had at this time. I was afflicted a long time with sore eyes and a stupid mind and i was tryed with the wicked. One day i toock the byble and opened to these words—he that toucheth you toucheth the apple of mine eye.[46] I was so fild with the love of jesus that these things did not now so much as touch my heart while i was feasting on the fatted calf. It lasted twoo or three days but alas i did not watch as i should and the cares of the world got hold on me. God withdrew from me and in this dark frame of mind i marryed to an unconverted man conterary i believe to the mind and will of god. Now i wanted god to teach me and i sought to him but then i went and toock my own counsil altho them words (in isaiah 8-21—and they shall pass through it hardly bestead and hungry and when they are hungry they shall fret themselves and curse their king and their god and loock upward). I say them words one day was like a clap of thunder to me. I was afrighted a little while but went forward. But o i soon see this scripture fulfilling. I was plunged into dreadful tryals of many kinds that i little expected twoo much to write. Now my husband and i lived in his fathers family a year after we was maried. Here my jonathan was born. I was in travel from the evening extreem bad till next day at noon. Almost was gone from all of my being delivered but them words come to me with power—i will be with you in six troubles yea in seven.[47] I spoke and told them of it but the midwife chid me and said she did not love to hear folks talk so or to that purpose. But soon after their fears and tears was done away for i was delivered. Now there was wicked practices in the family & i did reprove them and the family was soon set against me and we was turned out and no portion given us as the rest of the children had. We had not so much as a house to put our heads in. One of the neighbours toock us in and i a stranger away from all my freinds and o the false stories that was told about me i pray god to forgive them.

[*p. 21*]

Lord grant that my marrying may be a warning to all gods children. Remember what peter got by going to the high priests hall without warrent or call. God commanded the children of israel not to make marriages with other nations lest it was a snare to them god forbid.[48] Their communing with the nations and oulandish woman caused solomon to sin. They was a snare to him and drew him away after their gods. Ah what befel dynah for her visiting the daughters of the land.[49] A warning for all christians to be in the mind and will of god in visiting as well as marrying. Again god says be not unequally yocked with unbelievers for what communion hath light with darknes or what concord hath christ with beliel. And how can twoo walk together except they are agreed and agreed they cannot be while one loves christ and the other hates him and his. Read genesis 24 see how lovely isaac marryed abraham tells elezer god will send his angel before thee & see how he prospered being in the mind and will of god. I have often thot what running their was after eleazer met rebekah at the weell. Now rebekahs father & brother as well as abraham knew the mind and will of god as to this match and when eleazer had brought rebekah within sight of isaac lo he was not dressing up himself tho so rich nor fixing up his house but had been to the well lahairoi.[50] It seems his soul was so fild with the love of god he must go again into the field to meditate. It seems as if the eyes of his mind was shut as to all worldly things for it says verse 63 he lift up his eyes and saw the cammels were coming. It seems the bride was in the same temper of mind with the bridesgoom for she lift up her eyes and when she saw isaac she lighted down and covered herself with a vail to shew him reverence and subjection.[51] She was in her place & he in his for it says he loved her. Ah blessed couple this is the fruit of marrying in and for the lord. I believe isaac was a type of christ. Eleazer holds forth ministers of christ

[*p. 22*]

in his going to court a bride for his masters son but this is not all. I believe it is for a blessed pattern left for christians to observe in marrying. I believe abrahams marrying his sister and israels marrying their own relations holds forth to all believers now to marry none but believers & i believe such a thing may be known now as well as ever namely the perticuler persons that god has alotted for them. Isaac did not marry for fancy but was in the mind of god. It seems he never had seen his rebekah till she come to him. Read ruth 3 chapter in the 4 verce. Neomy tels ruth that boaz would tell her what she shall do and in the last verce she tells her sit still my daugher that is wait on god. Loock to him for the man will not be at rest untill he have performed the thing. And how was the mind and will of god uppermost in faithful eleazers heart. Altho he come from his journey weary & hungry and thirsty tis likely yet he would not eat nor drink till he had done his arrant. O be like him you that are godly parents. Be faithful and prayrful and i dout not you might know as weel as abraham and neomy who your children should marry. Much more might be said but i want time. I can truly say i love my husband nor would i if it were lawful change him for the best man on earth. But alas the yoak is unequal. But i hope for him from them words the lords hand is not shortned that it cant save & they come to me twise (once with power) when i had been begging for him. I believe now that i have prayd for him ayming wholly at the glory of God with thee i leave it. O my god yet make my house a house of god and every soul in it a tent for the holy ghost to dwell in for thy honour sake amen.

[*p. 23*]

But to return to what i was upon about my tryals. Soon after i married things was laid to my charge which i knew not god knows. The chief maker of these lies dyed senseles in an auful manner and was in dreadful horror and distres before he was senceles. O god let the rest be turned to god. And now i was exceeding weeakly in body under much pain but o my jesus was gone. I felt alone on the earth now. I had no husband that could know what aild my soul. There was none to take me by the hand no nehemiah[52] to go before me. I found one woman i could talk with but she was full of complainings twoo. Me thot i was like heman[53] while i suffer thy terrors i am distracted. I was so drowned in sorrow a great while i longed to loose my senses and prayed that i might: that i need not know what sorrow was. Ah wicked me and o my sins lookt like mountains twoo great to be forgiven. But o when all seemd to be lost & no helper found one morning just as it was light i was in bed this word was spoke to me with great power—father i will that those that thou has given me be with me where i am that they may behold my glory.[54] They was lovely words to me. Methot i knew jesus then prayed for me for i was with him and did behold his glory & his spirit was in me a comforter. I felt the power of his resurrection. All guilt & darknes vanisht. O how lovely then was a new pardon seald to my soul. O the smiles i see in jesus his face. I longed for the tongue of an angel that i might speak forth his praise. O this is bethel the gate of heaven god is in this place & i knew i knew it not. Ah what was all the reproach and lies of the wicked to me now when i knew the dreadful god of elijah was on my side. Glory glory to god who condecends to commune with me dust and ashes. Why me lord why me and leave so many. Why not devils rather than me who never abused the riches of free grace as i have done. But o if the lord will stoop to love a creature and commune with it who can hender him. Ah he loves because

[*p. 24*]

he will love o sweet jesus—and now o lord sanctify me by thy truth conform me to thy self let me live like one that has ben with jesus. I lived among cruel opposers but jesus was a friend that stuck closer than a brother while i was with him. I went in the strength of this meat i believe forty days and nights. O let me feed always till i get to horeb the mount of god. There i can sit down & rest me & dry me

Oations of sorrow can not quench: loves everlasting fire
tho sin oppose christ i have close: and i must still admire.

And now prepare o my soul for greater trials than ever. Now i felt after a while sensible withdrawings of the lord from me. He did not leave me at once but by degrees by little & little. Now when i had lost sight of christ many outward troubles come on me but my cries was to god. I use to sit up a nights after my family was abed & go away and on my knees on the ground pleading for mercy and sometimes i have lain on the ground wrastling & crying for help. I seemd as if i was cast out of gods sight and he in anger had shut out my cries. I had doleful thots cast into my mind twoo bad to be mentioned. One was that god was not a present help in time of trouble. One night as i was walking up and down without a glimps of comfort these words was some support to me—this is the time of jacobs trouble but he shall be saved out of it.[55] I believed i was a child of god a jacob and i would be delivered out of this dolesome condition.

Sometimes i could see a fountain open but i could not go to it. At times it would come into my mind how i knew the schriptures to be the word of god and my mind was so dark i could not tell. Then i was also tryed about christ springing from the loins of david as was promised for he was begotten by the

[*p.* 25]

holy ghost and i knew no account of his mother springing from davids loines. I cant tell to none what i underwent (i forget how long i waded with this trial i believe about 2 years and i was confirmed while at prayr but was confirmd at last by that in revelations 22-16 god had said it and it was anuf). I was tried also about what paul says in corinthians 1 epistle 7 chap i speak by permission not of commandment and i give my judgment. And i think i have the spirit of god and thot i how do i know but all he said was so. Ah he thot he had the spirit of god he was not sure of it. Alas how was my mind wayed. Ah now i had no husband nor friend to help my mind. I remember once them words was good to me thou god seest me. But then again unbelief and my wicked heart did say as gideon judges 6 chap 13 ver. Sometimes sin would rage in me like a lion ready to devour me. At this time i had an impression upon my mind of another great tryal that was coming on me. What it was i knew not. Now i was so unsubjected to god in the death of my children that i thot if they dyed i should dye twoo. But god soon shewed me otherwise for my babe was toock with the fever and flux and dyed in about 3 days. It was a daughter a year & half old wanting 8 days. When i see it was agoing i went away to the barn to pray with a trembling body and a terrifyed soul. O how hard i praid for the life of the child more than for a resind will to god. Ah poor me o how unfit was i now to meet god in such a tryal. Now my child dyed about noon on saturday and was buryed the same day about sundown i being persuaded to it by a neighbour that so i might not be hindred from going to meeting on the sabbath. But o how was i worried about it afterwards. It would dart into my soul your child was not cold it may be who knows but it might have come twoo again you have read & heard of such things but now it is twoo late

[*p.* 26]

ah it was a heavi tryal to me. And now i scrupled my intrest in christ because god did not hear my prayrs for the life of my child. But one night i was in bed it come to my mind how moses prayed (deuteronomy 3-25) to go into canaan but god would not let him so i thot i might yet be a child of god. But o i quarreled with god in my heart for taking away my child altho i believe i had a promise for its soul. Before it was born i was poor weeak and low full of fears. One night i was going into bed them words come to me sweetly—i will be a god to thee and to thy seed[56] and i believed it. And that day morning it dyed. That in the first of john 2 chap 12 verce it seemd good to me a few moments but my proud heart would not be still but ran further away from god. One great aggrevation more i had which added to my sorrow that day my child was taken sick. I washt its lower parts in almost cold water and it immediately was struck into a violent fever and when my adversary[57] was angry he would tell me i had murdered the child and my mind got so at last i thot so twoo. Here i sunk almost into dispair but all twoo late i mournd bitterly that i did it without consideration. Lord let me never do so no more. Now about 8 months after this child dyed i had a son born and when it was about 3 weeks old it dyed suddenly. At first i mournd bitterly but god soon stilld me by sending the ague in my head and teeth. I was in extreem misery and i believed it was for my wicked carriage under the rod. It lasted a long time till i knew not and they that was with me but thot i was a dying. Now i pleaded hard with God. I told him if he would

[*p. 27*]

give me ease and strength i would never mourn for my children no more and a little after the pain went of in a moment and i have not had it since.

"the ague in my head and teeth"

1750–1759

[*p. 27, cont.*]

It is now above a year and god gave me in some measure a thankful heart and has enabled me in some measure to perform the vow i in anguish made. Now after this i felt in great darknes. The terrors of god did drink up my spirits. I felt an absent god. I went one day to see a wicked man that lay a dying. His loocks was awful and terrifying. He seemd senseles and was speechles. Now as soon as i see him it struck me. The terrors of death get hold on me. It seemd as if satan told me the lord will leave you when death comes in my hands and then ile pay you for all. And sin and guilt and unbelief joynd with him. I cant tell with my tongue nor pen what i underwent for monthes together. It seemd as if i see death in everything what ever i did. Where ever i went death paines and dying groans did piearce my heart. It seemd sometimes as if i heard them. I was not then afraid to be dead but afraid to die. O that rent between soul and body was my torment and terror and i not knowing it was a temptation prayed for some time that i might never dye or that i might live till the day of judgment and be changed in a moment. I did not see then but that it was lawful to pray so my darknes and distres was so great. Now my body was weakned and i seemd almost fit for the grave. I had so lost the love of god that heaven seemd a mournful molloncolly place and god seemd to be a terrible angry majesty. Ah poor wicked me. Now when i was in this dolesome case no eye to pyty nor arm to help one morning as i awaked that word came with power to me glory seemd to be in it. A soul ravishing voice said (take thine only son isaack and go in to the land of moriah & there offer him up[1]). The words followed me & kept sounding in my mind and they was opened to me thus—that i was to take my child and go over to long island and see the christians and my parents

there that i had not seen in about eight years. I was much tryed about going. I feard we should be drowned agoing over
Year 1751

[p. 28]

the water or i thot it may be my onely son must dye there and so i am to offer him up but i dare do no other but go let the lord do what he would with me. Now my husband being very willing went with us to guilford and there left us. In a few days we went aboard the wind blew exceeding hard. My child cryed and was terrifyed with fear he was then 7 years old. But o what faith strength and courag did them words bring into my soul—he holds the wind in his fists and the waters in the hollow of his hand.[2] We got to shore twenty miles from my fathers house but o the kindnes them strangers shewed to me. Three invited us to stay at their houses. That night i hired a horse and we sat away next morning and got safe to my fathers before night. My parents seemd to rejoyce to see me in some measure like jacob to see his joseph genesis ch 46-29-30. I soon told father what distres i was in about dying and i told him if he heard i was dead to rejoyce that it was over. He said the devil often makes a hell of death and the grave to torment believers in. My parents talked to me but all to no purpose. Now gods work was going on at meacox on the sabbath. I went to meeting but o how grieved was the christians to see me in such a case. Many seemd to be fild with the love of god giving glory to his name. There was some christian indiens fild i believe with the love of god. O how lovely they prayed and praised god. Now while i was looking hearing & admiring this word toock fast hold of my heart with power—lift up your heads o ye gates & be ye lift up ye everlasting doors & the

[p. 29]

king of glory shall come in.[3] My soul now was like the chariots of aminadab. O how lovely now christ appeared to me lovely in himself lovely in his death & sufferings for my soul lovely at the right hand of his father crownd with glory & i believed i should in a little while be with him and be like him and see him as he is and o my ingratitude. I looked on him whoom i had pearsed and mourned and admired free mercy. Now methot i gave all up and made no reserve. Perfect love cast out fear[4] them terrible fears of death was gone. I longed to be with my god. Methot my heart was one of them gates the door was opened the king of glory was come in. O lovely jesus o free grace i cannot i cannot admire thee anuf.

After staying a month i toock my leave of my dear parents and part of the church after father had prayed and commended us to god: i felt an impression on my mind i should see them no more. That was fresh in my mind acts 20 chap 37–38 ver—they all wept sore and fell on pauls neck and kisst him sorrowing most of all because they should see his face no more. Now i felt hard parting. I felt such a unitednes to that people in the lord but that word was strength to me and helpt me such—god is in the midst of her she shall not be moved.[5] Ah how it melted my heart. I believed god was with me his spirit in me and so i come home to north haven with strength in the lord. Now i knew what i was bid to go to moriah for but oh what a dearth did i see in the world and security among my neighbors. I exammined and talked to them. Some would bare it but some would be offended and ridiculed me. But o how did i long for the salvation of their souls. Now god opened to me the 34 of ezekiel. Every word in that chapter seemd to go through my soul. I went and talked with mr stiles[6] our standing minister. He said if he was perswaded man could do nothing towards his own salvation he would never preach no more and many such like things he said. I thot i see him and

[*p. 30*]

his church and their conduct to be out of the rules in gods word as plain as i could see the sun in a clear day. Them words come to me—pertake not of her sins lest you pertake of her plauges.[7] And i knew to my sorrow that babylons plauges was hardnes of heart and blindnes of mind and so i dare not joyn with them no more but withdrew and never heard mr stiles no more.

This was in the year 1751. Now for a year and more i lived above the world in some measure. Many confirmations god gave me about the schriptures that they were his own unfailable unchangable words of truth. Now they was a lamp to my feet & a light to my path teaching and leading me into conformity to the mind and will of god. They was the men of my council. I felt the power of them from day to day. Ah in these days i use to keep the byble by me when i was at work so that i might often have a feast of reading. And o how did the spirit of god accompany it. I use to break forth often into praise thanksgiving and prayr. Me thot i see all the schriptures point to jesus christ and his kingdom. My god opened to me the 4 of mica and i see by many of the prophesies and by the manner of gods working and by his providences that the glorious day was near. That there shall be a new heaven and a new earth where in shall dweell righteousness.[8] In the 3 of zephaniah god says the preists have polluted the sanctuary they have done violence to the law but yet god promises to turn to the people a pure

[*p. 31*]

language and they shall all know him from the greatest to the least. In that chapter god promises to take away the dreadful judgment of false teachers. God says he will take away out of the midst of them that rejoice in thy pride. And o what glory then does he speak of. I will gather them that are sorrowful for the solemn assembly to whoom the reproach of it was a burden. Do read carefully the whole chapter. Also that it was sweet to me i will gather them like the sheaves in to the floor. Also that was powerfull in my mind i will be sought to by the house of israel for these things and i hant said to the house jacob seek ye me in vain (and to confirm all to me) he told me the word that was gone out of his mouth should not return void but accomplish the thing whereto he sent it by these schriptures. I see a work of god acoming in north haven and the lord gave me a great spirit of prayr for it at the same time so that my body would almost faint sometimes. After i had been wrastling with god to carry on his work here sometimes i use to go and ly on the bed till i recovered strength of body. Again that word in judges 5 & 7 verse use to run in my mind. The inhabitants of the viliges ceased. They ceased in israel until that i deborah[9] arose that i arose a mother in israel. I believed i should bring forth spiritual children in this place tho as to outward appearance all loockt dark as midnight But i see christs day as abraham and was glad with such joy i cant expres. When i was asleep some nights i had sensible communion with christ & when i awaked i was still with god. O what glories did i see in the upper world which made me cry out in my sleep sometimes o my god. And i use to sing praises in my sleep anights and awake my husband. Above a year i lived in the sight of christs coming to set up his kingdom in the world

[*p. 32*]

i see it by faith through the word. One lords day them words was life to my soul—they that wait for me shall never be ashamed and blessed is she that believed for she shall see a performance of those things which was told her from the lord.[10] Methot i sat in some measure on the watchtower a loocking for my god to carry on his work through the earth but especially in north haven[11]—by faith i see a saparate flock of god gathered and a little young teacher given them by god. I saw so plain i was not afraid to tell some of it at that time. I cried lord make me willing to wait thy time and not by unbelief and impatience cause thee to never let me see thy salvation as israel did who dyed in the wildernes and never went in to the promised land. About this time i was threatned by some of being had up for not going to meeting. But these words was good to me and comfortable—all men forsook me but thou did stand by me.[12] And o i see god was agoing to get to himself a great name and that was better to me than i can expres. I remember one sabbath god was pleased to lay a great croos upon me to try me & he withdrew his smiling face. O what sorrow i felt. None can tell but those that feel. I went crying and wringing my hands away into the woods alone. There on the ground did i wrastle in prayr. Methot i felt like israel by the red sea twoo mountains one on each side the enemy behind the sea before. O dreadful and o blessed methot like them the heavens was open to hear prayr. My sorrow lasted some hours then that word come softly—

[*p. 33*]

tho i sit in darknes i shall see light.[13] Then i believed i should have my jesus again and it was but a little while before that word in isaiah 53 took hold of my mind—he shall see the travil of his soul and be satisfied. Now i rejoyced again to vew the latter day glory that jesus himself should be satisfied. O how great then must that glory be. O with what pleasure could i think of the advancment of the redeemers kingdom and i knew it must be built on the ruins of ante christ. My soul and flesh felt the love of god. I longed to see jesus in heaven face to face. I said come lord jesus come quickly to judgment fold time together and come. But ah vile me and sweet jesus. O if thou wilt love a vile loathsome sinner who can hinder thee. O how comforting was that in isaiah 26-1—in that day shall this song be sung we have a strong city salvation will god appoint for walls and bullwarks also sing o earth rejoice & be glad for the lord will do great things.[14] About this time I was led to see converted ministers how they was turned back for christians had been making idols of them. Now me thinks they like acan have stole a weedge of gold and a babylonish garment and have dissembled (joshua 7). I mean joind with the world and preach for money. O lord uncover them to all mortals for thy honour sake. About this time the grand jureman come and told me if i did not go to meeting he would carry in the complaint to his meeting.[15] He asked me also how i knew mr stiles was not converted. I told him of them

Malachi 3 ?

words in malacai 3 chapter last verce. I told him my mind also about many things. I told him there was not a word in the byble to uphold him in having anybody up for not going to meeting. I told him my fears about his condition. Now this sent me to my god o how good a god have i glory to his name.

[*p. 34*]

North haven 1752 the coppy of a letter to a freind. Dear sir grace mercy and peace be multiplied to you through our dear lord jesus. Dear sir i feel my heart united to you in the bonds of love & fellowship of the spirit. Press forward dear brother. I believe the lord is with you. Keep cloose to your gloryous captain and nothing can hurt you. He will make your shoes iron & brass[16] &. O be conformed to god for these is none that goes a warefare for the lord upon his own charges. Fear not scoffing ishmaels. Remember abraham made a feast that day isaac was weeaned and your heavenly father will make you a feast of fat things when you are weand from the world and self. Although now gods children are devided and scattered by reason of false shepherds ezekial 34 but the lord will deliver his flock. O lift up your head for our redemption draws nigh and when a few more storms have passed over our weary souls we shall all be gathered at the mount of joy joy where the wicked cease from troubling. All sin will be gone all tears wipte away the weary will be at rest. O blessed rest when we shall see that blessed land that we have so often admired at a distance. We shall see that blessed face that was more merred than any mans. Here he had nowhere to lay his head. O sweet jesus the morning star of heaven—though wee seem black to strangers yet the kings daughter is all glorious within through christ. Yet a little while we shall assuredly enter into

[*p. 35*]

glory we shall be made white in the blood of the lamb. We shall have the society of angels prophets and dear myrters that seald the truth of this reproacht religion with their blood. Yet a little while and wee shall know that glory dwells in emanuels land. O astonishing that god should condecend to let us love him. Stand astonisht o heavens wonder o earth that he most of all (that god) should love us worms of the dust. O he loves because he will love & if he will love us who shall hinder him. Dear brother when you are near to god do not forget poor josephs and micaiahs[17] that are in prison and yet have great liberty when jesus is near. O that god would bring zion out of babylon and babylon out of zion. O when shall ante christ be destroyed and christs kingdom be built on its ruins. Let us not be ashamed to be called saparates. Moses begs for it exodus 33-16. There is glory and lovelynes in such a saparation. O brother comfort your soul jesus is a going to set up his kingdom. That little cloud like a mans hand that elijah see but quickly there was a sound of abundance of rain & in a little while the mountain of the lords house will be establisht in the tops of the mountains and exalted above the hills. The stone that the builders now refuse shall become the head of the corner. Christ is a going to reign king on earth. In the second of daniel we read of a stone that was cut out without hands and smote the great image and breake his feet to peices and the stone that smote the image became a great mountain and filled the whole earth. O make haste come lovely god is it not time for the work for the wicked

[p. 36]

have made void thy law and are setting up laws of their own inventing. O make haste gracious god and let us sing that song babylon the great is fallen is fallen that made all nations drink of the wine of her abominations. O jesus build up thy kingdom on her ruins and o that gods children might love god supreemly and one another fervently for love is the fulfilling of the law. By this says christ shall all men know that you are my disciples if you have love to one another. While i write i feel something of it as wee are members of one body one in Christ and one in one another. O let us abound in this love o that our souls may be fild with love to Christ for what he is in himself i beg we may see that gloryous house built the cyty whose name is the lord is there ezekiel last. Ah when shall the church be beautiful as tirza[18] as jerusalem and to loock forth like the morning fair as the moon clear as the sun terrible as an army with banners. O brother that you may travil in birth for souls and bring forth many spiritual children. I believe i have felt your prayrs for me since i was at your house. I am now more establisht in the truth than ever i was. O who could have thot it after so great shakings as you know i had. Now i can lean on jesus bosom and solace my soul in his love. O that he holds me up would make me to stand. O that i had an angels tongue to loudly sing his praise. But o the milk and honey is beyond this wilderness

[p. 37]

Sir i long to see you to tell you what glories i have seen in the 34 of ezekiel and in these words sing o earth rejoice and be glad for the lord will do great things and they that wait for me shall never be ashamed. I believe the lord will in his own time carry on a glorious work in north haven. It seems to me sometimes as if it would be the chief seat of christs kingdom. O methinks i rejoice and am glad like abraham to see christs day. Tho now all around are asleep and secure but i believe there will be a true church of christ here in this place. Dear brother remember afflictions are in the covenant. Yet a little while and we shant be put no more upon the exercise of patience but shall reap the blessed fruit. Then farewell faith and hope and welcome love love. You must not expect to fare better than your dear lord. He was belyed reproacht spit on mockt schorged crucifyed. O take follow this sweet pattarn of patience and love meekness humility resignation to his fathers will and remember brother it is said of christ that he shall see of the travil of his soul and shall be satisfyed. O what tongue can tell how great that glory will be that will satisfy jesus christ. Lord fold time together and come. O blessed redeemer gather in thy elect and o i can through free grace say come lord jesus to judgment. The very meditation of that day is anough to vanish ones heart. O i long to see that face that is white as the light a coming for our salvation. I see it a coming that christ will gather his floock like the sheaves in to the floor. O may god give me patience to wait yet i must say come lord jesus come quickly amen & amen. Hannah Heaton

[*p. 38*]

But to return. Now my heart went out to god in prayer for this poor grand jury man that he might be converted. Now i felt sweet peace and comfort in my soul for i know god had taught me to saparate from their idolitrous religion. I could now awake in the night and sing praise to my god and he that finds a place for the swallows to lay their young will find a place for me to meet with his children in his name and fear. For he has told me he will gather them like the sheaves in to the floor.[19] O how i long to meet with the assemblies of gods saints in his own appointed way. How much i have done in begging that mercy and how much time i have spent in pleading for it is not meet for me to declare. The day of the lord will declare that matter just as it is altho i am told often by the enemies of god they wonder i dont want to go to meeting if i am a good woman. Tho i tell them their gods are there my god is not there but they know not what i mean. It is just such a time now with christians as it was with israel when moses was gone up into the mount and delayed to come down.[20] Some were worshipping gods of silver and gold and some are worshipping ministers. O thou dreadful god of elijah that answerest by fire come with a spirit of burning as thou has promised and burn up all the idols that thy children have made. O let them no more play the harlot with other lovers. O come & thaw this frosen world amen.

[*p. 39*]

One morning my husband rode out to go to new haven expecting to be at home before or at night for he never use to stay out all night but now he did. Now i was in great distres concluding he was drownded acoming over the ferry. I kept watch at the door to see if somebody was not acoming to tell me he was dead. Now i went to my god and pleaded for a resind will let the case be how it would. I prayed also that if he was alive that god would tell me so i went to bed and went to sleep. I hope in some measure resind. In the morning i awaked with these words in my heart—when he see the wagons that joseph had sent the heart of jacob their father revived.[21] Then i told my little boy his father would come home again which he did a few hours after and brought the comforts of life home with him to his family. Now my heart went out to god with thankfulnes and praises glory to god he is a prayer hearing god. That word was lovely to me—i the lord water it and watch it every moment lest any hurt it.[22] O thy love and care to my soul. O let me now live to thee o sanctify me by thy truth o shall it not be jesus has prayed for it. Do lord pyty my dear husband and child. Pray give them an intrest in thy love. O that my house may yet have family prayr in it constantly. I must now practice family prayr whenever my husband is out of the way i mean not at home. Lord help me for christ sake. About this time i dreamd i was up a loft in a new building. I stood in great danger of falling but presently i spyed a great serpent swelld with spite and madnes against me and presently it jumpt at me and was fastned on my left hand.[23] I was extreamly terrifyed but i held fast with my right hand and tryed to

[*p. 40*]

fling it off with all my might but i could not. At last i see it had got no head. Its head was gone but i was still in a great fright and awaked with this word in my mouth—gracious is the lord and righteous yea our god is merciful.[24] Then i told my husband some great tryal was coming up on me which soon begun & now methinks i can interpret my dream thus. My being up in a new building is the grace of god in my heart but yet i am always in danger of falling into sin. The serpant that i saw is the devil swelld with spite at me. The serpant having no head why the seed of the woman has broke it. His sting is masht. My holding with my right hand means the hand of faith takes fast hold of christ and will never let go its hold and my trying to get rid of the snake but i could not teaches me i cant get rid of devil and sin in my own strength. And now i loock for trouble but hope i shall win through. O lord help me to honour thee. Now i soon began to feel sin rage in me. My sins lookt bigger than the sins of a whole country or nation. I was tempted to believe jesus never dyed for sinners and that god was not a prayr hearing god & how do you know says satan that you are guilty of adams sin altho i had then such a sence of the sin of my heart and nature. Now these tryals was so heavi upon me my body sometimes was almost overcome. Also it was prest upon my mind that my husband would die in a few days and whither this was from god or

[*p. 41*]

from satan i knew not but i was worried with it for some time. But i kept praying and crying to god for a resind will wholly to his will and that i might have his presence that was better than life and that i might know that christ was formd in his soul before he dyed. One day when i was under this trial i lit of a woman that i believed was a christian. I was sorely afraid to tell her but at last i did and it left me at once. Then i knew it was the devil but i was worried with this that there never was no jesus. I did sleep but little. My body was weeak but my cries was to god day and night. One day my distres was so great i cried out my father my heavenly father do help or i die. O help me and immediately the load was taken of. I felt easy & quiet. Ah good god to me but it was not long before the distres come again. I was so faint sometimes that i could not sit scarcely. I remember once i was wringing my hands groaning in anguish. Then words come to me—your eyes shall see your teachers.[25] Lord i thank thee o make me willing to wait thy time. Tho all looks dark to sense there is i believe not one minister within twenty miles of me that is settled in gospel order. They i think preach for money take christles persons into their churches of man.

September 1 1752. Now a good god often gives me an open door of exces to the throne of grace. I freely pour out my soul into his bosom. Now i kept a day of fasting and prayr alone for myself & my husband and child for north haven that god would destroy my heart sins. O sanctify me to god for many promises for north haven. God will yet work here and i hope my son will be an isaac go into the gospel field to meditate that word incouraged me once a few years ago for my husband. The lords hand is not shortend that it cant save nor his ear heavi that he cant hear.[26] That word come to me once when i had been praying.

[*p. 42*]

but o i know not but the lord will get glory in the damnation of my husband. But this i believe that i have prayed for him ayming wholly at the honour and glory of god saparate from self and selfishnes. Here i leave him lord save save his precious soul from the lowest hell. Lord i long to hear him praise thee for the abounding riches of thy free grace and mercy to him lord. I long to hear him tell his merry companions that thou art a lovely and a great forgiver—but alas his practice is worse & worse. He seems more & more hardned and careles. Lord dont forget the groans & cries i have been sending up to thee for him without ceasing this seven and twenty years. I have sometimes had this thot past through my mind that god cant convert him he is so stout and stubborn. Ah my god let me not dispair but keep praying. It may be thou wilt hear the last prayr i shall ever put up in time for him. I cant get a promise for him. O i think of the man that went to get loaves in the night & the woman that went to jesus for her daughter and toock no denial.[27]

1769 March [*sic*] (Now i sent another letter to my friends at the iseland. Lord do bles it or all is in vain). Once i was very poor and weeak. I beged god to send me somebody to help me and there soon come in a poor garl and offerd to help me and did a fortnit. Once i was very il i prayd for somebody to do for me. Presently

[*p. 43*]

there come in a woman and did for me o he is a prayr hearing god. Dost thou take knowledge of me seeing i am a stranger ruth 2-10. Now i sat apart another day for humiliation and prayr for myself my husband and child for north haven. To be secret i toock my byble went into the woods not far from the house. I pleaded with god that he lets the birds build their nests where they chuse even thine alters o god and thou dost allow thy enemies to meet where they chuse. I cryed how long lord wilt thou sit in heaven and see the purchase of thy own blood sit solitary as a widow alone. O how i longed for meetings with the saints in gods own appointed way. I had liberty in prayr but presently this was darted into my mind. Your prayrs are selfish god wont hear them. It struck me for a while but that come to my mind about nehemiah desiring of the king that he might go & build the broken walls of jerusalem. Gods cause lay so near his heart it made his countenance sad & he prayed to the god of heaven nehemiah 2. I thot of that in daniel 9 his setting his face to seek to god for jerusalem. I see i have at times been stript of self in praying for myself and mine and this place. The glory and honour of god was my all and my soul has been swallowed up in him and in him alone. Now at night that word was whispered in my ear ye did run well who did hinder you. Then i thot satan had been trying to beat me off from prayr. Ye did run well seemd to give me new courage to pres forward. He has said to me they that wait for me shal never be ashamed and i expect to see his salvation. I have been weeakly about ten years and when i am brought near to god in

[*p. 44*]

prayr i have sometimes been so overcome that i am fain to go and ly on the bed panting for breath ready to faint for some time and now like the men of succoth[28] i am taught with the thorns and bryers of the wildernes. O the snares that is layd for my soul but i am often struck silent I married not in and for the Lord but o a good god causes him to feed me as he did the ravens to feed elijah.[29] And i have thot it was as great a marricle but ah now i often look back to long iseland and think of them dear meetings and christians that i come away and left. But ah here i am alone like a sparrow on the house top.[30] Here is none to guide me of all the sons that i have brought forth in my spiritual whoredom. O the strong corruptions of my heart. And nature often sends me with great arrants to the throne of grace and cry as jehoshaphat did o my god wilt thou not judge them for we have no might against this great company that cometh against us neither know we what to do but our eyes are upon thee this day.[31] I longed incessently to take my farewel of this world by reason of in-dwelling sin in me. Indulged sin i see in others. O sin cursed sin through christ i have made war with thee & shall have the victory at last i trust. Upon the lords day i was low and full of sorrow. I could not pray. The heavens seemd shut up. This word come sweetly—ye are worth more than

[*p. 45*]

many sparrows the very heairs of your head are numbred.[32] O how good is gods word when he sends it. And now i met with a great tryal from a wicked person. It lasted for some time till i was just redy to give out and think god regarded not my cries. I went crying and wringing my hands. At last that word come to my mind—moses endured seeing him that is invisable.[33] Then i believed it was for tryal of my faith a rod to whip me to the throne of grace and i hope i shall endure to the end. And presently the person was turnd about and made a great acknowledgment to me and ownd he was greatly to blame. Ah it is just i should reap the same that i sowed o how still god sometimes makes me. This will be soon over it is not being in hell loct down in chains of eternal dispair. I deserve to be there true it is. Was i conformd wholly to the world i believe i could live quietly and very comfortably with my husband for i know he loves me but where the power of religion is in some measure kept up esau will be against jacob: ishmael mock isaac: cain hate abel: when enmity governs.[34] A few days ago my husband rode down to the ferry. There he met with a company of our neighbours and their wives who went down in a cannoo to see the fishing & to have a frollick. Poor man was one with them. When he came home at night o how his enmity raged at me. He said other men had wives to go with them but i have none

[*p.* 46]

and many wicked hard words which i forbair to write. But i held my peace. Then next morning i talked to him how i abhored such company where god was so dishonoured. I had rather be shut up in our dark celler all day alone. Ah there i could pray and meditate. But now he held his peace and heard me and said nothing. I have in cold weather gone and stayd in the celler to get rid of vain company when i knew i must not reprove these tryals. I write to shew what fruit my marrying brings forth. O take warning dear saints. Also tho my house is not so with god i trust he has made with me an everlasting covenant. Now i went to hear mr bird[35] of new haven but i thot there was an absent god.

1753. I sent a letter to mr frothingham of middletown[36] to ask the lord to let him come over to this macedonia and help us[37] for this people was wholly given up to idolatry. And a few months after he came and preacht at our house from matthew 5 begining at the 3 verce to the 12. I believe god helpt him and people seemd to hear. I believe he was the first ordained saparate minister that ever i heard. It was in the year 1753. Afterwards i had a sweet sence of the joys of heaven. Then i sung o sweet jesus them words came to me with power. For the devisions of rubin there were great thoughts of heart there were great searchings of heart judges 5 debora and barak sung these words after god had wrought deliverance for israel. I believe this was to

[p. 47]

shew me that great devisions & tryals was coming among gods children. But o i have laid up that promice more choise than a peice of gold. I will gather them & it will be fulfild throu the world and in perticuler in this place. I trust tho all is dark now but faith in gods word is light and life. Nehemiah vewed the walls in the night. Now some christians opposed me for not joyning with mr bird & going to hear him no more. But i thot how elial opposed david thot it was his pride and nautines altho he went in obedience to his father and was then in the mind and will of gods samuel 1 boock 17–28. He said is there not a cause that chapter has been sweet to me but i am in a hurre the lord is going to do great things i trust by striplings. For the solomons of this day are gone after other gods and i believe the lord is about to rend the kingdom from them and give it to their servants. Now in these days i got again into a very dull lifele frame. O my wicked heart. Now i went away to pray. O i found sin anuf to confes. O methot i was the worst sinner out of hell. O how i longed for free grace and presently the lord come near in these words and in that day they are often repeated in the short prophesies. O how lovely was them words and christ in them and in that day. O how clear i see now that christ should reign king on earth. In that day o come quickly lord take to thyself thy great power and reign. O how near faith brings it glory to god.

1753. I dreamed a vile snake was on my back. Methot i tryed to get it of but could not. Then i beged my husband to help me so i awaked in a fright. Then i thot a tryal

[p. 48]

was coming which soon came upon me. It was this i was extremely provoked by a wicked person. I bore it for some time but at last i got angry and spoke wickedly. O wo wo is me because i am one of unclean lips and i dwell with them of unclean lips. Now i prayed & cryed to god till my strength was almost gone. I found no where to go but to that god that jacob went to when esau vowed his death. Many nights i slept but little. Also i had a great longing for the conversion of the dear person[38] whose enmity raged against me. I got so near to god in one prayr that i thot i should have gotten a promise for the dear creature. Me thot i will not let thee go till i have a promise. I sayd lord it will be for thy glory thy justice will be satisfyed heaven and hearth will rejoice thou wilt rejoyce. Pray do that: that will make thyself glad. But by and by i was puld away as it ware from his mercy seat by my unbelieving heart. O sad sad i got no promise. Now this trouble was not toock of till i was willing to lye at their feet and be trampled on as it were. I see that fulfiled soft words turneth away wrath. God gave me so to condecend that i hugged the person in my arms and kist them tho they seemd then to hate me. Now when i was brought to this there was a sweet pease flowd into my soul. It seemd god was pleasd with me and forgave me. O sweet peace that god gives glory to his name.

[*p. 49*]

Now the persons anger was soon appeasd and acknowledged greatly their sin. I see much of god in this tryal. O my God wherein i have done amis help me to do so no more. Glory to god he doth all things weell. Now i think it is right for me to write down the remercable providences of god.

In the year 1753 the 22 of november. Just as the day broke this word was born into my mind trouble cometh out of the north and i told my husband of it but knew not what was coming. But a few hours after i heard an amazing loud noice in the north like a great gun. It lasted for some time before the sound was gone. People all round us heard it with surprise. Some said they see a streeam of fire some thot it was an earthquake and feard some place was sunk but i believe it was an alaram or sign of war with the northern french and indiens which soon began. But now how did i pray to my god to fit me for what was acoming. Prepare me o lord for tribulation make my shoes iron and brass. As my days are so let my strength be. Thy word and spirit has born witnes to me that bonds and afflictions abide me but these things dont move me when the blessed comforter is with me. O sanctify me make me holy then i shall be happy.

Now a little after i got into a stupid frame more to be dreaded by a christian than death itself. I could not meditate read nor pray with any delight nay i felt an avertion to it. O the athism i felt and enmity in my heart. I could not plead for the cause of god to none altho i believed all the while i should go to heaven. O i am no more happy than i am holy. Now one day i went to prayr and had two schriptures

[*p. 50*]

opened to me. One was concerning samsons taking honey out of the lions carcase his eating himself and giving of it to his father and mother.[39] This i believe holds forth to us that when christians have got through their lyon like tryals and begin to reap the peacyble fruit they will not only eat themselves but feed others. O now i hope the dear day is near. The other was about paul and silas being in the inner prison their feet fast in the stocks at midnight.[40] It seems the enimies of god thot now i have got them fast its all up with their religion now. But o how soon did god make them sing & bring them out and o methinks i see christians now in the same case in the nature of it. They are in the inner prison i say their inner man their souls are in prison by sin and unbelief gods enemies. And o methinks it is midnight. With a witnes o methinks i see their dear feet fast in the stocks. But o methot i had a lovely sight of the almighty power of god. He can make us all sing in this midnight and open the prison doors. And he can take of the oppressive laws of man in new england and will i believe in his own time. Come lord jesus come quickly.

Now about this time i heard the news of twoo poor indians that was in new haven prison for murdering an indian. I had a great sence of the distinguishing goodnes of god towards me. For some time i see i was as liable to commit murder as they if god did not keep me. O his tender love and care towards me a loathsome sinner. O this is free grace that seems to sweeten every cup put into my hand.

[*p. 51*]

again i had vew of my wicked heart. O my pride and my rechednes i am all sin. I toock my byble and went into the barn and daniel last last seemd to help me. O i hope at the end of these days of tribulation i shall rest and stand in my lot in christs church. Now my lot is prayr. It has seemd to me sometimes that i have stood in some measure as moses[41] did between the living god and dead sinners and he hant said to the seed of jacob seek ye me in vain. I hope to see the returns lord jesus come quickly. Now i was sorely afflicted with selfishnes o loathsome self that wont be content unles it rides sides by side with jesus christ. Saith a holy devine it is as easy to pull the marrow out of our bones as self out of our hearts. O dead dog self how are souls cheated with it. God has also of late been shewing me my ignorance of him. The more i see and know of god the more i see my ignorance of him. God is a sea without a shore or bottom which cannot be fathomed by men or angels. O lovely glorious pure blessed holy happy emence being. O when shall i be in heaven with thee and know thee even as i am known and see thee as thou art. Then o farewel farewel sin—forever. Yesterday i had a feeling sence of the captivity of zion. I had liberty to come near to god to plead. I loved and longed god would appear

& plead his own cause take his own work into his own hands. O thou that made time fold it together and come. Lazarus dyed when jesus was gone come lord and call thy lazaruses now and they will come forth they are bound hand and foot and their grave cloaths about them loose them and let them go.[42]

[*p.* 52]

1753. The day before christmas i had the impression of a tryal that was coming upon me a great wait was on my soul. I told my husband that some trouble was coming but what i knew not and that very night brother noah heatons house was burnt. Near neighbour to us ebenezer finch and his family lived in it. About 2 oclock in the night the woman awaked and the house was al on a flame. There was 5 of them they all got out alive tho some of their heads was singed as they got out of the window their bodies was almost naked and all their goods was burnt. And i knew not for an hour or two but that they was all burnt in the house for they went away to one of the neighbours. In the morning that word came to my mind—he that gives to the poor lends to the lord[43] and i was enabled freely to give them and o what a sence i had of the distinguishing goodnes of god to me and mine that we was spared when methot we deserved it more than they. And now the terror of fire got hold of my mind so that i could hardly sleep anights for sometime but the lord helpt me. Now i dreamd a few nights ago that these fiery tryals the saints are under now are the last they ever shall have here. But we have a surer word of prophesy that tells us the day is neer the glorious day. Now i had liberty in prayr and i had a sence of the tender love and care of the lord towards me which made me cry o ye angels and glorifyed saints praise him for me ah you can do it without sin

[*p.* 53]

while i can not. Glory to my god for free grace but soon after this impatience toock hold of my mind. I found it hard work to wait for the accomplishment of the promises that i have before written concerning the work of god in this place and my having gods appointed means of grace. I got so at last i thot i could wait no longer. I said within myself i must give out i must give out i must sink and die i fear if god dont appear. O thot i if i could but see or hear the least sign or motion that god had began his work o how my heart would leap for joy. But it is death within and death all around me. Now i cryed that i ever come in to north haven. Why did i leave my dear long isleland but thinking of one of my former promises stild my mind. It past through my mind thus that if i waited for god i should not be ashamed but if i did not wait he might justly make me ashamed. Then i thot i must wait still and the lord gave strength to my resolutions and got me in a waiting frame. Again o his goodnes to me a sinner o my ingratitude towards the lord. O i have been weary awaiting for god for 2 years. O how did he wait upon me almost 20 years before i would receive him for my saviour—

Now i met with a tryal. My husband heard the grand juryman would carry in a complaint against me for not going to meeting. He was bitter against me and thot it was my will that i could joyn with none of our standing ministers. But it sent me to my god that he would give me his own appointed means of grace & reading about the myrters helpt me.

[*p.* 54]

The copy of a letter which i writ to my three sisters at newark in the jerseys married there. 1753. Dear brethren and sisters. My hearts desire and prayr to god for you is that you all may know jesus christ and feel the power of his resurrection. Remember death is coming as fast as time can bring. It dont rest with the form of religion. If you do you will sertainly be lost eternally as sure as god is in heaven. Perhaps you may think you dont rest in a form for you are concernd and distrest for your sins. Ah remember so was esau and sought carefully. Pray dont rest here. Remember what the angel said to mary he is not here but is risen. O seek you a risen jesus and when you have found him if you live to him you will like the samaritan woman[44] invite your neighbours to this lovely jesus. O that you may know now in time the worth of a christ. O that the everlasting arms may take hold of your souls for himself. Abraham made a feast when isaac was weand and when you are weand from self and converted to christ your heavenly father will make you a feast of fat things. Do think of the prodigal when he returned. The father ran to meet him to put the ring of everlasting love upon him and the best robe the righteousnes of christ. O amazing love. He also fell on his neck and kist him a seal to the day of redemtion and o what musick was there in his fathers house

[*p.* 55]

joy in heaven and we may well think in his heart twoo now a tent for the holy ghost to dwell in: dear sisters if you will come to christ no doubt you will have as kind a welcome as he for the spirit and the bride says

come. Ah my dear sisters i write to you all as one. If you are on a self righteous foundation you are in the most danger of perishing of any in the world. Christ come not to call the righteous but sinners and they that can heal themselves will not go to the physition. O think and tremble o that jesus that made clay and opened blind eyes would bles these lines to you tho written with much indisposition of body. Perhaps this is the last time i shal ever write to you who have reason to think i am near to the grave and heaven. There i shall see him that had nowhere to lay his head when here ah it was crownd with thorns for sinners. O sinners o heaven my happy home which is my country dear. What cause have i to long for thee & beg with many a tear. Dear hearts i want to tell you but cant what god had done for my soul of late. Could i see you here methinks it would rejoice me like as josephs wagons did jacob. I see a rain of righteousnes acoming it looks near. I see it by the spirit of god opening his word to me. God says his secret is with them that fear him. The blessed jesus has promised to send the spirit of truth and shew them things to come. Abraham rejoysed to see christs day tho many hundred years before jesus come yet he saw it and was glad. O my dear sisters i pray you may be brought to have communion with christ here that i may meet you in heaven at gods right hand never to part there. All sin will be gone every vessel will be full of love and glory there. I shall have no weeakly body to clog the soul there. I shall see jesus who groand and dyed for us. We shal be like him & see him as he is.
Hannah Heaton

[*p. 56*]

Now i had sight of the sins of my heart and nature. I felt sin anuf in me to make a hell. I felt pride rage. I felt tuche apt to be angry. I felt bound hand and foot and thrown into the prison of sin. The schriptures are seald & loct up from me. I feel a heart against prayr. Now i wanted always to be alone mourning and lamenting for my sins. Sometimes i wonder i am not in hell then. I am still but o this is a hell in the soul to be disserted of god. But one day i thot i could live no longer. I cryed out with a flood of tears. O great jehovah has thou disserted thine own cause that thou has left me to be trampled upon by my sins and carryed captive. Lord art thou a man of war then do appear and fight for me. Now when i had pourd out my soul to him my burden was lightened. But alas it soon came on me again and i felt no might against the great company that was come against me. Now the wicked say so would we have it. I thot i was so wicked that god would not forgive me now without bringing some dreadful thing upon me. That word use to strike me to the heart—tho moses and samuel stand before me yet my heart can not be towards you. I thot god was weary of forgiving me so many times as he had done. Horror overwhelmed me gods arrows stuck fast in me the poison therof drink up my spirits. Altho i then believed god would bring me to heaven but that alone dont make me happy for i am no more happy than i am holy. I wanted to have sin destroyed in me & to have a new pardon seald to my soul by the blood of jesus.

[*p.* 57]

Now i was in this dolesome horrible pit and miery clay a great while but one night i went to prayr but found an absent god. I could not pray at all. I grew worse and worse. When i come out of my closet my child come to me with a book in his hand sent to me by one of the neighbors. I felt my heart affected. I begged god to bless it to me. I opened it and found it to be mr jeames jeneways[45] heaven upon earth or the best freind in the worst of times. I had not read long before i felt a heaven in my soul. Methinks i come near to god that night in prayr on my knees in the cow yard. On the ground i had liberty to beg against my predominant sin. O the love of christ how it fild my soul. I long to say farewel world and sin. O the raptures of joy i felt. I come near to plead for my husband. I told the lord i could not be denied a blessing i must have. I had leave to beg for my child. I begged he might be like jonathan[46] of old. O let him taste of the honey that his eyes may be inlightned. Glory to god i believe he will be saved. Now i have writ on page 27 how god bid me offer him up at long iseland. Now my understanding was so unfruitful that i did not know that i offerd him up then. I did not think of it altho i knew i offerd up myself and all that i had without the least reserve (o here i must hope for my husband twoo) but alas i knew not in what way god bid me offer my son up. But about half a year after i had been behind the barn pleading for my child. When i was going to the house i thot i knew not whither the lord would have mercy on him or not but in a moment i heard as it were a voice saying to me (if i did not intend to have mercy on your child why did i bid you offer it up). There seemd to be a power and glory in these words oh what peace and satisfaction i had. O unmeasurable goodness that thou should be a god to me & to my seed. In the night i awaked with heaven in my soul. That pashige was sweet to me in samuel 2 book 19 ch

[*p. 57 sic*]

about berzilai and chimhams being invited to go to jerusalem with king david.[47] O shall my young chimham go with me to live with king jesus. And o it must be mercy itself that can pardon such sins as mine. O it is love with a witnes that can imbrace such a loathsome sinner as i am. O what a wonder it is that the earth has not opened and swallowed me up. Ah vile me and sweet jesus. Now i longed for the day of judgment a joyful day to the saints o lord make me holy then i shall be always happy. Now i see a glory in the 31 of proverbs from the 10 ver to the end. O the glories of the church when they live to god and trade with heaven. Then her husband jesus will be known in the gates when he sitteth with the elders of the land. I see such a fulnes in this chapter it seems whole voloms might be written from it but i must forbaer. I can get but little time to write having no woman kind with me in my family. At this time I was athreatned by the grand juryman for not going to meting no more but it sent me to god at this time. One day all at once i was fild with a great longing for the convertion of my husband. I was constrained to pray loud for him. It seemd as if i took hold of the lord for him but he laught at me and said he beleived i was bewitcht. Presently i began to listen to him and it withdrew. But god was gracious. I did not feel guilt for it but after a while i was very ill in body and the king of terrors got hold on me. Again i think i am one of them that through the fear of

[*p. 58*]

death are all their life subject to bondage. O jesus fill me with thy love then i can go over dry shod. O my god how the devil threattens to pay me for all at death. Lord i deserve to be left in his hands then but has thou not said i will not leave nor forsake you. Lord let this petition cry to thee that i have now writ to my last breath. O my doleful wicked unbelieving heart that cant trust god to deal out to me as he pleases at death when i have given up all to him many and many a time. I trust god be mercyful to me a sinner. Now some sabbaths i feel lonesome at home & i cant get my heart joined to mr bird. I fear god hant sent him forth to preach. O lord make good thy word to me on which thou has caused me to hope. When lord shall i hear the sound of thy goings as on the tops of the mulberry trees.[48] O that i could hear the voice of the angel among the myrtle trees. O how it loocks to me to see sinners going so merrily in the road to hell. How do they incourage one another. How they sing laugh &

dance as if they had the best company while i sit mourning alone for myself and them lord help me to [illeg.]. O let my life tell them to their faces they lie that i have the best company to make one chearful in life and at death chearful and glad i can witnes to it. True religion is not a maloncolly thing. O sinners what made the dear myrters sing in the fire holy caynam when his arms & legs was half burnt said he felt no more pain than if he was in a bed of roses. What made paul and silas sing in prison when their backs was raw. What made stevens face shine like an angel.[49] O sinners it is i it is i that sin and force the lord to

[p. 59]

withdraw the comforter and then i go mourning. But i find a pleasure in mourning for sin. But this i know that all the comforts and pleasures on earth are just nothing compared to that of walking in the path of holines. The way is a way of pleasentnes & all her oaths are peace. I tryed the ways of sin nineteen year before i knew the lord and i found every peasure had its poison twoo & every sweet a snare. O blessed jesus come quickly & give the kingdom to the saints. O i long i long for that day. Now one night i dreamed there was a terrible earthquake and the house was aturning over and i cryed to god for mercy for me and mine and so awaked in a fright. Again i dreamd i see a snake i had a weapon and went to kill it but the serpant come at me with great rage. But i see it had no power to hurt me. Soon after them words in joshua 10 come to me—sun stand thou still. Then i believed god was agoing to withdraw and great tryals was coming on me but i hope to win threw. My dreams seemd a little to help me for the earthquake nor snake want alowed to hurt me but dissertion and death soon come on apace. My sins and corruptions ralled up their forces. Now i was afraid i should be destroyed. The devil and unbelief worryed me greatly to give out. I was in a kind of a dispair. O says satan & unbelief you have been praying & striving against sin these many years & you hant got the victory over one corruption yet. Nay the power

[p. 60]

of sin is stronger than ever. It is vain for you to seek the lord & walk mournfully before god give out give out. I say this was so powerful for some time on my mind i was at the point to give out. None but god and my own soul knows what i underwent. I could not hide it from my poor husband. O he is stumbled at it. O i am now apt to be angry lord help me or i shall sink. O i think now god is agoing to wean me from the milk and draw me from the breast it is hard work. O when i come to heaven i shall be as holy as god can make me but o i groan i cry i long to be holy here if there never was no heaven for me to go to. O destroy my sins or take me out of the way of sinners by death. Ah how fraid have i been of dying but now my heart sins are worse to me than death. O my god i have been crying 13 years to thee and hant got the victory yet over one corruption. O thou man of war great captain help help me my heart akes. I kept a day of fasting and prayr alone lord give me patience & submission to wait thy time for santification. Now about this time i went to a neighbor in travel. In her distres she said lord have mercy on me a sinner i deserve no mercy but said the midwife yes you do. I asked her if she thot there was a creature on earth that deserved mercy but she said nothing. Lord pyty make all the midwifes like the hebrew ones. This day i had a little glimps of my righteousnes being in heaven. I see cause of joy that it was not in my own hands but it was like peters sheet[50] soon gone. Now i lit of john bunyans grace abounding[51] & it afected me much to think i had been weary awaiting for god 2 years when he lay in prison 12 years and a half in gods cause.

[*p. 61*]

I had fellowship with him in his tryals & temptations altho not exactly like mine yet in the nature alike. One temptation now comes to my mind. In years back i use to be worryed to make away with myself when sudden trouble come on me and when my little daughter dyed satans tone was go hang yourself go hang yourself and when i saw a convenient place i could hardly tell how to keep from it sometimes. But o i use to faall to prayr when it come and in that spot i believe the snare was broke and i delivered glory to god pray pray o saints when tempted resist the devil & he will flee from you. Also when my little daughter dyed i thot i had murdered it as much as if i had cut off its head for it was my sins that caused the lord to take it away. So it was i from first to last and my washing of it as you may read (in page 26) without consideration was my sin & it dyed in 3 days and a half. One day i remember i was lamenting roaring & crying. I wondred in my heart why the athority did not come and take me to prison and put me to death for murder for i thot i deserved it as much as ever any did in this world nay i thot some times i was willing. Now i lay bound under this tryal about nine months till i thot i must have lost my sences.[52] I had none on earth i durst complain to with freedom and ah my godly parents now towards a hundred miles off & the sea between.

[*p. 62*]

I had no nehemiah to go before me but o then a kind god saw me when sinking in dispair and his own arm of power toock off the load and sent the devil away i believe a shamed and relieved me in body and mind at once (loock page 26). And afterwards i see it was my looking in time of tryal to second causes and not eying the hand of god here satan got advantage over me. Take warning all afflicted ones labour faithfully to see gods hand and subject to it lest you smart like me. And now again i was sorely afflicted with carnal worldly thots in prayr. This was the tone leave of leave of says satan you have prayed anuf your worldly conserns suffers while you pray so much pray no longer. But now the truth and faithfulnes of a god that cannot lye bears me up from sinking when storms are all around me. But i fear i am now agetting into a legal spirit for sometime i feel sin lie heavi on me and i feel guilt. Ah the thunder & lightning of sinai made a holy moses excedingly to fear & quake. I have got a little help from hebrews 12—read from the 18 verce to the 25—o you that are out of christ what will you do when you shall feel to your eternal amazement and astonishment & confusion the breach of this holy law. Lord shew the world what a holy law thou hast made & they have broken it for cursed is he that continues not in all things written in the boook of the law to do them. O god grant that through the law they may be dead to the law that they may live to god do get and read john bunyans book on the twoo covenants[53] a choise book.

One day there come an old man a church member of mr stileses and talked severely to me for not going to meeting. He said i must not expect a

[*p. 63*]

blessing in the chimney corner at meeting was the place and much more he said. I begged him not to rest in the condition that he was in. I told him he would never be paid for going to meeting. I believed the gospel was not preacht there. I believed mr stiles was a blind guide leading the blind.[54] He said he believed mr stiles was a good man for in many a drouth he would pray earnestly for rayn and it would sertainly come. I told him of hager and ishmael[55] & that god sometimes grants the desires of the wicked to ripen them for judgment. He gave them their desire but sent leannes into their souls. O god do convert him altho old in sin & unbelief. But alas i fear to hear any of our standing ministers for they have joynd a confederacy with the world when god says there shall be none.

Now i think of dyner what befell her for visiting the daughters of the land so i must stay at home.[56] Yet i believe god is truth itself and will be faithful to his promises. When his time is come i shall have his own appointed means of grace i believe. O god come quickly o how i long for meetings. I am learning now to live by faith. My sweet feelings seems to be going but i believe my righteousnes is in heaven perfect and compleat before god. And so my goodnes and good frames cant make my righteous no better nor my wickednes & bad frames cant make my righteousnes no worse. I have been twoo much like a little child when learning to go its father leads it by the hand then it can go and loves its father but when its father lets go its hand

[*p. 64*]

then down if falls & cries and is offended and thinks its father loves it not. O how hard it is for a child to be weand from the milk and drawn from the breast. O how will it run to its father and cry dadde dadde. Lord let me live by faith and not by sence.

But now i hant seen sun nor stars in many days. I think often of pauls voige at sea acts 27. I believe some since paul have taken just such a voige in a spiritual sence been in the same dangers and escaped safe to land: this day god gave me to spread the doleful state of Zion before him. O how i longed for their deliverance out of babylon. O methinks i with nehemiah can vew the broken walls in the night. Lord stir up the spirit of a cyrus o send a nehemiah to build the wall altho wee are far one from another and there is much rubbish. The strength of the barers of burdens is decayed so that we are not able to build the wall. Lord come quickly. O there is death all around me and much within but i hope against hope for the outmaking of gods promises. Now i find my way from the cyty of destruction to the new jerusalem is right against the world the flesh and the devil. I think often on the two milk kine that was to carry the ark of god into another country and leave their calves shut up at home they went loing as they went. O Lord appear for thy honour sake—

Now me thinks i must give a short account of a tryal I had for leaning on creatures. I had a neighbour with whoom for many years we toock sweet council

[*p. 65*]

together in discoursing of religion. I believe she is a christian. We seemd to see eye to eye in the things of the kingdom. I loved her & she seemd to recken of me but alas she lift up her heal against me. It was thus. My husband had a brother that was a minister setled in the standing order & by chance i had heard in the family that this minister as he had dône before was now agoing to send to another minister (living 56 miles from horse neck) to hire a great pack of notes to read over to his people and they call it preaching & as i have been informed since it is a custom among blind guides. But now i felt burdned with this private cheating soules to death and instead of being faithful to this minister about this thing i went and told my neighbour charging her to tell none of it. She promised me not to tell none of it not so much as her own husband but about 5 years after she told her husband of it and broke her promise and he went directly and told the minister of it and now my husband the minister & others was very angry at me. But i think i had some measure of meeknes and i was in some measure faithful to my neighbour. She owned her fault for braking her promis and being unfriendly to me but never asked me to forgive her. And this was a trial to me and by turns a bar between my soul & hers i cant feel free towards her as i use to. O this sent me

[*p. 66*]

many times to the throne of grace with groans & cries that god would make my proud heart to overlook it and unite me to her by his spirit. Ah gods word is a brother offended is harder to be won than a strong cyty. O my god let me never lean on creatures no more. Oh help me to lean on thee to trust in thee for in thee i can not be straitned and thou wilt never deceive me. O sanctify me by all thy providences—Now about this time i saw my dream fulfilled about my house turning over (in page 59). It seemd to be imprest on my mind for some time that i should hear of my father Cooks death or some of his family. I told my husband what i expected. Now it drove me to god many a time begging of him to prepare me to hear the news that i might lay my head in his bosom who sits on the waves. And soon after i had a letter that brought news my father was dead he dyed in fabruary 1754. He was from home took in an instant with the num palsey. He never spoke one word more he lived a week & dyed. I had the news in the morning & i spent part of the day in secret prayr for myself my mother & brothers & sisters & the place where my father had lived that god would raise up some there to stand as faithful witnesses in my fathers room. O that the spirit of elijah might rest on elisha.[57] I thot god gave me freedom to come near to plead that day with him. Them words was upon my mind i will sing of mercy & judgment to thee o god will i sing.

[*p. 67*]

Now i see i had been leaning and depending twoo much upon my father his prayrs his spiritual letters he use to send me & he was i believe the instrument of my conversion. Ah methinks god is taking away all my props that i have leand on. O lord bless my mother with the choisest of heavens blessings. O that she may now stand in fathers room and be leader & teacher in her family. O make her house still a house of prayr make her a mother in israel and o god be the widows husband & the orphans father for jesus sake. Lord i have given them all to thee let me leave them with thee. Now about 2 days after i had heard of my fathers death the devil come upon me and fastned this in my mind that there was no heaven and that my fathers body and soul was now in the grave together. I cant tell the distress i was in my body was so weeak i was ready to faint sometimes. But them words come in to my mind but i wanted faith to believe them—because i live you shall live also & i am the god of abraham the god of isaac & the god of jacob he is not a god of the dead but of the living & that about moses & elias being with jesus at his transfiguration.[58] But then this was cast into my mind how do you know that the byble is the word of god. Then i said thus have i not felt the power of gods word in my heart many a time. But ah says the temptor

[*p. 68*]

should you not feel the power of a wonderful history that told of great and terrible things done in forrain countries & then the king of terrors death was set before me naked. O how was i worried night and day. Now says the devil why will you sleep or eat or injoy any other comfort of life when you know in a short time you will feel with a witnes what it is to die. I say my very soul was prest under these temptations. Another i had also at this time that i shall not now mention. One day there come in one of my neighbours and seeing me sad she asked me what aild me. I told her my father was dead. I had heard this woman had formerly known him. O says she how happy he is to what wee be. Ah thot i—i can see nothing but that his soul and body is in the grave together and jesus and heaven was now wholly hid from me. The word of god was seald up from me. I had no faith to take hold of one word in the byble. I felt just ready to give up taking any care of my family and ah i had now no companion to tell these tryals to none to take me by the hand altho they might easyly see some great wait hung on my mind and by turns my nature works strongly about my fathers roting in the grave. O i think of them dear hands and that sweet smiling face that use to be so often reaching up to heaven in prayr & praise is now consuming in the ground. Ah must worms eat that dear body here. I cry & think of josephs falling on his fathers face & he kist him when dead & now i am so fild with unbelief and athism i cannot see & believe that god loves and will keep the dust of his saints and make up that very body at the resurrection day. It was often cast into my mind that god is a deceiver he will never

[*p. 69*]

do so weel for his children as he has promised it raily takes hold of my mind. O the anguish of my mind o the anguish of my soul god is hid from me. It seems my case is singuler & i feel helples. I cannot get out of this horrible pit & miery clay. Ah my godly father has done writing to me. Now i cant no more write my tryals to him that he may pray for me. I have heard my father tell that he was converted when young & seald in the year 1741 in the time of that great shower. He seemd to have the treasures of gods word opened to him and he and others thot he had a call to preach. But tryals coming on he did not come up to the help of the lord in that way but use to exhort & expound upon the schripturs & one day while he was aspeaking from the 7 chapter of john he was took speechles and never spoke no more. This was a sore tryal to me. God grant this may be a warning to all christians to come forward and improve the perticuler gift that god has given to them for the building up of zion.

But to return now i found i could not give out but i followed god day by night altho sometimes i could not pray nor feel any desire to be delivered out of this dolesome spot. Sometimes i was in such anguish i thot i would have changed with almost any creature. I thot o that i had never been born woo is me my mother that thou has born me. Now i wanted always to be alone crying groaning & ringing my hands and the wicked seemd now to be stired up to afflict me. One

[*p. 70*]

night i took the byble and lit on the 45 of isaiah. I thot it was a wonderful chapter especially the 17 verce. I had a little glimps of the everlasting love of god in them words ye shall not be ashamed nor confounded world without end. I went quietly to sleep that night but in the morning i felt myself much as i did before. Now i dont make dreams a foundation of my faith but we read in the word that some was to dream dreams and to see visions in the latter days & in job 33-15 in a dream in a vision of the night when deep sleep falleth upon man. In slumbrings upon the bed then he openeth the ears of man & sealeth their instruction. One night this verce past through my mind at once numbers 24-24 and ships shall come from chittimes coast that shall anoy asher & when theave fild their measure up ile ante christ destroy. And the next night i dreamd this methot i see a company of poor people & that they come up from egypt and come the way by the red sea. And i dreamd i looked beyond them in the road where they come and it was a strait and narrow road & methot the sand by the red sea loockt white like snow & that this company was the children of israel. O how lovely they lookt and while i was gazing on them my soul was fild with glory & methot i went to prayr with them and while i was giving glory to god for that he had brought them safely up from egypt by way of the red sea i awaked and hoped i should see the glory of god in the land of the living. Now after this i felt again like one shut up in prison and left alone. I felt such a weit on my soul i was ready to sink but are i was aware

[*p. 71*]

there come a shower of longing down into my soul for christs coming. I was out of doors but o how did i look up to heaven with these words come lord jesus come quickly come lord jesus come quickly to judgment. I cant expres how i longed to see them lovely chariots come rolling down the skies. But at once this was dashed into my mind somebody will see or hear you and think you are beside yourself. But glory to god it did not stop my mind for i longed to see my jesus. Ah now i knew there was a heaven & a jesus glory to my god for this revelation of jesus and heaven to my soul o let me live thy praise. Now i was taken ill but i felt willing to die of this ilnes if it was gods will. Methot i was willing to trust god to do with me just as he pleasd.

And now i think i must write a line or twoo of my tryals when i have been like to have my children for i always then was exceedingly weeakly & almost over born with vapours.[59] And now this was always satans time to try to tempt me to sin against god by holding up gastly death before me and on purpose twoo afflict me. This always has been the devils tone at such times—you deserve to undergo more than ever any woman did and to die in your dificult hour when it is come & it is the hardest of all deaths. And my unbelieving heart use to say amen to it twoo. I was afraid to trust my god. I am one of them that through fear

[*p. 72*]

of death am subject to bondage. Now i say at these times i could scarcely ever eat or sleep or injoy any of the comforts of life but satan would bring it fresh to my mind what i had to undergo that he might mar all my comfors. Now i have written in the 20tieth page a little concerning my first child: and when i went with my second child whose name was hannah i may say as david my tears was my meat day & night. But that word was good to me i will be a god to thee and to thy seed see page 26. O it made me love and admire and i see this word fulfild in the time of my distres which was not much above an hour. And i see gods salvation & expect to meet that babe in glory and when i went with my third child i was weeak full of pain in body and soul & i found i could not trust god and he justly hid his face from me. But when my hour of distres was come them words come with power—god is in this place and i knew it not.[60] Altho my sorrow was long yet salvation come. O how good was god to me and my ingratitude towards him. He has given me a new life. O let me always live to thee i beg i beg.

After a while i went to meeting to new haven where i heard mr whealock[61] of labanon preach. He spoke clear truths but methot he wanted the spirit of christ. But when he prayed for gods children

[*p. 73*]

that was under persecution whose necks was under the iron yoke o how my soul went out to god in prayr that i could hardly contain myself. But while i was at meeting i was afflicted with vain and athistical thots. I felt the most of the day bowd down with sorrow of heart. I see that man had the least reason to be proud of all creatures on earth. O what a sence i had of the shortnes of time & the sertainty of death. When i looked round the meeting house and see the white wigs o i thot how dare you adorn your poor heads thus when in a few moments more they must be buried under the clods to rot. I felt so bowd i thot i wanted to creep under the seats and lie there where no eye could see me and so i come home sad. O let me lord feel the power of thy resurrection to raise my sinking soul o once more let me love and praise thee

Now i felt still benumd in soul and body especially on the lords days. Now the payns of death keep taking hold me. I am just like the man in the fifth of mark i am among the tombs crying and cutting myself with stones.[62] I feel like one just decending into the pit. I am limiting god &

saying can god furnish a table in the wildernes can god comfort me. O lord is thy mercy clean gone forever has thou forgotten to be gracious. O lord if thou hast any mercy in thy heart bestow it on me for thou has sore broken me in the place of dragons & covered me with the shadow of death psalms 43-19.[63]

[*p.* 74]

Ah here my wound is that by turns i am so distrest about the pain of dying and rotting in the grave. It is self love. I love myself then better than i do jesus christ and i believe the devil and the law of moses has a great hand in it and o that i could get rid of this hateful self love. O that i could fly out of this blind withered naked sore stinking rotten proud selfish self love into the arms of a lovely jesus. O then how pleasant & desireable death would always be to me this my soul knows right well. O come to me for i am a sinful woman o lord.

Now some of my neighbours cant tell me how to bare it that i visit them no more but now i think of david how he kept himself close for fear of saul and at another time he could go out against a great goliah without fear. Ah just a little while ago what glories did i see in god and christ & heaven. The seald book of god was in some measure opened to me but in my attainments i got spoild with pride and self confidence and i forst the lord to depart. Ah here I am. The whole head is sick my heart is faint my weakly body is ready to sink under this load i am dayly looking for some terrible outward calamity. I feel peevish and impatient i cant bare with my husband. Now sin rages in me and like jordan overfloos all its banks. Now i envi them that got ashore to glory and can say farewell sin forever. Now i lit of john bunyans book the travel of true godliness.[64] I think the 100 page he says—this ads to my joy in the midst of my sorrow this longed for day is

[*p.* 75]

near and hastens greatly. It is but a little while and there will be a great alteration in the world. The angel who inlitens the earth with his glory will suddenly cry babylon the great is fallen is fallen. I say this for a few moments sat my soul alonging & praying but on a sudden like peters sheet i was caught up again. Ah my soul mourns. Now i went to see my christian neighbour that i have spoke of before but o what complaints did she make. She cryed and said she was so disserted of god she could not read nor pray. She said it is midnight and i thot so twoo and so i come home in destres. It seemd as if god had forsook the earth. I can hear of no breath nor stirring among the dry bones[65] nowhere. I went away to prayr to god. I cryed to him in anguish of soul for help but in a moment there was the form of a mans face sat before my mind that took it & it was darted into my mind that jesus lookt just like that man that i had a ideah of here. I was stopt then i see the temptor was at work. He cares not how foolish his temptations are if he can but obtain his end. O my god my god why hast thou forsaken me.

May 1754. It being sabbath i stayd at home alone and i felt alone and altho i am in bondage now yet i have a little reviving in it. Sometimes i long to appear before

[*p. 76*]

god in the eternal world & sometimes i long for the day of judgment but now these are short blinks to me and then sin will rage in me as if it would drive me down to hell. And the wicked are joyning with sin and devil to ruin me when christ is gone but they dont know it. Father forgive them.
In 1754 there was a terrible thunder storm. I know one that trembled at the majesty of god. But the storm is over & we are all alive and out of hell. It is a wonder lord i thank thee o god remember thy sons prayr. O santify me by thy truth for thy honour sake.

This morning i got up with a great load upon my soul. I felt pevish but when i could hold in no longer i cryed out to god to come & help me. I cryed with a bitter cry and god litened my burden. I find sin wont die easily but it will up and fight for its life. O it seems to me that if the lord dont kill the power of sin in me i shall dye. And yet i see i secretly am crying to make up a righteousnes of my own altho i know & feel i can not help myself no more than dead lazarus could come out of his grave. O if my heart sins was but distroyed i should be more happy this day than if i was made queen of the world. Lord help for i have no might against this great company that cometh against me. Neither know i what to do but my eyes & cryes are to thee. O give patience to wait thy time for deliverance. I firmly believe i shall go to heven when i die but i cant sit down and rest there for am plauged with a wicked nature.[66] I want sin to be killed that i may injoy my god & live to him here and honour his holy

[*p. 77*]

name. If there was no heaven i long to be conformed to god and made holy here but i am in the furnace. O stand by great refiner and purify me.

Purge out the droos take away the tin make me come forth like gold tried seven times more fit for my masters use. Lord pyty me a poor sinner.

Now i see it is the greatest thing in the world to believe gods word. Sometimes i read gods word but i dont believe one word truly & realizingly. I find athism in my heart for sometimes i see no difference between praying to god or praying to the stones as to being helpt or heard. My past experience affords me no present comfort i cant live upon the food i eat last week. Now all that ever i have done in religion as it come from me looks as black as the evil one and now i am fearing gods heavier rod. Destruction from god is a terror to me & by reason of his highnes i cannot indure Job 31-23. Lord i want a new pardon seald by thy spirit to my soul i want armour out of heaven to war against sin. I want to see god in christ smiling i long for the death of sin in me. O subject me to thee to take thy own way to sanctify me. One day i was very sad

[*p. 78*]

walking alone in anguish of soul. I thot god dont hear my cryes. I am a good mind to never pray again. O did hearing of my fathers death strike death into me and are i was aware that word come like a shower in job 19. I know that my redeemer liveth and tho after my skin worms destroy this body yet in my flesh shall i see god. O how was my hard heart led to think i should yet see god in the flesh altho the grave and worms must destroy my weakly sinful vile body. I knew i should at the great day see god to be my god my redeemer through christ. O how i loved & longed for the hour. Also that word was sweet to me—the god of peace shall bruise satan under your feet shortly.[67] O these are gods words let satan say what he will he is a lier o satan you will not only be bound but cast into the bottomles pit and a seal sat upon you.[68] O my god hasten the day glory to thee for this help.

Now it was not long before all was out of sight again but i felt my soul supported against the fears of death and the grave for some time. Lord i thank thy holy name god my redeemer lives and often from the skies he will look down and watch my dust till he shall bid it rise. Corruption earth and worms shall but refine this flesh till my triumphant soul shall come to put it on afresh arayd in glorious grace shall this vile body shine and i shall be christs & he forever mine. Now i had a little vew of them christians that like the jews of old are in their hearts gone back again to

[*p. 79*]

egypt alas i fear they will die in the wildernes—numbers 14:23. But those christians that are carried captives into babylon i hope will come back like nehemiah and others & help to build the broken walls of jerusalem. For my own part as ezekiel and daniel i am among the captives but god can give revivings in bondage and shew his glories to his children in captivity as he did to ezekiel and others. I think of jacobs waiting 7 years for his beloved rachel altho he served how patient. And is it jesus i am waiting for o let me remember he that believes should not make haste.

Now in these days i go not to any meeting but sildom and they tell me it is a duty to go to meeting to hear the best preachers if god is not with them. But i think of that in samuel first book 15 chap how saul and them people spared the best of the amalikites cattle and sheep to do sacrifice withal when it was gods command they all should be slain. O if people did but see the ministers in the standing constitution o how would they exclaim against their conduct and in that way slay them or hew them in peices as samuel did agog.[69] Lord made the world to know: that to obey thee is better than sacrifice & to harken than the fat of rams. O what ailes gods children to be so content in their graves with lazarus their eyes closed their mouths shut up. Lord call them then they will come forth.

[*p. 80*]

Now i feel strong corruptions working. I feel discontent because i have no more of god and yet i abuse what i have. I cant perform no duty but my corruptions will thrust in themselves. I cant present myself before the lord but satan will come also. And sometimes o how shamd i feel to go to god to pray when i have been sinning against him. But i continue to go tho it is against sin death & hell and if i get nothing yet i must go still & wait the returns in gods time. O that god would make me like davids worthies first book of chronicles 11 chap. There we read of two that slew each of them 3 hundred enemies at once. Benajah slew 2 lionlike men & he slew a lion in a pit & an egyptian five cubits high & in the egyptians hand was a spear like a weavers beam. Ah he went down to him with a staff and pluct the spear out of his hand and slew him with his own spear. And when david longed for water of bethlehems well by the gate three worthies brake through the host of the philistines & fetcht water to him & he pourd it out to the lord.[70] Ah so does every christian when they have a vew of christ the water of life once at bethlehem now at heavens gate. Ah methinks these worthies hold forth to us noble victorieus faith. O that like these heroes i could kill or get victory over 3 hundred sins & corruptions at once by faith in christ. Lord kill my lionlike corruption that has got a worse spear in its hand & a greater than the egyptian had. O methinks much might be written from these schriptures but i hant time no woman kind lives with me in the family. There may be many improvements or instructions taken from one place in schripture and yet

[*p. 81*]

the schripture properly mean but one thing. Now 3 times this week i got liberty to come near to god in prayr once to beg for help & strength against the sins of my heart & nature once for the glorious day of christ to build up zion. Methot i by faith see glorious jesus lovely jesus altho satan often tells me there is none but i apprehended him and long for the down pourings of his spirit upon the world once with groanings that cant be uttered. Every limb knee rib & joynt & sinnew of body & soul was in prayr. Lord thou hast prepared my heart to pray i beg. Cause thine ear to hear for thy honour sake.

One morning i got up a wait was on my mind and a temptation took hold of me and i was angry and spoke rashly & wickedly but in a few moments i was sorely greived.[71] I went alone & confest it to god then i went & sat down by him that i spoke wickedly to. I told him i was sorry i asked forgivenes & my tears ran freely. I told him that god had said he would not forgive them that would not forgive others & he was mild & spoke lovingly to me. Ah i am ashamd to write what i said in anger. I am ashamed to have it brought into judgment at the great day. Ah i cant forgive myself altho i believe god has forgiven me for. This same day i went to pray again & o free grace o free grace god gave me liberty to come near to him altho i had just been sinning against

[*p. 82*]

him. O how my sins makes free grace look. I had freedom in wrastling that ante christ might be destroyed in the hearts & lives of all flesh but especially in gods children & that the long loockt for day might come that jesus shall reign king on earth. But now god seems to be withdrawn from the earth and sin rides on horseback & jesus goes on foot to alude to that eclesiastes 10-7. O lord do send thy nehemiahs in this night to vew the broken waals of jerusalem & to build them. Ah they are broken down to the ground and the gates thereof are burnt with fire. But o i see a great harvest acoming o lord send laborers into thy harvest. I can say sin & satan has thrust sore at me that i might fall but the lord has helped me. Thou rulest the raging of the sea when the waves thereof arise thou stillest them. In these leaves are some of the spoils won in battel which i have dedicated to maintain the house of the lord.

August 4-1754. I have been tryd in my mind many times about the schriptures of the old testament that was written so long before our day whither god spoke to us now in perticuler by them as he did to the old testament people. But that word in hosea 12-4 helped my mind much namely—he found him in bethel & there he spake with us. I see that what god said to his people in old times in the history now is held forth to us in the mistery or spiritual sence. Now one morning as soon as i awaked that word come to me softly he that is slow to anger is better than the mighty & he that ruleth his spirit than he that

[*p. 83*]

taketh a city. Now i soon had a temptation to be angry but a good god caused it to pass away glory to his name. O my bace wretched hart if it be good for nothing else yet serves to commend & set off the unserchable riches of free grace. O that the lord should take so vile a loathsome creature into nearnes to himself. O let me lie always as in the dust. O my god keep me at thy feet always admiring the riches of free grace in christ. Now i lit of mr flavels book of keeping the heart.[72] His sweet schripture arguments against the fears of death has got hold of my mind. I am comforted glory to god glory to god.

Now this many years before trouble & tryals came upon me i have an impression of it upon my mind that sends me to god for help and strength that i may safely win through & give glory to god altho i dont know then what sort of trouble it is that is coming. Ah the secret of the lord is with them that fear him & he will shew them things to come. Imaginations & dreams are not a foundation of my faith but jesus christ is my foundation stone. O let me let me live to thee.

Now again i was afflected with wandering thots in time of prayr. But altho my sins & corruptions & the world flesh & devil have pitcht battle against me yet god helps me along & maintains full assurance

[*p. 84*]

in my soul of going to heaven. But i see i am apt to have a legal spirit. O my pride of heart that costs me a sea of sorrow. I find i am more willing to take christ for my prophet & my priest than i am to take him for my king to rule & govern me by his dyvine laws rules & institutions. Alas i often conclude against myself that my case is singular no heart like my heart. I remember i asked my father once if it could not be that i had more of adams sin than others. He said we all pertook equally alike. Lord my soul mourns for sin o wash me in thyne own blood.

In the year 1754 in september. I was much excercised by hearing the indians had done much mischief in the upper towns had killd some & carried some captive. I had a sence of the justice of god if he should give all up into the hands of these cruel barbarous enemies. O the back slidings of Gods children call for great judgments. That in amos come to me—amos 6-12 shall horses run upon the rock will one plow there with oxen for ye have turned juddment into gall and the fruit of righteousnes into hemlock. God has tryed us with outward peace & plenty & that wont do. Who knows but god will not redeem zion with judgment. Now i thot of that word that come to me when the noise was heard in the air namely trouble cometh out of the north.[73] Once while i had such a sence of gods justice in destroying new england that my mouth was stopt i could not pray. But one night i was in bed

[p. 85]

then my soul cryed to go with this language o my god didst not thou tell me that thou wouldst be gracious & carry on thy work in this land and now art thou about to give it up into the hands of these cruel enemies. Lord i have looked for peace and glory but behold trouble & destruction i fear is coming. Then i toock mr flavels book & read how to keep the heart under outward straits. Then i was stild & went quietly to sleep & since i feel willing that god should take his own way to redeem his people the affairs of zion are all in a good hand. Flavel says it is better to be as low as hell with a promise than in paradice without one. O my god come quickly.

Now again in the week past i have had sore tryals. Lord forgive them for they know not what they do. God gave me a heart to pray for this dear creature with groaning unutterable. O that i may be at his feet & be willing to be trampled on & be called anything. O give patience o give me charity to suffer long and be kind if it is thy will. O how tribulations indears christ and heaven. I long to leave sin darknes & ignorance & begone to heaven where the wicked cease from troubling & the weary are at rest. I spent some part of last night in prayr. I had some liberty in wrastling for my husbands soul. Them words darted into my mind the lords hand is not shortned that it cannot save neither his ear heavi that it cannot hear. This was a good incouragement to me for a few moments

[*p. 86*]

but it was caught away from me again. Then i prayed to god that if he intended to have mercy on my husband he would be pleased to speak that word to me again & a few days after the same words past through my mind with power. Then i thanked the lord but soon unbelief began & said this word is not an absolute promise it wont do to depend upon it for his conversion. But o i cant quite let it go no no. But now sin began to rage in me twoo hard for me. I could not lay nor conquer it it was my master. O my weeak body totters under the load o i fear gods judgments ah my legal hart wont look to christ. I went to meeting to new haven. I hear mr troop[74] of southold but i felt nothing but death. At noon i went to captain dogesters. There mr edenbourrous prayd with some nearnes. We converst & sung. I would have them have the meeting there in the afternoon but i believe the fear of man prevaild. Here once i felt my iron heart long after christ but wee went to the meetinghouse again. At night i come home heavi anuf. O my soul mourns for sin. I can say as bunyan o my sins my sins they kill me with care they daunt me with dread they consume me with sorrow they pine me with pain they eat me with grief they overcome me with bignes they pres me with wait they overlay me with load. O the multitudes of them o the deep dye scarlet & crimson o heavy anger the goliah of all when the least grain of it comes into the heart christ goes out.

[*p. 87*]

Altho others might think my provocations are great but o there is no excuse for sin. Lord do take away this adament that wont be refind. Ah for sin i roar i groan i cry i pray but cant get constant strength against it. I feel at present a load of sin & guilt & my poor weeakly body hangs like a wait or clog to my soul. Lord help help this once.

And now again about a week i was afflicted with angry resentments for past ingeries. This made vile me many an arrant to the throne of grace that god would give me a christlike forgiving temper of mind & that i may bind these reproaches as an ornament about my neck. They belong to my christian profession. Moses counted reproach for christ greater riches than the treasures of egypt. But on the lords day morning before i was up that come—lift up the hands that hang down & strengthen the feble knees.[75] I use formerly to fling away schriptures sometimes when they did not come to me with great power & count myself unworthy. But now the word of the lord is precious in these days. Now i can pres towards the word & cetch hold of half a word if it be but whispered to me. A horse that is in the mire will flownce towards sound ground altho he sticks in the mire still. But the last cyted schripture made me hope help was

[*p. 88*]

coming. But now i can write but little by reason of darknes i cant put my experiences into language. On sabbath morning that word come to me—rent your heart and not your garments & turn to the lord lest he brek out like fire in the house of joseph & there be none to quench it.[76] I got up and got my byble but o that day the distres i felt in my soul. O dear me fraid i was of gods rod & instead of fleeing to christ i ran to sinai. There i stood & see the lightning & heard the thundrings of gods law and them words did run over and over in my mind with great fear and terror. The next day it began to ware of till it left me just so as i was before. A day or twoo after i had an impression of trouble being near & presently there come in twoo carnal woman to see me. I felt dolefully. I was soon weary of their company but o like saul when he forst himself to offer sacrifice i forst myself to tell them the certainty of death and the dangers of not being prepared. O such company is a torment to me wo is me that i sojourn in meshek and dweel in the tents of keder.[77] O when shall i go from them that are not my people to them that are my people where the wicked cease from troubling & o methinks i feel like a marinor that has been out in a long storm his sails torn to peices he mast split his vessel full of leeaks his heart full of fears. O how he longs to get on shore to his own home and yet thanks be to god my ancer is fast. Lord bid me come to thee on these waters of affliction.

[*p. 89*]

One day there come one to my house that had ronged me with their tongues. I talked to the person about it. When they was gone i went to prayr. As i was kneeling down that in 2 of samuel 28 come to me. What right have i to cry any more to the king. Now i hoped god was going to deliever me from that prejudise or resentment in my heart for past injuries and so i had no right to cry for that which was done for me. Already i feel delivered at present glory to god.

This week i had an impression on my mind of trouble coming. It sent me to god for help. In a few hours it was upon me but wherein the wicked dealt proudly god was above them. Lord blessed be thy name.

Not long after this before i was seisd with dispair and molloncolly i did not want to see nobody but wanted to be alone grieving & mourning for my sins and backslidings. I have many a time when my family has been gone shut up the doors & when people have knocked i said nothing & let them go away. O the unbelief and athism i felt. I was afraid i was going to give out. All seemd to be out of sight. One day i asked my child if he believed that there was a christ poor thing said yes. Ah thot i—i dont believe it. O it is the power of god to make a creature believe there is a god.

[*p. 90*]

Now i had some sence of what unconverted ministers are adoing. I believe they are adoing the work of the devil leading souls down to hell. This sat me to crying to god for help for mercy that he would convert them or thrust them out of the ministry & set faithful ones in their room shepherds after gods own heart that know how to divide the word of truth and give to each one his portion. I have a hope i shall yet see that word fulfilled. Your eyes shall see your teachers & as a young man marieth a virgin so thy sons shall marry thee.[78] Now i hope yet i shall see gods promises fulfilled and rejoyce in his salvation. It seems to me i shall not die till i have seen christs kingdom built up but o it seems to me i feel more of adams sin in my heart than ever mortal did. I think my body would be weel if my soul was at rest in god. O it seems to me there is not one in hell that ever had such a heart as i have. Sabbaths now are a burden to me. I dread when they are near. O vile heart what cant be willing to serve god becaus thou hant the appointed means of grace. One morning i had a wait on my soul. It seemd pincht & straitned. After a while i had liberty to pour it out before the lord. My soul cryed for christ to come & set up his kingdom on earth. But ah i soon was in bondage again. Next day i went to east haven and heard mr robbins[79] of branford preach. He preacht well but i got nothing as i know of. This week i spent part of a day in prayr alone but i got nothing in hand.

[*p. 91*]

Now i met with a sore affliction and i fainted to see i dispised the chastisement of the lord & fainted when rebuked by him the breach of a plain command. O my sins my cruel sins i think exceeds manassas.[80] His was done in ignorance mary magdalen & saul sind ignorantly the angels when they fell knew not the way of salvation by christ. It seems to me there is but one sin that exceeds mine & now the temptor is telling me that god is and will be true to his threatnings but not his promises. This has got hold of my mind. I can now say as heman while i suffer thy terrors i am distracted.[81] Now satan insults over me because i am a stranger alone none to take me by the hand in soul matters & my dear poor husband knows not what i ail. Poor heart he has no other breath but threatnings and slaughter against christs religion. Also i was now very ill & weeak in body & now i could not by reason of darknes see it just for god to hide his face from me because jesus has fully paid the det satisfyed gods justice i believe for me and now his grace is free without money or price. Then why will he not come with his love to my soul—here i blunder founder cry & quarrel—yesterday when i could hold in no longer i got me away & roared out with a flood of tears. O god if i am cast out of thy sight do aniolate me to nothing. This i repeated over and over with a heart ready to split and when i had vented out my sorrow to god my soul was easd. O god i thank thee that i am not in

[*p. 92*]

hell where i could never pray. But o i am here where i may beg for mercy. Just now i could not see gods justice in his hiding his face from me but ah altho christ has payed god to a tittle for my breach of the moral law & has justifyed me i believe freely without money or price i mean anything in me but mark i have sind against the law of christ and broken covenant on my part & must be punished with [stripes?]. Much might be written here but i want time and o when i look around me i see the tender care of god towards me. He feeds me to the full with the good things of this life when many of gods saints have wanted the same samuel 1-25-8—give me i pray thee said david to nabel whatsoever cometh to thine hand. Also dear renound musculus[82] was forst to dig in the town ditch for a maintainence. Famous ainsworth[83] was forst to sell the bed he lay on to buy bread but what speak i of these. Behold a jesus that was lord of all had not where to lay his lovely head and being a hungry he looked on the barran fig tree if happly he might find anything thereon mark 11-12. Lord dont let my hard proud heart never overlook thy dear mercies again. O how apt have i been to abuse what i have & quarrel in my heart because i hant greater mercies. O my sins how long shall i be plagued with thee come lord & deliver me. One night after i was in bed that word was good to me—better is the day of a mans death than the day of his birth.[84] I see with comfort that it was a goodly time when the righteous

[*p. 93*]

dyed. O to be uncloathed that we may be cloathed. O my soul haste thee away from the lions dens & the mountains of the leopards. O this day i long i long i long to be with my jesus to say fareweel sin i am now where you can not come. O now i hope the lord is subjecting me to death and the grave. Now i see the grave is a refining pot a sweet bed of rest to saints. O take the sting of death away. Let perfect love cast out all fear. O let me lean on jesus bosom then i can die sweetly grant this for christ sake.

Yesterday there come a young woman to see me from a neighboring town. Her father was dead and her brothers was wicked. She cryed and told me her great tryals & what promises had supported. I felt a simpathy with her in her trouble. I told her afflictions was in the covenant. We must not halve christ. We must take croos and all and follow him. Gold must be tryed for the fire to purge away the droos. We sat up late and talked of the things of god. She seemd to fear before that in samuel 1-3-14 where it says elies house shall not be purged &. The next morning she sat away to go home but her talk was profitable to me lord pyty her and bless her.

Soon after i was taken with extreem pain in my head & teeth again. I cryed to god for mercy. It was cold weather the beginning of winter still the pain continued. Some fits was so hard it seemd as if i must lose my sences. My head was so mazed

[*p. 94*]

and benumd with the extremity of misery i endured & now this added to my grief that i was a stranger alone away from all my relations. Ah now i wanted friends but o i was diserted of god and that made all my circomstances seem so hard. I was much fild with dispair & sunk down many days & long cold winter nights. Now everything lookt dolesome but i still followed god in prayr and o one morning before i was up them words was whispered in my ear (the sea parted away and jordan was driven back). Then i believed god would heal me but i got up and the pain still continued hard. Then i sunk again but now & then them words would come and lift me up a little & i did say to myself i do rayly believe god will heal me and soon after i was easy and comfortable. Glory to god he is the god of pain and the god of ease. Now after i was easy i douted whither i was heald or not so i thot i must wait & see for the pain use to come by fits. Then them words come to me with power & sweetnes. He wist not that it was true that was done to him by the angel.[85] O the tender care & love i then saw christ had for his chosen ones. O did he send an angel to heal me & yet must i dout. O to be easd when in great pain is a great thing but o here is a greater

[*p.* 95]

that when god had but one son he gave him to die for sinners. Oh i have done anuf to provoke him to cast me out of his sight & forget to be gratious. O let me live to thee who has wrought a new salvation for me. O when shall i be with him and leave sin behind. Hold out o my soul a little longer. The fight will be over the warefare ended & thou shalt flee away and be at rest and sit down with abraham isaac and jacob and all the prophets in the kingdom of heaven amen. Even so come lord jesus.

Alas it was not long before i got into a dark low frame. I could not pray only with words. This day i see that god makes the sins of his children profitable to them for he says all things shall work together for good to them their sins makes them abhor themselves and learns them not to trust in themselves but in god alone & in the righteousnes of his dear son. Ah they dont sin that grace may abound. God forbid for whenever true grace is in excercise it sets a kean edge against sin. Sin it is that makes saints go groaning & crying through this wilderns and if we grow not in self abasement we grow not at all.

Now i dreamd mr whitefield was come and while he was preaching he stopt and spoke

[*p.* 96]

in perticular to me. Now about 2 days after i heard he was come & agoing to preach at new haven. There was a cause why i did not go to hear him. I had twise a spirit of prayr for new haven and a few days after there come one & told me several sweet pasages in mr whitefields sermon that was very precious to me. So i see my dream fulfilled then. Them words was renewed to me again blessed are they that wait for me.[86] Methinks i see god has not forgot his sweet promises. O i believe he will come he will come and carry on his work.

Now this same day there come a little gairl to my house. I asked her if she thot she was fit to dye. I told her she must come to christ & be converted or lie in hell forever. I told her christ was willing for he says suffer little children to come to me and as i spoke them words i felt them. I see the willingnes of christ to save lord purify my heart.

Now i had troublesome visators. I could not be faithful to them. If i had i supposed they would soon have left me but my soul was in prison. But when they was going i went into the barn. There i poured out my whole heart to god. Now i was afflicted again with a carnal mind. When i first wake in the morning & in time of prayr my thots are with the fools eyes. O my sins my sins makes me long to go home altho i fear the passage o the time looks long. One day i felt an impression of trouble coming. In a few hours

[*p. 97*]

it come. Wo wo to that child of god that dont marry in & for the lord. And altho i am now under so much trouble yet i cant pray because of my minds being fild with distracted thots. O the distres of an absent god. O i feel the very venom of hell in my heart. O how impatient i feel. Alas now i feel the power of sin so in me it is worse than death itself. Lord help me if thou see fit that i must drag the power of sin always with me here through this vail of tears. O let me see i deserve it o make me thankful that thou hant reserved it for me to feel it in hell. I will hope in that prayr john 17—sanctify them through thy truth. O come lord jesus in infinite compassion. Lord to my complaint give ear whole troops of sorrows bair me down o when wilt thou appear.

Now i had a remarkable dream but i must not write it. I believe dreams & imaginations & impressions are a foundation of sand to be depended on. But yet they do good when they drive or lead the soul to god & his word which is the shure foundation stone which god in zion lays to build our heavenly hopes upon to his eternal praise.

But o how many years have i been groaning crying praying over this doleful heart but yet have got no cure no absolute deliverance. Still my sore runneth and ceaceth not. I have been in travil but have brought

[*p. 98*]

forth nothing but wind at preasent but o must dispair of ever being delivered from sin till death. Then i have cause to say come o my god & eternally deliver me. Now the word is a seald book to me. I feel an unthankful hard heart. O how ignorent stupid & born down with guilt & fear of gods heavier rod & besides a weeakly body to carry through this wildernes. And now there is the sound of war[87] against new england & a wonder it will be if we are not given into the hands of our enemies for dispising an offered christ. And alas methinks the church is with david driven into the wildernes while a hitophel and absolem[88] is plotting its utter ruin. I say these things way down my mind o could i but trust god for all methinks it would make soul and body strong. But o i have got to sinai and stand trembling there. O how worn out i feel with sin & sorrow but i think of israel when they complained god was angry & consumed them—numbers 11-1. Lord i beg give me patience to indure like a good soldier. O help me to stand in this evil day and having done all to stand this day i see the same language in my heart as israel had when they had been to search canaan—numbers 13. I am fearing before my giant corruptions. I believe i shall go to heaven when i die but here is my tryal i fear my sins will master me while i live

[*p. 99*]

& that i shall be a reproach to true religion. O what a monstrous wicked heart i feel &. O the horrible nature of unbelief. Them men that went to serch canaan and brought up an evil report dyed of the plague. Now i bring up an evil report before my family & others & spiritual death is upon me. O could i believe like caleb and joshua[89] but for want of this i go bowed like the woman and cannot lift up myelf. Lord help me.

Now i met with sore trouble from a wicked person but that word come into my mind—i wot brethren that what ye did ye did ignorant.[90] Then i see i had the more reason to be patient. Lord give me a double portion of patience.

But alas now i am in a continual storm not a day but i have some gyent corruption to mourn over & groan under. O if my god should take away restraining grace from me what better than a devil should i be & o my poor trembling week body. Anights i am tossed to and fro till the dawning of the day & sometimes when i first wake i have a corruption to fight against. But o are the wicked to feel the weit of sin in hell & not here. One morning this word come to me in psalms 76-5—the stout hearted are spoiled they have

[*p. 100*]

slept their sleep and none of the men of might have found their hands. Ah did jesus at his death spoil stout herted sin for his. As to its damning & reigning power & altho sin & corruption looks like men of might yet they hant found their hands to prevail or overcome wholly. As paul says—i am afflicted but not forsaken cast down but not destroyed[91] so a good god helps me along here a little & there a little.

But now my corruptions soon was in arms again. O how selfish i felt. I see my prayrs was selfish. I am now looking within me and the more i serch the more i see. I find in my heart one abomination under another & another under that & i am afighting against them but cant lay them. I am now just like a silly maid that sees her house defiled & goes to sweeping of it and the more she sweeps the more the dust flies & all because she forgot to sprincle it to lay the dust. Lord help me to look by faith to the blood of sprinkling to lay the dust and smudder of sin. O is there a time acoming when i shall have a better heart when i shall be as holy as god can make me although now my soul is in anguish while i am writing because of sin.

Self in myself i hate this matter of my groans
nor can i rid me of the mate that causes me to moan.

Now the fears of death seem to be taken from me but o i have now another fear for i am like to have another child and i thot a great while some great tryal was coming

[*p. 101*]

& now them words come afrest to me that i have mentioned before— turn to the lord lest he break out like fire in the house of joseph &. I was terrifyed because of these words and the hour that was coming. O how often did i weep & cry before god. I wrastled with them words o lord who shall live when thou dost this to break out like fire. O turn me to thee & all wil be wel.

Now i heard that the woman that i have spoken of before in page 65 that was unfriendly to me was in travil. I went to god for her i had a great spirit of prayr & liberty. Next day i heard she was delivered safely glory to god for salvation to her.

Now there come a woman and staid at my house 4 days. O the carnal ungodly talk i heard. If i reproved her i had as good speak to the stones for she regarded not gods truths. But i kept pleading with god to send her away i groand to god to deliver me from her. At last by a providence she was fetcht away. Then i thanked god but my mind was confused for a while but i got calm still and composed. Lord help me to improve better the lonely hours i have.

Lord fit me for the next tryal. O let me lay my head in thy bosom for thou dost sit on the waves. O make haste accomplish thy furnace work upon my soul. O bring me forth like gold tryed seven times more fit for thy use. O purge out the droos take away the tin subject me to thee.

[*p. 102*]

At this day i was weeak in body and mind and a wicked person got angry at me without a cause and abused me with the tongue. I cryed & grieved till i was quite sick. O wicked me i ought to endure hardnes like a good souldier but o the athism that now seisd my mind. Gods providences seemd all to work against the promises. When saterday night was come i read a story of a childs conversion how she spoke of the love of christ & longed to die to be with him. This gave me a little relish. I told my husband i thot i could go a great many miles to see a christian fild with the love of god. On the lords day morning when my husband was gone to meeting i went to prayr with my child of ten years old. God gave me freedom to pray for myself for zion for my husband & child that have no christ but are dispising the greatest offer that ever was made to mortals. Lord compel them to come to thee i beg i beg. Not long before i was troubled with distractions in prayr and meditation. O how am i straitned in myself. O dear could i now fly to him that i cant be straitned in.

February 1755. The week past i was stupid a hart as hard as a stone. On the lords day i was a little moved by reading a letter that my father had formerly sent me. O that god would accomplish his furnace work upon zion & appear for her up building and glory.

[*p. 103*]

Last night i had a temptation to walk in the ways of sin as if that was the best life. Lord help me to fight in thy strength against sin as long as i have one breath. Now i dreamed one night that i was witnising for christ before a number of people. My soul was fild with love to god & his word but not long after we suffered a loos by a neighbor. Now i felt a covetous revengeful heart. I see the nature of murder was in my heart. I cryed lord hold me dont let me go for if thou dost i shall be a devil incarnate. Lord help lord help me. My lusts did rage corruptions swell as if theyd drive me down to hell. Also i am now fearing the hour that is coming which i cant escape. Now i lit of the 7 of romans it was precious to me from the 7 verce to the end & altho i have strong corruptions within trying to get the victory & opposition from without & cruelty they that should be best friends to the stranger are joyning with sin & devil & are accusing of me before god day & night. But altho my feet has been hurt with fetters & i have often been in prison laid in irons & altho the archers have sorely grieved me and stot at me & hated me but my bow abides in strength. I believe i shall win threw and get to heaven. This hope has been anchor when i have rid it threw hard storms & lost my sayls & mast. Bless the lord o my soul.

[*p. 104*]

Once when i was in a dark doleful frame one night that word come to me when i was asleep i have heard i have heard thy groanings in egypt and am come down to deliver thee. I was fild so with the love of christ that i sung when asleep and my husband heard me. O methot i was far away above all these things. O methot i see jesus appearing for the deliverance of his chosen ones out of bondage. I awaked and felt sweetly o is god contriving in this dark dolesome day to bring on zions glory. O come lord jesus come quickly lest the spirit fails before thee and the souls that thou hast made. And now again i got a distance from god and like them discyples i toyld all night and caught nothing. O that the day might break.

It was in April 1755. I was under much trouble in body and mind and one night them words was something to me. Zion heard and was glad and the children of judah rejoyced.[92] O i hope god will appear ere long for poor dear zion whose eyes are now fild with tears because thy beloved is gone. But ere long you will lift up your victorious heads and rejoyce and be glad over babylons ruins altho she now bears you down and your souls cleaves to the dust. Read revelations 18. On the next lords morning i had a great weight on my soul. I went loaded with sin and sorrow for several hours. At length it drove me to god.

[*p. 105*]

But o how my soul went now to god up with groaning for my poor family that jesus might come in to my house and dwell here that it may be a house of god and every soul in it a tent for the holy ghost to dwell in. O jesus i dont remember that ever any in thy word invited thee into their houses and didst refuse when thou was here on earth. Then dont let me be the onely one that thou wilt refuse the invitation o come in dear lord here is work for a god to do. Sin abounds here o make thy grace much more to abound as it has been wont to do. O let us all yet live and sing thy praises and all the glory will be due to thee and to thee alone world without end. One day that schripture was on my mind in number 27 concerning zelophehads daughters pleading for their inheritance & god heard & told moses to give it to them. And doth it not hold forth this to us that god will hear the cries of his children that are pleading for his spirit presence and glory to be powerd down upon the world which is their promised inheritance in this latter day. O come quickly lord and get to thyself a great name.

[*p. 106*]

Now i felt sin rage again i felt a quarreling mind. I was weakly and had much bodyly pain. I had a terryfyed mind about the hour that is acoming which i cant escape. Now i had many tryals and o my doleful heart. I wanted to get my neck out of the iron yoke. When sabbath was come in the morning i could not pray but i opened the byble to the 17 of samuel how david went against goliath in the strength of god alone and boasted of no other weapon and got the victory. But for want of faith in god i sink i faint i tire amidst the race. O them words in samuel 15-23 for rebellion is as the sin of witchcraft. O my stubborn heart art thou adoing that to thy god that is as bad as witchcraft. And now the lying tempter would persuade there is no relief in heaven and all my swelling sins appear twoo big to be forgiven. And now it pleasd the lord to take away by death one of my husbands brothers. He was about 40 years old. It was evident to all that he brought himself to the grave by excessive drinking & he dyed stupid and senceles. Now it seemd so awful to me i thot i would not go to the funeral. But that word ran in my mind—it is better to go to the house of mourning than to the house of feasting.[93] Then i went.

[*p. 107*]

Now there come a turn of extreem fear and terror upon me about the hour that i cannot escape and now it draws near. I got into a fit of extream crying. I was alone begging for mercy. O how fraid i am of the pain tho i believe i shall not die till i have seen them promises fulfild in the building up of zion. But o for want of faith in god as to this hour how i tremble for fear. Now when i had cryed to god and come out of my closet this word come to my mind cast your burthen on the lord and he will sustain thee. But then i found & felt i could not do it no more than i could make a world. Them words did nothing to my iron heart. O lord do make me cast all into thy dear hands. O what will become of me if thou dont appear and help me. O god is not thy grace as free now as ever it was. O hast thou forgot who it was that left thy bosom & come down and took my nature that groand and sweat as it were great drops of blood in the garden. O hast thou gracious god forgot who it was that hung on the cross between

[*p. 108*]

twoo theives at mount calvery. There he groand out his dear soul. Ah well might the sun hide well might the earth tremble when the blood of the son of god was asoaking into it. And after he said it was finisht he ascended up to thy bosom again. O my god think on this o hear his intercessions o let me once more feel the power of his resurrection. But o now gods blessed word that once was my souls delight is now in a great measure a seald book to me. I dont see believe & feel with comfort & glory the lovely truths containd therein altho i cant yet quite let go some promises. But o i feel a great deal of athism in my heart and alas now here i am alone one sabbath after another. Ah doubly alone. The comforter is gone not a christian can i see that i dare be free with and if i was well i dare not go to the standing meeting. And ah my vile heart is unwilling to serve god because i hant his appointed means of grace when it is a wonder that such a creature as i am am not in hell. Lord have mercy on me. One day there come a wicked man & raged at me for not going to meeting. He said i staid at home everlastingly & pretended to be good or to that purpose. I saw he was angry & i said but little to him but it stuck in my proud heart several days that such a wicked man thot he reproved me.

[*p. 109*]

Then i thot of that word samuel 2:3 chap last ver i am this day weeak the sons of zurviah do twoo hard for me & whosoever killeth you will think he doeth god service. Lord pyty me for jesus sake. Now i am expecting the hour of sorrow o do come lord to my soul do send the comforter. I am terrified with fear and i kept praying for an absolute promice. O let me not go through such a great work with an absent god. O have mercy on my babes soul & body. And one morning i was in the garden that word come with a melting power in isaias 63-6—in all their afflictions he was afflicted & the angel of his presence saved them.[94] O said i lord dost thou simpathise with me what in all my afflictions and shall jesus the angel of thy presence save them both mother & child. Methot i did believe o it was refreshing. The next lords day as i wake them words was with me—the mountains shall depart & the hills shall be removed but my loving kindnes will i not take away nor cause my faithfulnes to fail.[95] O the love in these words. This schripture confirmed the other they was a great support to me glory to god. Now i spent part of a day in prayr alone for myself and mine. Lord let my groans come up to thee.

[*p. 110*]

Now one evening there was a hard thunder storm. My husband seemd much terrifyed. I went alone & prayed for him once more & in the morning that word come to me—they were meet together with one accord. I had raisd hopes that we should yet be one in the lord altho i often fear i fear god has said he is joyned to his idols let him alone. O jehovah pyty him. One day i met with sudden trouble & a rash word broke from my lips but in a moment or twoo i was grieved. I burst out with a shower of tears that i was so wicked. Methot i longed to break my fetters & begone to that place where no sin cant come & where my jesus is. This sin was the means to indear christ & heaven to me. O hast thou loockt on me to repentance as peter o make me watch better for time to come.

But now in these days my comforts are short lived almost all my time is spent in wars & combats. O how often i see & feel my heart a hater of god by nature. O the athism that is in it. How do you know says satan that there is a jesus or a heaven. My weeak faint body totters under it how full of pain i am how hard i fetch my breath. One morning i sat down to eat my breakfast & a sorrowful soul i had. But are i was aware that word took hold of my mind—tho he cause grief yet will he have compasion.[96] I left eating for a while

[*p. 111*]

and my heart gave glory to god. Methot them words was the sweetest in all the byble. O will god have compassion on me soul & body. O will he help me vile me in the hour i have feard. O let me never distrust thee again. O let me lay my head in thy bosom for thou dost sit on the waves. Now this was comforting to me while it was passing threw my mind. But last night i waked in a terrible fright as tho the hour was come but i fell to prayr & the fear went off quick. Glory to god that helped me to resist the devil and he fled from me. I believe the devil and sin is the authors of all this fear. Satan keeps throwing my past sins at me & says you deserve to go threw a hell & ah i know this is true. Sometimes i am so terrifyed with fear i feel all over as weeak as water i am fild with self love. O did i love christ above all i should love to trust him with my all but for want of this i sigh i groan i cry alone. And now at this time one of my poor bours was stird up by satan i believe to bely & reproach me at a hard rate. But i felt now as if i could bare that for my mind was under the power of fear of my sorrowful hour but i kept pleading with god for mercy. But that word come to me—his flesh upon him shall have pain & his soul within him shall mourn.[97]

[*p. 112*]

Then i thot great sorrows i must bear gods will must be done let me be as fraid as i will. Now the time being come i was about 7 hours i was in as great distres i think as i could be in & live. I kept praying god gave me wonderful courage & patience. They that were with me admired god is a prayer hearing god. Now towards the last i was so bad some feard how it would be with me. But a good god sent salvation. Calvin was born and in a moment my soul winged away to glory. I felt far away above all these things. It seemd as if my soul joynd with saints & angels round the throne. I cryed out glory glory glory to god the father son & holy ghost. Some bade me hold my peace keep my strength but one woman said if she dies apraising god let her die. But o i was constraind to praise god not only for what he was in himself but my eyes had seen his salvation. I saw his word fulfilled on which I had hope. O he is a god of truth. Now for some time i felt a calm sweet peace in my soul. But alas a treacherous heart & carnal company stole me away & instead of performing the vows i in anguish made i ungratefully departed in a great measure from my god. Wreached one that i am because of this body of death. O my jesus when shall i be free from sin and live with thee.

[*p. 113*]

Now it being in september 1755 about a month after i lay in one saturday night about midnight i was waked out of my sleep by a larrom of war. One rid by and cryed wake wake wake. The drum was beating guns was shooting the bell aringing. Now my first thots was that the french & indiens was just at my door acoming in to kill me for our new england army was gone to ground point[98] and we was fearing dayly how it would be with them. But i had heard nothing what the matter was. Now i was so surprised my joints seemd to be loosed. I got up out of my bed terrifyed for i was weeak. That in jeremiah 4-19 come to my mind—my bowels my bowels i am pained at my very heart my heart maketh a noise in me i cannot hold my peace because thou has heard o my soul the sound of the trumpet the alarm of war. Also that word—cursed is man that trusteth in man and that maketh flesh his arm.[99] Now i see if god did not help no deliverance nor help could we have cut off we must all be. I see we all deserved it as a people. Many sins lookt god provoking but o poor new england has dispised an offered christ. Now this sin lookt the most dreadful of any. Ah in the year 1741 and down since when god sent forth his imbasendors by a special

[*p. 114*]

mission from heaven o how few has imbraced the gospel in comparison of the refusers. Dear ministers scornd & dispised and some put into prison & them christians that lived to god mocked hated & reproached. O the cruel laws of man to hinder the progress of the gospel of the dear jesus. Now my husband was going out of the door some hours before day to go to the neighbours to see what the matter was but i begged him to come back. I wanted to go to prayr & he did willingly when in fear and i prayed but methinks my soul stretched itself to heaven for mercy for new england. My soul cryed for free mercy in christ jesus. When i had prayed my husband went and brought me news that it was supposed all our army was cut off & more men must go. O methot that night i sometimes got near to god to plead for poor sinful new england. I prayed that night till i was so weeak that i could but just speak or go alone. I being so spent i layd me down on the bed & dropt to sleep. Now about sunrise i waked & reacht out my hand to the table that stood near the bed and took the byble and opened it. I cast my eyes first on jeremiah 5-1—run ye to and fro through the street of jerusalem and see now and know and seek in the broad places there of if ye can find a man if there be any that execute judgment that seeketh the truth and i will pardon it.

[*p. 115*]

O i cant write the glory that i see in those words. O i see that god made me to seek the truth for christ is the way truth and life so he would pardon it. O it was for his own name sake. O them words—i will pardon it—i was almost overcome for a while. I cryed out o is he gracious o is he gracious is he gracious o the love i see in the heart of christ the pyty and mercy altho just before all was justise. O how happy i felt that word come—the solitary places of the earth shall rejoyce for thee.[100] O methot god helped me to wrastle till i got a blessing. Glory glory to god in the hiest. O let a day of thy power make us all willing subjects to thee. Now we had soon sertain news from our camp there had been a fight at ground point and some of our men was killed & o how terrified everybody was with fear that new england was agoing to be destroyed. Christians would say we all deserved it and the french are gaining upon us but my soul was comforted. Gods word was secret strength to my soul all that war. Bless the lord o my soul. Now about this time there was an earthquake one night i heard it. It was terrible i felt something surprised at first but this went off my mind without any remarcable impression. Only i believe gods word is fulfilling the end ahastening

[*p. 116*]

again. I was afflicted with a hard heart. I was under great dissertion the comforter was gone sin & devil has carryed me away captive into babylon. There i am bound with fetters but i have sometimes a little reviving in my bondage. But ah i am not onely in babylon but i feel babylon in me. O how busy satan is. He darts such blasphemy through my mind. I dare not write it it may hurt some & besides these thots are darted into my mind that there is no heaven nor no jesus o the secret groans the anguish i felt a long time.

In the year 1758 a young man one of my neighbors i believe was converted. He seemd to have god with him and wee began to have weeakday meetings. And some brethren from new haven did sometimes meet with us and here and there come a few from wallingford and some here that durst did come to meetings. We had also night meetings. Some young people was under great consern acrying out what shall i do to be saved. There was several hopefully converted agiving glory to god & some old christians seemd revived. This work of god was heard of to distant places and the dear saparate ministers did come and preach to us sometimes bringing their lord & master along with them. But o i had more comfort in vewing this work of god by faith 7 years

[*p. 117*]

before it come than i had after it was come for i was worried day & night in my mind thus it may be there is no heaven nor no jesus that dyed for sinners. It lies so heavy on my mind it seemd as if i should sink under it. I could but seldom pray & when i did it was thus o god i am at infynite uncertaintyes about a jesus & heaven & thy word is seald up from me. O pyty me pyty me have mercy on me & when i have cryed out my anguish to god i feel something released but soon worried again anew. It may be there is no heaven nor no jesus. O you that think you believe gods word i tell you it is the hardest thing in this world to believe gods word. Truly i find it so. As to these doubts thrown into my mind ah poor me i thot when this work of god come i should have nothing to do but go right into it but o how i want sanctifycation. I have had faith but o how little am i subjected to god. Ah i think of balam read numbers 24. O my fears are great i hant given up all my idols. And o methink i am like the man on whose hand the king leand he see a pleanty with his eyes but did not eat thereof.[101] Now i can see others fild with gods love while i am under the very feet of satan and a weeakly body & much pain. O come to me o lord for i am a sinner a great sinner o purify my soul.

[*p. 118*]

Ah i fear i have been making an idol of my faith and intended to have no croos but all comfort & joy when gods work come but o how am i disappointed and o lord how many times do i abuse thee by sheltring my sins under thy croos. But i cant stay but a few moments upon one thing but must hasten along for want of time. O thou that did eat & drink & company with sinners pyty me make me as truly willing to have my sins mortifyed as to have them pardoned altho with the publican i fear to lift up mine eyes to heaven. Yet let me with the humble mary magdalene come behind thee weeping and wash thy feet with tears and kiss them.

Now for a long time i was more weeakly than common and went to meting nowhere on the sabbath. And now the opposers of gods work that was among us was inraged against us & our meetings altho as yet only on the weeaktime i did use to go sometimes to mr birds meeting which was 7 miles off. But i now being poorly my dear neighbour took the advantage of me & complaind of the stranger to the grand jury man and he come to talk with me. He told me knew our meeting and this work was the work of the devil. I asked him if he

[*p. 119*]

ever felt the power of christs resurrection and beheld his glory. He answered there is no such thing. I told him if he was a good man he would not dare to talk so. I think i begged him to get an intrest in christ for if he dyed as he was i could have no hope of him i believe. He took it as a hard saying and was offended and went his way. Now when my husband come home he said he would pay nothing for me i should go to prison and he seemd cruel against me. Now i had more room for many arrants to the throne of grace but here was my great tryal to see gods providences work right against the promises. And were it convenient i could write much about my tryals now but i forbair.

It being about the middle of may 1758.[102] The grand jury man carryed the presentment to justice davenport brother to mr james davenport that went about preaching in the time of that great work of god that began in 41. Now there was twoo women my neighbours that go to our metings and i believe christians

[*p. 120*]

happened to be at mr davenports & he sent word to me by them of what was acoming but upon inquiery i found these friends did not plead for the cause of god nor for the afflicted stranger. O how it made the tears run. Now i think of jesus when he was betrayed into the hands of his enemies. The discyples all forsook him and fled. Now soon after the officer was sent and served his writ upon me. I was poor in body and had just been sweeating for the ague in my head. I was took away forthwith in the rain it thundered and lightned and raind all the way. Soon after we got there the court was set. I was cald to speak but to no purpose. Once i attempted to speak but the justice bid me hold my tongue.

[*p. 121*]

There was twoo women that spoke for me & told how weakly i was but they was not regarded. And there was twoo women that was summonsed against me. One womans concience smote her so that all she said was that i had been exceeding poorly. The other told the justice that she did not remember that she had seen me to meeting lately. The justice said there was no need of the evidences being swoarn if they was creadible persons. See there was no evidence against me nor was i allowed so much liberty as a criminal for he may speak for himself and paul was allowed to speak for himself when accused altho he was before the greatest ones on earth.

Now this court being soon over i spoke & told the justice there was a day acoming when justise would be done. I told him god had said he would execute righteousness & judgment for all that was oppressed. I said there is a dreadful day acoming upon them that have no christ. The justice

[*p. 122*]

told me i talked sasse and he lookt with a lofty angry countenance. And now my husband being scarred he without consideration payd twelve shillings lawful money to the justice. It was contrary to my mind. Now i sat out to go home. It raind i was wett & that night i had a fit of the ague and was poorly a long time and sometimes almost overboarn with vapours and weekness. And now my husband was much offended with me because he was forst to pay so much money for me & i had disgrased him and the children. Dear man i think felt his enmity against me more than against the dear neighbours that was the means of it all. But now gods word was sometimes refreshing to me that about jonahs being in the whales belly[103] altho he had sind yet when he cryed to god he heard him. O i wondred at it this morning. Lord humble me down at thy feet & keep me here o sanctify me by thy truth. O god thy son has prayed for it thou mayst deny us but him thou canst not deny. O come quickly and distroy these antichristian laws and the power that lords it over souls. O when shall

[*p. 123*]

the first dominion come to zion o when shall officers be peace & exactors righteousnes o when shall violence be no more heard in the land o when shall our walls be salvation and gates praise. And o my god forgive my dear neighbours for i believe they dont know what they have been doing. Altho they never said one word to me nor took gods word for their rule in their practice altho profest church members but what they did was in the dark privately. O lord grant that they and i may meet in heaven at thy right hand never to part there will be all love & halalujahs. This week i run away from my family and spent half a day in prayr. Methot god helped me wonderfully. I thot i got near to plead for holiness for myself & for salvation for my family lord let me see the returns of this day.

Now after i was had up christians seemd to forsake my house. I believe they feard to shew me kindnes they was afraid they should meet with persecution twoo. O how am i now drove to god but i hope the wall is abuilding the christians are far one from another.

[*p. 124*]

I think now if i had time i could write much of what i see in the 4 chapter of nehemiah. O when shal the rubbish that hinders the building be clensed away o when shall they that should bair burdens have new strength—verce 10. O when shall christians that are far one from another be gathered like the sheaves into the floor.

O how little love do i see. I have much hardnes to endure especially from one that should be my best friend and now i am tempted hard to fling up all them promises i thot god gave me because providences workt right against them. Satan is trying to get me under his feet because of this and for what i deserve for sin. Now my cry was continually at the throne of grace.

Once i thot i could say in this tryal glory to god for what i see in ezekiel 16th & from the 60eth verce to the end. Ah will my god yet remember his covenant. I want to write what i see in this chapter but must hasten.
One day i had been to prayr. Methot i see why god suffered his children to be parted. I believe they have been leaning on each other & it seems to be of necessity that they should be parted

[*p. 125*]

and all the old building thrown down & not one stone left upon another that they may be builded anew into that glorious church that i hope god is about to build in this latter day. Soon after i suffered by our ante christian laws i see a dear saparate minister that lived about 20 miles off. He told me that he felt it with me. It affected my heart. He told me he had been in prison twice once for his rate & once for improving his gift that god had given him for the good of souls. Lord do give him benjamins mess.[104]

Now one night i went to bed sad and heavy. I dreamd that a christian woman was with me one that goes to our meetings and she sayd to me bind the tears in bundels and burn them. Now as soon as she had said that i thot i heard the great god speak thus let them alone for in a little while i will send my angel that shal gather the wheat into my garner & the chaff shal be burnt with unquenchable fire (luke 3-17). Good part of the night i watcht for fear i should forget this lovely dream. In the morning how refresht i felt while i was making my bed. Glory to god i am full of hopes

[*p. 126*]

that he will send the angel of the covenant in a very little while to gather his flock that have been scatterd in the dark & cloudy day. Lord grant that the discord & contentions that are among thy children may be done away. Now I know what that means in the 5 of judges 15-16 verses for the devisions of reubin—there were great thoughts of heart for the devisions of reubin there were great serchings of heart. O when shall thou lord be thy childrens all above all with all who shall hereafter be all without all. Ah it is good to be stript from creatures for there is a worm at the root of these gourds. O let me be laid at thy door alone for help i beg o lord.

Lately i have sundry times set apart several hours in a day to pray for myself & family. I had a heart enlarged in prayr for my own sanctifycation. O how i longed to be made holy that gods image might be perfectly restored to my soul. Ah my family is a little one o spare this little zoer[105] and my soul shall live tho i am as a publican & harlot. Yet pray come into my house & dine here. O make it a bethel a house of god.

[*p. 127*]

Now of late my husband has been seemingly unconcerned about his soul. One night there come to see me twoo young women we talked of religion & sung himns & read wattses vision. As soon as they was gone my husband went to bed but he soon cryed out o lord have mercy on me i am asinking into hell. With his hands lifted up he cryed o eternity eternity how can i lie in hell to eternity. How shall i answer at the bar of him who is most pure. I cannot answer for myself myself i cant endure. He mentioned some that had just dyed that he thot was gone to hell. He cryed out about an hour and was so spent that he lay still but he said he slept none that night. O now i had raised hopes of his conversion but presently it was darted into me that this was the effects of my prayrs and it was but a few days before his concern was almost all gone. O alas how tryed i then was for fear i had hindred a blessing and my mind about it was so unfruitful i could not tell whither it was from satan or my proud heart. Lord have mercy on us.

[*p. 128*]

One morning i awaked with them words—enoch walked with god.[106] O how i longed to walk with god & be like god to be holy to honour god to be subjected to god o to be holy was the all i wanted. And i see i never could be happy in this world nor in the next if i was not holy. O god make me holy and then i am made for time and for eternity. O for the compleat restoration of the blessed image of my god. I want time to write lord give me a patience to wait thy time to be made holy. I hope quickly i shall get what i long for.

Now this week i spent part of twoo days alone methot god helped me to pray for the church of christ and for my family. I longed for every soul in my house and can do nothing for them. I can add nothing to free grace but o i long for them with groanings unutterable. O in these days of sorrow & exile i feel a travil on my soul for my family. O god pyty pyty them i beg for jesus sake.

Now i lit of mr brainers diery[107] it was the means to strengthen and invigorate me in my journey. I see many of his tryals to be exactly like mine. He tells of his soul being often sunk

[*p. 129*]

down with dissertion i say under the hidings of gods face & his comforts and enlargements was good to me. Ah he had a weeakly body that hung as a wait on his soul. And he tells of his being much alone when with the indiens but few to converce with that caused his tryals to settle and penatrate into the inmost secret recesses of his soul. Ah methinks i had fellowship with him but he was carried threw all and made more than a conquerrer and dyed triumphing but dear man did not see into this saparation but writ against it. Lord open the eyes of all to see that it is a latter days reformation from the mistery of sin.

Mr frothingham come and preacht to us upon infant baptism. He held it up but my mind not being establisht about it i was flung into great distres and confusion. My mind seemd turnd as it were upside down. I got so at last i was in such darknes that i felt sensibly shut out from god. I felt as it were inclosed in hell itself. This i think was horror nearest akin to the torments of the damned that ever i endured. I felt almost overcome ready to faint & dye. One evening i was crying to god for

[*p. 130*]

mercy by the side of the barn and i wondred why the cattle did not fly upon me & tear me to peices and vindicate their masters honour that was by me abused. At another time it being late in the night my family being all in bed i went to pray in the same place but i could say but little for like david i roared by reason of the disquietnes of my heart. I could not tell what i wanted but i cryed till i was much spent looking up to heaven. Then i came in and went to bed & had a distressing night.

This was in 1759. I was still low and sorely distrest till one day i was talking to my husband beging him to live a better life. I pleaded with many tears my heart seemd melted down i felt constrained to pray out let who would hear. I cryed o lord do come into my house and dwell here. O make it a bethel a house of god. O let me see thy footsteps here & thy salvation. O that every soul may be devoted to god. My soul longed for god for the living god. Afterwards i felt sweet peace quiet and composed. That night went to bed slept comfortable in the morning awaked singing praises to god. Now i know god can give songs in the night to the tempted distrest & afflicted. Glory to my god. Now i had raised hopes that god will yet show mercy to my house tho a house of beliel. O father look upon thy sons blood & the tide will be turned.

[*p. 131*]

Now i think i must remark the threatning providences or judgments of god that soon come upon them that was the instruments of taking the law to punish me for not going to meeting with them. The grand jury-man a few days after was at the raising of a house and he fell from the chamber down into the celler among the stones the men said it was a wonder he was not killd. A young man that lived at his house at that time was seized with a hard fit of sicknes seemd nigh to death. One of the women that the justice took for an evidence against me had a child took sick for some time exceeding bad. The other womans son had a child layd to him by a young woman and he married her but it greived the woman his mother so people thot it was the means to bring on her an uncom-mon illnes and was likely to die and was sorely afraid of death. And a few days ago this womans son was working their only oxen and one fell and broke his thigh and they was forst to kill him and borrow one to work for they are poor. Also a man that wee suppose was in the private affair of complaining of me fell from his horse & bruised his flesh and had a great sore was under the docters hands lay a great while dangerous and hardly recovered. And a few days after mr devenport had me up he went to the grave with his wife & with a negro

[*p. 132*]

man & not long after he dyed himself with a canser on his mouth that eat down into his throat so that he could not scarcely speak or eat for some time before he dyed (i do desire and do believe i write in some measure in the fear of god without reflecting not knowing how i am to die myself nor how soon). I say o that mouth as it is creadibly reported that use to lie cheat & extort from the poor and others to inlarge his own estate. Who knows but his soul is now with the dives in hell abegging for one drop of water to cool his tongue and like king jehoram he departed without be-ing desired 2 of chronicles 21-20. Again i got low and much discouraged about my seeing the glorious day because israel dyed in the wilderness & josiah died in battle. One night i was tormented with my own thots that i slept but little. It seemd as if the devil was all night by the bedside. In the morning i was so burdened & boarn down with sorrow till i could stay no longer. I went away behind the barn and pourd out my soul to god with groans cries & tears for help and deliverance from sin & devil. And afterwards i felt relieved but not delivered. I felt my soul in some measure humbled for sin a longing after holines.

[*p. 133*]

February 1759. Mr morse[108] of new london came and preacht to us from revelations 18-4 come out of her my people that ye be not partakers of her sins & that ye receive not of her plauges. He shewed who and what babylon was in the history & now in the mystery and who and what gods people was and what they must come out from what babylons sins was & what her plauges was at large. The power of gods word seemd to seize peoples minds. I thot never man spake like this man. I believe god was with him my soul groand and longed after god. I felt new strength and courage to pres forward. Next day methot i wanted to spend my life and strength for god a lovely holy god. Conform me to thee bless the lord o my soul.

But alas it was but a few days before there came a great fear upon me that the byble was not the word of god & that this work among us is a delusion. On the lords day i felt as if i should die with anguish of soul. I thot i could change

[*p. 134*]

conditions with any creture that had no soul for i had lost the sence of the very being of a god. I could not pray it was strange to me i was tempted to say as they in malachi 3-14-15. I felt sensibly shut out from god the world lookt like a dolesome place. I said can god make me happy can he make me to rejoyce in himself. O i felt comfortles on gods earth like cain. One day i prepared my families dinner. I eat none myself but took my byble and slipt out and run away into the woods and thot to try to spend the afternoon in prayr. But o how many things pulled me back. Methot i went like the milks kine that left their calves at home & went loing as they went. Then it come to my mind how that jesus thot nothing too much that he could do for sinners. Now god helped me to pray. I see much of the vileness of my heart and nature. I see i had been building babel & had joined with herod and his men of war and had set christ at nought a hundred times by sin & unbelief. O it was a time of mourning and repentance. Methot i looked on him whom i had peirced & mourned and was in bitternes as one that mournd for a first born. O god be merciful to me a sinner.

[*p. 135*]

Now the next night but one after my fast i dreamd there was a terrible earthquake and thunder and lightning. I thot the earth shook. Me thought i was sorely surprised and fell down on my face on the ground to prayr for mercy and so awakened. That word come to my mind revelations 16-18. O god prepare me for every storm that shall be in my face here on earth.

In 1759 there was a separate meeting begun on the sabbath but i did not go for 3 or 4 sabbaths after it begun for i was waiting for something new and special to go out upon. Altho i had been praying for many years and see it clear duty but at length i went having a mind in some measure established from gods word. Now i have reason to think that the darknes and distres and temtations that i have just written was then brought on me for my not coming forward to witness for god in going to the meeting when it first began. But the first sabbath that i went to the saparate meeting eight of us was complaind

[*p. 136*]

of to squire sacket and when he had threatned us & see we did not intend to pay our fines he went & put it into squire davenports hands thinking it seems that cruel govenor he will do the work. And now wee was looking every day for a fortnit or more to be carryed to prison or to the whipping post or stocks or be sold to pay the charges & fines. I felt in a measure resind. I had a prayrful mind & so i must beleave it not knowing what the event will be.

Now after a while wee heard that mr davenport went to the great men at new haven but they would not help him. Then he went to the assembly at hartford but they would not do anything about it so it droppt threw. The hearts of all are in gods hands lord destroy ante christ by the spirit of thy mouth and the brightnes of thy coming.

Now for about half a year i was in great pain by turns in my head and teeth. I was up many nights asweating groaning & crying sometimes vomiting means seemd inafectual. I seemd

[*p. 137*]

at last almost to forget what ease was. One evening i was in great misery. I felt alone none to pyty me. I thot i was forsaken on earth but when i was sunk down in dispair in a flood of tears and anguish left alone o thou my god didst think on me. That word come to me judges 8-17 he beat down the tower of penuel. This word come to me in the evening and i could not let it go altho i had a distressing night. Now as soon as it was light i went out away alone and cryed to god to fulfil that word that come last night. I begged o break down this tower o give ease or i die dost thou not delight in mercy. And it was but a little while before i was easy and have been so a month. As to that pain glory to god o that thou shouldest take knowledge of me seeing i am a stranger. I was fetcht to a woman in travil there was not a word spoke for god. I felt so bound up in my mind that i could say nothing. It being over i hurried home. Then i felt like one got out of prison. O how my heart went out

[*p. 138*]

to god. I got alone and cryed lord give me not my portion in the friendship of this world nor in the riches nor honours of this world. Ah i had a sweet time of prayr & praise and love glory to god that has caused my bonds to be a means of my liberty and brought good out of evil.

Now mr marshel[109] of cantrebury come and preacht to us. God was with him and my heart seemd fild with love to jesus. And in one of his sermons he cryed out thus o saints there is a soul intangled in the net. Lift up your cries to god help me draw the net ashore. And before sermon was done one man i believe was converted. He manifested it then and shews the fruit since. He preacht one sermon in my house. He said he see mercy hovering round this family. He believed mercy was coming into this family. Methot i see the love of christ ran freer than the water in the river. This word was powerful on my mind i am not worthy that thou shouldst come under my roof. I believe the spirit of god was here. O what can a worm say glory to god.

Now elder morse of new london come. He preacht from samuel 1 book 4 chap 21 ver. He shewed that when christians got away from god it might be said the ark of god is taken. He shewed how tender god always was of his

[*p. 139*]

ark how he defended it. He kild uzza[110] for touching of it as you may read in the history and he opened the mistery. He shewed how god loved his children how tender he was of them and he charged the opposing world to let them alone. I heard gods word with delight that night. I thot i as it were sat at jesus feet and washt them with tears as it were wiped them with the hairs of my head. I went away alone & had a sweet feast of love. This verce was sweet to me

Those soft those blessed feet of his
that once rude iron tore
High on a throne of light they stand
and al the saints adore.

Also i had an impression on my mind at this time of great tryals a coming which soon come on for the camp distemper began amung us. Twoo of our neighbours have buried seven children and it spreads daily. I expect it every hour to begin in my family. I felt a travil in my soul for my husband and twoo children and for myself. I pray & fasted yea a week almost without ceasing. I longed that like noah and his family wee might get into the ark. O the storms of gods wrath that i see a coming on the wicked world. How many times did i now lie on the ground on my hands & knees nights & days

[*p. 140*]

crying for mercy mercy o spare thy people lord o try us once more with a day of the down powring of thy spirit and condecending grace. Will it not do more than all these sore judgments have or can do in turning souls to thee.

At this time mr miner[111] of lime come and preacht at our house. I believe god was with him. He seemd to come like david 1 of samuel 17-17-18 to see how his brethren fared in the war & to bring us parched corn. When he had done the meeting the enemy raged & mocked but we went to singing praises. Next day he prayed before he went away. He seemd to get hold of god for my husband and my soul joind with him. He seemd to have faith to believe he would be converted. When had done he turned to me and said i tell you sister your prayers shall not be lost. Only believe & you shall see the glory of god. When he went away he went with his hand up towards heaven. Now when mr miner come to my house i was perswaded he was a messenger of the lord come to warn us and i was sorely afraid we should not take the warning. That word was heavy on my mind that night so that i could not sleep quiet. It was about davids being affraid to bring the ark home because god had kild uzza. I was in distres in my mind for god had just taken away 3

[*p. 141*]

children from the family that we use to meet at on the sabbath and the fourth lay just adying and did dye soon after and o the warnings that they had one messenger after another. O how fraid was i of the lord. That day i said in my heart how shall the ark of the lord come to me 2 of samuel 6-9. Ah what a dreadful thing i see it to be for the great god to send a prophet to my house to warn us to fly to christ and for us not to regard it. My soul cryed to god that day for my family. Methot i could die if they might live to god. I longed to hear them praise my god before i die.

In 1759. I think there is nineteen that have dyed here within a little compas in this sickness. I think now of what i see in my sleep writen in page 135. Now at the beginning of winter i was taken ill. I was by turns in as much pain as i thot as i could live under. For 3 months i was sometimes up night after night in pain and weaknes and now my neighbours contrived to have me up again for not going to meeting. The grand juryman come to talk with me. I read to him the 2 of collosians 16 i begged him to think of it. Lord open thy word to his soul do convert him to thyself for jesus sake.

1760–1769

[*p.* 142]

Now this last complaint was dropt threw i believe gods hand was in it. But now i think the devil is come down in great wrath against me for the place was fild with bad stories about me. Report it say they & we will report it. Now the terrors of hell got hold of me & the floods of ungodly men made me affraid. It seems god is withdrawing from this place. Now my sleep went from me. I watered my couch with tears my tears were my meat and o my weeakly body.[1] Now i have none i dare open my mind to. Christians are parted. Now the best comfort i have from them is they will come and tell me what the opposers say. I feel sometimes as much alone as if there was not another person on earth. Now i feel as if i was forsook of god & man. I feel as if i shall never get to heaven. I see the righteous are scarcely saved by reason of the oppositions they meet with here. Lord help me to persevere.

1761 february 13. I kept alone with fasting and pryr for myself & family. I read in jeremiah & lamentions it was a day of mourning. I longed greatly to be made holy. God be merciful to me.

[*p.* 143]

Now mr stiles being dead this people have hired and setled another teacher. Now he can walk with this church and many of a corrupt life are indulged in it without being dealt with according to gospel rules. I find no heart to joyn with him.

In 1761 the rain of heaven was suspended. Every green thing was almost dryed up crops much cut off. And in 1762 the drouth was harder still poor cattle ready to starve for grass. God threatens us with a famine still i see no turning to god. O then what must we expect lord have mercy on us.

In 1762 there was a woman that was a saparate.[2] She come to live in this neighborhood. She was a cunning creature & had great knowledge in the schriptures. She was much for talking and in all her discourse with us she would pretend to know and see something clear away beyond us all. Christians got intangled by her and was led into darknes. Her talk got such hold of my mind about 3 days that i was almost in dispair. I tryed to fling away all my former experiences and get something more wonderful. I cant tell the distres i was in. At length i was helped to vew my past experiences and to

[*p.* 144]

believe it was gods work in my heart and his teaching. And in vewing this i got an establisht mind and then i see this woman standing plain. I always before found her talk led my mind into distres there was no relish in it. She was a person of good behavior but fild with pride. She caused devisions among us and was a means to break up our meetings i believe. But not long after she went out in her head & was in a dispairing way saying often i fear i must sertainly perish. And after some time she got well in her sences and said that she had been the means to break up our meetings. After a while she joind to the separate church at middletown and there for her contending and wanting to govern she was dealt with by the church. And now 2 or 3 years has she seems to have no regret for what she has done but stays at home and will not go to meeting.

In 1762 in the winter the young man that was leader in our meetings got discouraged and moved the meetings to wallingsford 6 or 7 miles off for there was 2 men there gifted to help carry on in the meetings. Now i knew it not till the next sabbath. I went to the usual place but i soon understood that they were all gone to wallingsford

[*p.* 145]

to meeting. Now i soon returned home with a heart full of sorrow. O methinks i must mourn now with nehemiah ezra and jeremiah when they seeth walls of zion broken down and her gates burnt with fire. O methinks zions breach is great like the sea who can heal thee lamentations 2-13. O now i had many arrants to the throne of grace. My tears were my meat. Ah moving the meetings i believe is not the mind and will of god. Ah me the ark is carried away. Ah how do things work against the promises. Ah how did jacob feel when his benjamin must be took from him and go away to egypt.[3] O lord god hold me fast in thy hand or now i shall let go all thy promises. And now my husband keeps me at home ah userped athority.

1762 lazarus ives of this time got up in the night and left his wife and eight children in bed with nothing on but his shirt. He went about a quarter of a mile on the snow barefoot a cold night

[*p.* 146]

to a lonesome barn and cut his throat & dyed with the knife in his hand. His blood flew on the timber of the barn. This took hold of my mind for a long time. It made me cry often to god to keep me watch and preserve me. I see if god left me i should do so too. It seemd to seize my mind with anguish and terror but it made me pray hard. And in 1762 i think in may our companies was training and enoch ives a young man was shot down dead. He had not time to pray or speak one word. Now again through gods grace this quickened me. O lord grant that i may be actually ready when death comes. Also in 62 john pain was drownded in the river. O my god dont let none of these sudden deaths be lost to this people i pray. In october 7 i kept a secret fast for my self and family. I bewaild my own sins & the sins of the family. My heart seemd enlarged in prayr for my family and for my own sanctifycation & purging for gods own appointed means of grace in this place. O lord hasten the

[*p.* 147]

day thou did give me to see 11 years ago when thou would gather thy flock like the sheaves into the floor. O send out the hunters to hunt them out of the holes of the rocks.[4] O come quickly & gather the outcasts. I am one of them. My meetings are moved away i am left behind. O methinks i lament with israel after the ark. Lord make me truly willing to love & serve thee alone til i die if it is best for me. Dear elijah was willing to serve god alone and joseph in his prison cell tho his feet was hurt with fetters and he laid in iron just like poor me.[5] Lord come quickly.

Now a wicked person ronged me much. I cryed but did not feel revenge full for i see i deserved it from the lord. I went away and prayed with a longing heart for the convertion of the dear persons soul. Shew pity o lord forgive.

1762 november 7. That word in isaiah 37-36 took hold of my mind concerning the angel distroying an hundred and four score and five thousand of the churches enemies and has not god said that babylon shall fall like a milstone cast into the sea shall be found no more at all.[6] Lord hasten the day.

[*p.* 148]

I have been to meeting a few times to our standing meeting but the abomination that makes desolate is set up there. Mr trumble[7] preacht 2 sermons from isaiah 2 chap 2 and 3 verces. He put a temporal meaning to his text and said that meeting house was built for prayr & praise & he said it was zion and they that did not come there to meeting no place was for them but hell meaning the saparates i suppose. Lord shew him by thy word what a zion of god is. Ah will he not then purge out hypocrites & scandalous persons out of his church. Ah now i am shut out from hearing him any more. I know not but i must stay at home till i am took to prison.

1762. Twoo children dyed suddenly one was drownded the other choaked and a woman dyed suddenly all in this place and in this town a child was burnt to death. O what work sin has made. I am now almost 42 years old but o what a monster of ingratitude do i still continue after resolutions and hopes of doing better. O how dull & wandering is my heart o lord is not quickning clensing strengthning promised in thy covenant. I feel and know there is nothing in this world can satisfy my

[*p.* 149]

soul and yet a little disappointment in creatures sometimes will discompose my mind. O how hard am i to forgive past injuries. O jesus wash me in thy blood. Lord thou knowest i long to be made holy. If i never go to heaven more than ever i longed for anything in this world. These ten years have been years of prayer with me to let me see the returns. Last week heard mr miner preach. I went out and returned and o the opposition in my family. But my heart went out to god for help and strength i gave freely to the dear minister. Now i spent part of a day alone in prayr for hopes and expectations are sometimes raised. I hope to see souls yet born in my house. I hope yet to see god at work among my neighbors. I hope yet to have the appointed means of grace and to see god build up zion. One day my dear husband got in a passion and talked against me extreamly to a sad man one of our neighbours. Ah if it was an enemy i could have born it but i fear i was now more touched for my own honour

[*p.* 150]

than i was for gods. Now it was but a few days before he acknowledged his fault to me & went and took the man alone and acknowledged his fault to him and desired him not to tell of it. Dear man he has sometimes tryed a long time with his abusive talk to provoke me to be angry and has tyred himself out and confest he could not. It was nothing in me it was gods mercy and free grace to me. I have felt sometimes my mind so ingaged in the things of god that i have scarcely heard what was said and sometimes sin has sat heavi on my soul and then the cross has sat light again. I have felt sweet peace and strength not to open my mouth. Ah my greatest tryals i write not. O tis a heavi but just affliction for i knew he was an enemy to god when i marryed him and gods word forbid it. O how many times has that word been heavy on my mind—o assyria the rod of my anger the staff in their hand is my indignation.[8] Ah it is gods rod for my sins and i hope god is killing my sins every day. The lord has done right he hant ronged me. I it is that have sind but thou o lord doest all things weel—Hannah Heaton

[*p.* 151]

Last sabbath i was mourning at home alone for the sins of the family. While i was praying that word raisd my hope again—in the habitation of dragons[9] where each lay shall be grass with reeds and rushes. Lord is not my family a little one. O spare this little zoar and my soul shall live. O say to me as thou did to lot genesis 19-21.

1763. Now my poor husband seems more prejudist than common. He will not let me have the horse to go to the saparate meeting at wallingsford. I hant been a great while. Alas i confess one day my spirit did rise and i forgot my principle of lying at his feet. This i must confes to god & to him my sins has cost me more sorrow than anything i ever felt in this world for in my nature is a continual spring of sin. Lord clense me o when shall it once be.

Again i was weeak & fild with pain & i was tortured for some time with my own thots things that were meere trifels of no consequence. I was worried night and day. Alas flies or lice was twoo hard for a faroah when god sends them.[10] This week i spent some hours in prayr alone one day i had some freedom. O lord let me see the returns i beg i beg.

[*p.* 152]

One morning i awaked with this in my mind—he that has known my name on earth ile honour him in heaven. There my salvation shall be shown & all thy sins forgiven. This verce helpt me glory to god. I have writ in page 114 how my mind was exercised in time of that war and how i had faith in gods word that he would spare new england. The war began in 1755 & it held till 1763 & then peace was proclaimed and a day of publick thanksgiving held throughout our colony. But o vile me i did not give god the thanks the mercy called for when i had seen gods word fulfild.

1763. I think about a year and a half after our brother benjamin moved the meeting to wallingsford he left it & goes now to the standing meeting here constantly and was not willing to be subject to the church at middletown and so was turned out of it. O lord let thy promises hold me up that i have laid up and pondred this 12 years. O lord let me yet hear the sound of thy goings. Pray sanctify my soul prepare me to serve thee with my whole heart day by day till the last moment of life.

1763. In september caleb todd of this place was killd in a moment by a timber falling on his head. O god dont let this young mans death be a lost warning.

[*p.* 153]

1764. O now my sins are gone over mine head as a heavi burden twoo heavi for me to bear. My stroak is heavier than my groaning. I feel the power of sin strong in me it drives me to god with lamentable tears sighs & groans & sometimes get a little ease again. What a hard heart do i feel. Ah my sin is anuf to bring a curse on all i do. I am under the feet of satan in respect of it. Lord i know not whither ever such a sinner as i come to thee for mercy whither ever such a work was done to any poor wretch as the saving of my soul must be. If god helps me now it must be a creating work. God is angry at my sin o i have done very wickedly i am in the furnace of affliction but unhumbled i have great mercies but unthankful. I am now 43 years old but days and years hant made me holy. This week i spent part of a day alone in reading gods word & prayr. My mind was in some measure engaged some lively sighs and groans to heaven. I had a sence of the mortality & nothingnes of poor fallen man. O god pyty me.

1764 may. I hear gods work agoing on at long island wonderfully in many places. O arise here sun of righteousnes lord come quickly.

[*p.* 154]

1764. In july my jonathan was sick with the putrid fever. Now my cryes was to god for him. Now he would let me talk to him. He is now 20 years old but i believe he lives wholly without prayr and but little regard if any to the sabbath and but seldom reads & goes often to frolicks disobedient to parents. Ah i am often reproved in that spot. O how he is to his tender mother that bore him and brought him up with sorrow. He in pride joins with his father against my going to the dispised saparate meeting. But o i have a fixt hope for his soul that he will yet be an isaac and go out into the gospel field to meditate. Ah it is better to build by faith on gods word than to trust to signs & tokens. O god i beg do convert his soul. Dont let him hate thee & thy cause & thy children no longer. Dont let him spend no more of his precious youthful days in sin and vanity. O thou that made him this soul—glory to god that has heald him of his sicknes. O let me live thy praises. Ah me i hant returnd him the thanks the mercy called for. O fit me for thy whole will.

May 5 1765. This morning i felt prest in my mind till i pourd out my soul to god. I told the lord ah my soul cryed lord thou dost visit the earth and bid the grass appear and wilt thou forget the purchase of thine own blood. O look on zion & see her down in the dust & sinners

[*p.* 155]

saying so would wee have it. Methot now i felt resolved to give god no rest day nor night till he shall make zion a praise in the earth. Methot if i never see it i will go mourning down to my grave for zion.

Twoo nights ago i thot i heard them words spoke to me—i will avenge thee of thy adversary.[11] I cryed so and made such ado i awaked my husband & myself. I hoped god was going to destroy sin in me my worst & greatest adversary. O sin that enemy to gods honour & glory. I felt afterwards comforted and strengthened & hoped in his mercy.

July 5. I felt a load of distres upon my soul. I got me away to prayr. I thot i see these words was in the covenant—in the habitation of dragons where each lay shall be grass with reeds and rushes.[12] It seemd as if i come to god with a warrant for mercy for my family. These words have been words of hope to me these many years. Methot i see the awful case of my family. I see them ondone & lost forever forever without the power of almighty god. My soul wrastled hard for them with tears & groans unutterable & with god. I must leave them i can do nothing for them.

[*p.* 156]

1765. After much consern in my mind about the sabbath being changed from the seventh day to the first i was established from hebrews 4.

1765. In july my calvin narrowly escaped death. He was riding & the horse took a fright by a dog and he fell and was stuned. One of the neighbours saw him lie on the ground senceless & spechles and carried him in using proper means. He recovered. Glory to god for his mercy to him and me. O give him an intrest in thy love and favour o sanctify it to us all.

1765 August. Now my husband was taken sick with a malignant slow fever. I tended him alone a fortnit day & night with watching & carefulnes. But o how impatient & fretful he was altho i waited on him as well as i could for my life. One day my heart was ready to break to see his ingratitude. I went away into the barn and lay down on the straw. I thot i could do no more for him let what would come out. I cryed & grieved & prayed. I thot of running away into the woods. I thot of davids caves & dens o that i had one now. I knew not what to do. At length my judgment began to work the governing faculty of the soul. I thot altho i felt as i did yet i knew to take christ for my pattern that was right. Then i thot how he did good for

[p. 157]

evil he gave blessing for cursing so then i got up and went in. Now i felt it to be a heavi cross but i went to waiting upon him again as well as ever i could (now he was so weeak he could not sit up but a few moments at a time). And now my burden fell of i felt peace in my soul i could do for him freely and pleasantly and in a little time he grew better and was well. Glory to thee o god for thy salvation to my soul and for healing mercy to his body. O god pyty his soul o make him thine by converting grace o make him thine in that day when thou wilt make up thy jewels & o let me live thy praises. O my god keep me at thy feet.

1765 november 15. I spent part of this day in fasting and prayr for myself and family in the woods not far from the house it was cold. Now i had some freedom in prayr some mournings to my god some lively sighs to him to let me see the returns of this and all such days i have spent in secret.

On the lords day morning i had a mind to go to meeting to wallingsford but many deficulties presenting i stayed at home. I was tempted to wish that god had not shewed me the standing

[p. 158]

meetings was rong. Methinks i felt like israel when groaning under their bondage & wisht moses had not spoken to them but afterwards i was sorry & afraid.

O methinks what a pyty it is that god that made the world should have so little of its service & satan that undid mankind should be so willingly and expensively adored. O how long shall he have that userped tytle. The god of this world help me. Lord i beg i beg to plead for thee by my practice and pray and believe and live suffer & hope and repent thyne & my enemies down lord help me.

1766. This winter has been a time of confinement to me and a time of prayr repentance & mourning for sin. God gives me to often pray in bitternes of soul and with great desires for my own sanctifycation and for poor christless ones and for the appointed means of grace in this place. I feel my hopes raised by that precious word—thou wilt prepare our harts to pray and cause thine ear to hear.[13] Ah lord thou canst save to the uttermost of our thots desires prayrs tears groans. Lord i long to see thy lovely arms spread out here to pull in souls to thyself like him that spreadeth out his

[*p.* 159]

arms to swim. Lord here sin abounds. O let thy grace much more abound as it has been wont to do. O jesus i invite thee into my house today. I dont read that ever thou did deny any when here on this earth but did go in and eat & drink with publicans & sinners & must i be the first that ever was denied. O let me see thy lovely footsteps in my house & my soul shall live. Alas the wicked has seen many of my miscariges & stumbled at them but they hant recovered by my secret mournings sobings pantings & persuings after gods forgiving love again.

February 6 1766. I see and felt myself to be a great sinner. I went & confest it to god with heart groans cries and tears. I see how little i have loved holines. O jesus thou hast taken our nature into union with the divine & dost now wair it in heaven. Then why wilt thou not help me purge me purify refine wash me seal me shalt thou not be more admired and loved for such a matchles mercy than for all thy works of creation or acts of vengeance & will it affect me only. O will it not run threw heaven. Likely many will admire more for thy grace to me than to themselves. Lord help me.

[*p.* 160]

1766. In june my little calvin was very sick with a fever and flux. My soul was in distres for the child. I went to the lord for him and god suffered me to plead hard for him & if it was for his glory that he might live. And now when i come in there was a strange alteration. I se the child was better & he soon grew well. O god i thank thee that thou art a prayr hearing god. O pity his soul do convert him. O now let me live thy praise. Now twoo mornings as soon as i awaked these twoo verses was in my mind

With dreadful glory god fulfils
what his afflicted saints request
And with almighty power reveals

his love to give his children rest
Then shall the flocking nations run
to zions hill & own their god
The rising and the setting sun
shall spread the saviours name abroad.

Now i had raised hopes of gods appearing to build up his own cause. O come lord jesus come quickly in thine own time i beg i beg.

1766. In the fall i met with long & sore trials & besides i had teagious boiles under both my armholes for about six weeks. I suffered much for want of friends but o i felt a backsliden hert from god the worst of all. Oh methinks the agonies i felt in the first pangs of my repentance are not worthy to be compared with that of a backslider in heart when holden with the cords of their sins & fild with their own ways & o to see them i toock to be great—proverbs 5-22—

[*p.* 161]

christians turnd back and crying out by their practice there is no heaven no jesus that dyed on the croos at mount calvery. Ah i see sinners secure & saints asleep. I hant been to meeting about 3 months but god was kind & merciful to me a sinner. I lit of a book that helpt my soul. It was harrisons spiritual pleadings & expostulations.[14] I was in great darknes fearing the fruits of grace was extinguished in me. I felt so much of the power of sin but saith he every creature convays its sound its tone & tune to the young ones & none of gods children are stillborn. The spirit unties their tongues & sets them acrying aba father. He knows thou dost cry sometimes not as a thief at the bar but as a poor child when in distres who daily asks his fathers blessing. Thy heart is against a covenant of works but for all the works of the covenant he knows that nothing in the world so wounded thee or went so near thy heart as thy tempting & grieving him hath done. Thou art resolved never never to forgive thyself tho he does. Here we are full of hindrences & incumbrences which makes a hell above ground to an honest heart. O then how intolarable is the nether most hell for there is never a nooch nor corner in it where a poor sinner maight weep eternally without blaspheming without hearing blasphemies

[*p.* 162]

without hating god without sinning against god saith he. The cananites are left in land to keep down pride to try whither wee will follow the lord or our lusts to teach us war & to exercise our graces to make us keep more above upon the mountain god makes. Our corruptions do us some service which our grades can not do without them—judges 1-34.

1766. Sin in my family grows worse & worse wo is me that i sojourn in meshek. Now i see i have done the same in the nature of it as israel did in having a king. They would be like other people altho samuel told them from god the cruel tyranny & sorrow they should be under by reason of their going contrary to the mind of god—1 of samuel 8 chap. Read from the 11 ver to the 19 and ye shall cry out in that day because of your king that ye have chosen you and the lord will not hear you in that day. O wo wo is me for not harkning to gods word when i married o that word in isaiah 8-21. O christians take warning by me. Ah i have lain under this assirian rod 23 years but blessed by the lord it has sent me often with arrants & gods promises to the throne of grace. O my god grant that out of the eater may come forth meat & out of the strong sweetness.

[*p.* 163]

O let my petetions cry before thee while i live and after i am under the clods much much have i lost on earth. O lord let me gain it in heaven and sing louder than any there. O how would the extent of free grace be ever known if it was not for such superlitive sinners as i am[15]—Hannah Heaton.

I being much kept under at home & often shamed for going to my meetings and sometimes shame goes with me to meeting and there devil sin & shame will plague me so i cant hear o my jesus who bore the shame deliver me.

The lords day being come i was under tryals i mourned & prayed till my spirit tired. I staid at home methinks i felt like them that could not go over the brook besor they were so faint.[16]

October 20. This sabbath my husband rid to his meeting so i staid at home. I prayed all day for sanctifying grace. I pleaded hard methot i felt my famine. Lord help me to make haste away like jacobs sons to my brother joseph to get bread that my soul may live. Ah he is not only lord of egypt but of goshan canaan labanon. O that cluster of grapes which the spies carried from the brook of ashcol to kadesh

[*p.* 164]

they bare it between twoo upon a staff—numbers 13-23. Ah methinks it held forth christ crucifyed between twoo on a croos of wood.

November 21 1766. I spent part of this day alone in prayr in the woods not far of for my own sanctifycation for gods help on all accounts. When i am shut up this winter in my prison it may be not hear the name of god seldom mentioned for months together unles it be in a way of prophanenes. It was so last winter. Now i begged for the souls in my family o spare zoar & my soul shall live and o dear redeemr come to zion. The next sabbath i went to meeting to wallingsford. I felt my mind helped & comforted while one improved from psalms 11-1 & 2 ver & from s: song 4 last. I felt my mind more calm and quiet all the week than usual. The next sabbath my husband would not consent to have me go. Ah poor week me i got overcome with grief. I thot how hard it lookt a good horse to lye in the pasture all day and how kind i have been to help him to go to his meeting. O help help help me lord thy hand has permitted it. O give me and keep me in a forgiveing frame & father forgive him for he knows not what he does.

[*p.* 165]

Now one night these words seemd to take my mind in isaiah 5—i will throw down the waal and take away the hedg it shall not be digged nor be pruned. Now i thot this was the case in this place for god suffered mr beech to remove our meeting to wallingsford 6 miles of and the glory seemd to go up from us to heaven when our meeting was gone. For i believe god calld him out to improve here and was with him. But now he has left that meeting & fell from his standing & has for a year & half gone to the standing meting in this place & is shut out by the church at middletown & gods heavy hand fell upon him & his wife. They were both seized with great ilnes of body & weeaknes and was under confinement the bigest part of a year. Now in this affliction he seemd to have some regret of mind acknowledged his having lost the love of god & now his mighty hand was out against him.

1767. I was taken again with extreem pain in my head and teeth i cant express what i endured for about 3 months. O how many times was i up anights sweating groaning crying praying. I thot sometimes it was like hell torments. I suffered much for

[*p.* 166]

want of kindnes but now when gods time of mercy was come one day i thot i got hold of god in prayr. I thot my cryes reacht heaven & soon after i grew easier day by day tho weak & faint. O god let me never forget to thank thee for ease. Thou hast remembred me in my low estate when no eye pitied nor arm could help thou did bring salvation and take knowledge of me a stranger. O let me live thy praise lord i beg it of thee. O christless sinners what torments i felt when god touched me as it were with one of his fingers. Ah what will you do when he will press you with both his hands & pour on you full vials of his wrath without mixture of mercy & no deliverer there. Now christ is offered except & be forever blest.

September 1767. This sabbath was a sweet sunshiny morning a sweet sabbath a pity this day should be lost. Ah cruel faroah would not let me go to serve the lord & when i go how am i thrust out in haste. But ah he is my dear husband still he is a good provider. Methinks god as raily now makes a raven to feed me as he did elijah. I think of late i have seen that i have an intrest in them gifts jesus has received of his father for men even the rebellious & he must give

[*p.* 167]

them out. Lord what will thou do with thy gifts if thou dont bestow them here in my house the most kneedy. Now i often think of that word 2 of samuel 6: 12—altho his sin was put away yet god says by the prophet the sword shall never depart from thy house because thou hast despised me.[17] O lord grant that through the croos i may be crucified to the world and the world crucifyed to me.

Again on the sabbath my husband was not willing to let me have the horse to go to my meeting. Ah i think of them words if it was an enemy i could have born it but it is my familiar friend in whom i trusted and them words i come to set a house at varience & your chief foes are them of your own house hold.[18] But o my stroakes are fewer than my crimes and lighter than my gilt. I am yet here where i may pray i am not in hell and i sometimes see a lovely sweetnes in gods word glory to his name. But o alas who could believe me if i should rite the sorrows & sufferings i have endured this summer and fall. Lord have mercy on us.

[*p.* 168]

This week one morning i was troubled with vain company till my soul was burdned and pinched. I went out to the back side of the lot and there i pourd out my soul to god with great freedom. I with a free soul confessed my sins and wandrings from god. I thot i was glad that god knew all my sins. I prayed hard for myself & others. At length i held up this schripture before the lord which i thot i had many years ago for a promice for north haven. It was this: viz there shall be boys & garls yet playing in the streets of jerusalem and old men leaning on their staves for very age[19] but now all at once this dout was dasht into my mind that this promise was to the jews only. And i got to thinking about it and so lost my hold of god but when i come in i had great peace in my soul my burden was gone.

November 1767. A woman come to see me & told me she had been in such distres in her mind she could not sleep nor pray it seemd like hell. Now her talk cetcht my mind. I thot i felt just like her i was in such distres i could hardly speak to her. Now she said twice she was fild with comfort and felt happy but i felt so i could not see where she was nor help her. Now when she was gone i went to god for help help my soul cryed for mercy mercy.

[*p.* 169]

I see that instead of my being in the work of god i had got into the horrible pit & miery clay. I thot that in joel first 17 was my case (the seed is rotten under the clods). The word and grace of god that was sown in my heart i held it down under the clods of sin like rotten seed that dont spring and grow. I fear i fear but i thot i had great liberty to confes sin and beg for mercy. And now all at once i felt my burden took off. I was calm & quiet and felt peace. Then i thot i see the womans case that she had got false comfort no conformity to god. I must be faithful to her the first oppertunity. Lord help me to be more faithful prayrful watchful i beg it of thee.

In november 1767 ah poor me with an anctious mind i sat away to go to new haven. The tide being very high the crosway all over covered for about half a mile when my horse had carried me about half way it stumbled & fell of the crossway into the mire and so i waded out about up to my knees. I ringed my cloaths rid a mile got to a fire & got dry things. Lord i thank thee for preserving me from such an awfull death and preserving me from taking cold. O god let me always be in an act of readiness for death. O let me never again be anccious about anything beneath the clouds. O break down the power of sin in me for his sake whose holy humanity was arraignd to dye as a criminal hurried away to execution & hammered to the gibbit.

[*p.* 170]

1767 november 29. I heard mr miner of lime preach from mark 16-9 & 10 concerning jesus appearing first to mary magdalene and she went and told it to them that had been with him as they mourned and wept. He shewed where there was much for forgiven they would love much & what was the causes of god childrens mourning in many perticulers. One was their mourning for want of sanctifycation & conformity to god. O said he you will get your case. Jesus christ is set for it twoo. It seemd as if i got new strength & courage and raisd hopes for my family. God was gracious to me. Oh let me never dispair again.

February 1-1768. This day i saw a man extremly angry for a meer trifle. I went away into the cow house with a soul full of sorrow. There i pourd out my heart to god for him with great freedom. I thot i was willing to die if i stood in the way of his conversion or if it might be a means of it. Strong & powerful schripture arguments come to my mind to plead for myself and family. O god of prayr hear i beg i beg & answer. O wilt thou not when thou has prepared the heart to pray. O have mercy on us for his sake who had his nerves racked on calvery with exquisite pain and his blood inflamed by a raging fever cryed i thirst & o unparralled hardship was denied the poor refreshment of a single drop of water. O god look on him for us who had nowhere to lay his head till laid in the silent grave.

[*p.* 171]

1768. Mr beech[20] was excepted again by the church at middletown and not long after began to preach. A few days ago my jonathan went to the funeral of a lad & come home light & airry & played with his brother. Ah poor child turns a deaf ear to gods calls and he feels a bitter enmity against his mothers religion & minds me not about frollicking but i have layd up that word in my bosom—genesis 22-2. I hope i hope he will yet be an isaack o god bless my seed while sun & moon endureth.

Now one morning i awaked with this verce in my mind—why hidest thou thy face when foes stand threatning round & sad calamities appear & not a helper found. It ran in my mind in the morning & about noon a temptation come heavy upon me. My mind was sunk indeed i felt god hide his face. I cant tell the anguish my soul felt & so i forbair. I dreamd i see it writ on paper that without conformity to god i could never be happy.

1768 aprel 6. I kept a day of secret fasting and prayr. I had some liberty in praying for sinners. I pleaded hard that of the unjust judge helping the widow & will not god do much more. Lord i invite thee to come into my house & dwell here. Yet dont let me be the first that ever was denied. O god must i be tryed every way. O bring me forth like gold tryed more fit for thy use.

A tabletop monument, symbol of wealth and status, marks the grave of the Reverend Isaac Stiles, Congregational minister of North Haven. During Stiles's ministry, Heaton disapproved of his uninspiring sermons, criticized his opposition to the Separates, and finally withdrew from the legally established church. Courtesy of the North Haven Historical Society

A

SERMON,

Preached at the EXECUTION of

Moſes Paul,

An INDIAN,

Who was executed at *New-Haven*, on the 2d of *September* 1772,

FOR THE

MURDER

OF

Mr. *MOSES COOK*,

Late of *Waterbury*, on the 7th of *December* 1771.

Preached at the Deſire of ſaid PAUL,

By SAMSON OCCOM,

Miniſter of the Goſpel, and Miſſionary to the *Indians*.

THE THIRD EDITION.

NEW-LONDON:

Printed and ſold by T. GREEN, 1772.

Heaton was impressed when she heard the sermon delivered by the Mohegan Indian preacher, the Reverend Samson Occom, on the occasion of the execution of Moses Paul, an Indian who had murdered a white man. Occom's sermon, on the evils of alcohol, first was published with memento mori imagery and later reprinted by temperence societies. Samson Occom, *A Sermon Preached at the Execution of Moses Paul* (New London, Conn.: T. Green, 1772), title page. Courtesy of the American Antiquarian Society

4

pray i haue some times been so ouer come that i am fain
to go and ly on the bed panting for breath ready to fai
nt for some time and now like the men of succoth i am
taught with the thorns and bryers of the wilderness
o the snares that is layd for my soul but i am often
struck silent i marvied not in and for the lord
but o a good god causes him to feed me as he did the ra
uens to feed eliah & i haue thot it was as great a
marvicle but ah now i often look back to long island and
think of them dear meetings and christians that i com
away and left but ah here i am alone like a sparrow
on the house top here is none to guide me of all the
sons that i haue brought forth in my spiritual whore
dom o the strong corruptions of my heart & nature
often sends me with great arvants to the throne of
grace and cry as iehoshaphat did o my god wilt thou
not iudge them for we haue no might against this gre
at company that cometh against us neither know
we what to do but our eyes are upon the this day i long
ed incessently to take my farewel of this world by
4 reason of in dwelling sin in me and indulged sin i see
in others o sin cursed sin through christ i haue made
war with the & shall haue the victory at last i
44 trust) up on the lords day i was low and full of sor
row i could not pray the heauens seemd shut up
this word come sweetly ye are more worth than

Heaton gives expression to her sorrowful emotions, remembering happier days spent on Long Island, in contrast to her situation in North Haven, where she is "alone like a sparrow on the housetop." Written in 1753, when she was in her early thirties, the handwriting was crisp and clear, despite her despondent mood [p. 44]. Courtesy of the New Haven Colony Historical Society

261

was taught that she was a sinner and a
bout wor ship twoo and how he shewed her
that he was christ and he is now wont to tea
ch his children i had some mel tings of soul
while i heard the gos pel preacht when i come
home there was car nal com pany in ter taind
that night and i was sore ly dis trest with sin
and de vil my sleep dis tur bed so i almost
lost my met ting but i plead yet o god be mer
ci ful to me a sin ner o pyty my dear hus ba
nd that would not go with me to this meting
pray bless thy word

397 aprel 22 1775 there was a gre at a lar rom
threw all this coun try on hear ing that
gen neral gage es ar my and the boston men
was fighting and great num bers was kill
ed on both sides now great num bers of our
men are gone and go ing to help them o god
have mercy on our na ti on o bring them
in to near nes to thee and to one a nother
o the sin of this land in re fu sing and
trampling on an of fer ed christ im priso
ning christs mi nis ters that he has sent
forth to preach i beg lord give our rulers
to con si der while they are fight ing for
li berty what cruel bonds they have laid
upon thy saints

398 a las it is a try ing day with new engla
nd i know if god doos by us as wee deserve
wee shall all be dis troyed in this af fair
my soul has been in an guist for this land
but last thurs day mor ning i a wa ked
ear ly with them words - the angel of the
lord encamp peth a bout them that fear
him and de li ver eth them o i hope and be
lieve god will take care of his children let
what will come al so on sater mor ning i a
wa ked with them words the wil der nes
and so li tary place shall be glad for them

Heaton reports that British general Thomas Gage had arrived in Boston, where many men had been killed. Written in 1775, at age 54, the entry [p. 261] shows her anxiety about the impending Revolution, which also is evident in the tightly bunched lines of script. Courtesy of the New Haven Colony Historical Society

424 June 1-1790 i was refresht by the [illegible]
that the work of god was begun at
west feild some hopefully conuerted
and many crying out what shall i
do to be saued i cryed lord make thy
dominion spread from sea to sea and
from the riuer to the ends of the
earth come lord iesus come quickly

1790 i haue lately heard from longise
land that the work of god is begun in ma
ny places on the east end o lord grant
that the showr may reach poor north
hauen that seems to sleep in security
here the righteous are sad the wicked mad
while thou withholdst thy spirit good
spread out thy hands of mercy here like
a man that spreads out his hands to
swim to pull in souls to thy self i beg

now i was poor and weak distrest with
pain in my back these words come and
run in my mind take my yoke upon
you learn of me o methot i felt more wil
ling to take up my croos lord help me to
liue and walk in thy spirit that i may
not fulfill the lusts of the flesh i beg i beg

1790 now aboue a year i haue had sore eyes
and in aprel there was in the night a
thunder storm there was one very hard

Despite physical ailments, at age 69 Heaton continued to record news of spiritual matters, including the occurrence of revivals. Poor eyesight made it difficult for her to read and write, but this page [p. 424], dating from 1790, composed in a handwriting larger than usual, is still quite legible. Courtesy of the New Haven Colony Historical Society

[*p.* 172]

I lit of a myrtyr book[21] of the percecucions in king charls sewards reign. Ah the sweet words of the dear myrtyrs. Young ezabel allason & marion hervi sung at the gallows. Not long after i got another myrter book of the heathens before the comeing of christ & after his coming also of the papists in queen maryes reign. The religion of christ is the same now as it was & the enemies of god have the same opposing spirit as ever they had.

1768 may 8. I was unwatchful & so was holden with the cords of my sins proverbs 5-22. I felt such anguish of soul a week i could hardly sleep or eat. Sometimes the mouth of my soul was stopt up i could not pray. Sometimes i could vent out my anguish with tears cries & groans to god for help help or i die. O god thou has stiled thyself a god of mercy. O how shall i know it except i have an experiment of it. O let me not think that thou art as the waters that fail. One day i got some help in a way of prayr but it comes upon me again. Now i thot my case was written in job 19eth from first to 22 verce. My family was strangers to my inward soul anguish and they was bitter and cruel to me & (my lusts did rage corruptions sweel as if theyd drive me down to hell). I thot i see & felt sin anuf in me to kindle a hell. O god dont let it break out. I thot of exodus 23-5 there we are told to help up the ass of one that hates us

[*p.* 173]

if he lie under a burden. Lord my sins are enemies to thy honour & glory as well as to my peace & comfort. O lord i am sunk under the burden may not i hope in thy mercy. O god dont deliver me up to the power of sin. Samson was more affraid to be in the hands of his breethren than to be delivered into the hands of the philistians—judges 15-12. Now when i had been tryed about a fortnit i got a little help to my mind a little reviving in my bondage bless the lord o my soul.

In may 1768 mr miner come to preach in the place but my husband could not be willing to let me have the horse to go to meeting. But the same day offered to go & cetch the horse for jonathan if he would go to a frolick. Ah he is my dear husband still father forgive him for he knows not what he is doing. My poor jonathans mind is after fine cloaths & vanity but i hope he will yet be an isaac & go into the gospel field to meditate. Lord let him tell his mates thou art a great forgiver o make him to love thee much.

June 13 1768. Last night & this morning as soon as i wakt that in isaiah

33-20 was upon my mind. I hope yet in his mercy that can & i trust will give me meetings in this neighborhood. My eyes & cries are to him that hears the ravens & lions when they cry and shall the purchase of thyne

[*p.* 174]

own blood sit solitary as a widow alone. I beg give resignation to thy will to wait thy time. O god sanctify me bring me forth like gold tryed &. O make my rod like aarons to bud and blossom[22] and bring forth the peacible fruits of righteousness for jesus sake. I awaked one morning with this in my mind thy sins are perdoned for the dear sake of christ that dyed. It was some help to my mind but i was soon plunged into the ceanest distres i think i ever felt. It was from a tryal that come upon me. God hid his face. Mourn over me angels and men my tears are my meat day & night.[23] Earth does oppres who heaven has forsook. Ah my sins my sins oh that i could run away to some lodging place in the wildernes where no mortal eye can see me until my calamity be over & past. And now i was week and a pain in my breast o if i am saved it will be scarcely. O the traps snares & sins laid on every side to cetch me and o my weeknes. Sin in myself and others made me long for the grave where the wicked cease from troubling and the weary are at rest. The lord has told us that twoo shall be against three and three against twoo but here is three against one.[24] O jesus give me a large intrest in thy prayr in john 17 chap-15-17-19.

[*p.* 175]

1768 september 11. This sabbath i stayed at home. I had uncommon liberty in prayr with my twoo children. Methot i pourd my complaint into the bosom of the lord for myself my family the church this place the world & with god i leave my petitions hoping to reap and see his salvation.

In november 18 i spent part of the day in prayr for myself & family. I read daniel 9 & isaiah 37. All that i can do is but my reasonable service but lord give me a blessing. O let me like hannah go my way & be no more sad.[25]

1769 may 15. I heard mr frothingham preach from this word awake thou that sleepest & arise from the dead & christ shall give thee life.[26] All meeting time i felt bound & my soul shut up. I could hardly look at the minister. When meeting was done the power of god seemd greatly to seize my mind. I wanted a corner to pour out my soul to god. I hastned me home crying from the heart root. Lord search me try me to the bottom o wash me clense me o bring me into subjectednes to thy kingly athority. O make me love thee supremely to honour thee in all my practis and in heart lip and life. I longed for sinners lord pyty north haven all souls are alike precious. I prayed till it was dark o my god crush all sin in me dstroy all my idols and altho

[*p.* 176]

thy enemies are so bitter to me father forgive them for they know &. My sins are against light i do worse have done worse by thee than ever they have done by me. Groans & tears of repentance now ran freely and my bonds was the means of my liberty. O let my cries come up before thy throne o fit me to have the iron yoke taken off it gals my neck o my father fit me to have my sore feet like josephs took out of the fetters and my irons took off. I prayed hard for death if i may have jesus with me through the valley if i may not live to god. Lord i want meeknes humility watchfulnes patience self-denial the greatest of all supreme love to god charity the fountain of all graces then i shall love my neighbour as myself. O let me honour thee o god more with my substance. Our beginnings was small but god has made us to increase in estate. O make me a good steward and give me a wider heart to give when god calls for it for all is thine. I firmly believe lord thou art just & holy & righteous in all that has befallen me and in my being so holden with the cords of my sins—proverbs 5-22. I believe the punishment of my sin has been worse than the punishment of the sin of sodom. Lord help

[*p.* 177]

me into subjectednes to thee. O pluck me as a firebrand out of the burning and i will be thine forever. O help me this once to break threw the host and fetch water out of the well of bethlehem springing up always to eternal life.

Next day it began to wair away trials came on again. One morning this word come to me—give & it shall be given to you good measure prest down & shaken together.[27] Then i thot it may be somebody is coming that i must shew kindness to & before night a poor widow woman come to see me and i gave her.

1769 i think in june. This word ran in my mind—the inhabitants shall not say i am sick.[28] Ah i hoped a long lookt for shower was acoming about 4 days after mr miner come and preacht at mr beeches. The power of god was visable sinners crying what shall i do to be saved. Now in 1769 nathaniel & elias beech was brought out and set up weekday meetings.

Bless the lord o my soul that i have in some measure seen thy word full-filled on which i have hoped to perfect thy begun work.

Now i am praying for someplace to run to where i may live ah i think of davids strong-holds & caves where he

[*p. 178*]

hid from saul. O that i had in the wildernes a lodging place but o cant the mighty god in whom i trust yet save me here as weel. O the opposition within and without. If i am saved it will be scarcely. I am in great distres of soul but that runs in my mind christ my lord is conquerer still threw all the wars that devils wage and this is some help to me now—to think of the salvations god has wrought for my soul in years back my cries are still to god. O how satan rages o lord let his time be short. Lord help help i am like one in the briers and cant get out. O sin and devil is never tired. Ah how many times has my poor weeak knees and feet gone afoot to meeting when i must not ride and the horse may be rid to frolicking with freedom. I am now in the slow of dispond with a burden on my back.[29] O it makes me pray o thou great help take me by the hand and my burden will fall off. Now i often give a little hint of my trials in the family hoping it may be a warning to christians to marry in & for the lord. O god have mercy on prayrles families.

[*p.* 179]

September 9-1769. A violent temptation come upon me all of a sudden and i did a wicked action[30] for which my heart immediately smote me. I asked everyone in my family to forgive me. I pleaded with them to do it. I told them over & over again i had done rong i was sorry. Now it was not long before i hastned away to a private place in the bushes. There on my knees i acknowledged it to god that i had sind against him. Tears of repentance ran freely. My soul cryed lord forgive me keep me hold me in thy hand let me never do so no more. I pleaded hard for death or to have sin killd in me. I thot i could say without reserve i had rather die than to sin and dishonour god as i have done. This sin sent me many times to the throne of grace. My cry was lord do give me the grace of watchfulnes. Now i see the mercy of god that kept me from doing worse than i did no thanks to me but to a good god be all the glory world without end. In the mount of the lord it is to be seen. Ah i cant lay my sin to the devil nor to the temptation. I have no excuse for my sin against a good god i have done it.

[*p.* 180]

Now i was so fraid i should be provoked again to do rong. I thot i would get my things ready & run away but this flowed into my mind vain hopes & vain inventions all to scape the rage of hell. The mighty god on whom i call [illeg.] save me here as well [illeg.] i flung it all up and o the mercy of a good god to me. I felt my burden took of and a still mind. I hope god has forgiven me.

Not long after this i was tried about going to the saparate meeting. The sins failings and imperfections of gods children was held right up before me so that i could not see one of their graces. It come so hard upon me i felt as if i never desired to see one of them again. I wanted my husband to sell what we had & go away into some other place to live. My mind was so seized i could hardly think of anything else all day a saterday but i kept praying to god for help. On sabbath morning i thot i would try to go to meeting altho i felt no heart. I begged for the horse and my husband let me have it. I prayed by the way it was 6 or 7 miles and when i come there it was all gone. Now i know it was the devil & sin. Lord help me forever to go forward

[*p.* 181]

in that i have seen to be duty from thy word altho i should be beset by a legion.

1769 october. I writ another letter to my mother o god bless it i beg i beg.

October 7-1769. On sabbath morning i begged to go to meting but [illeg.] went to meeting. Now when i come home sin & satan raged at me. Should i tell the sorrow i met with i should hardly be beleived so i forbair. O father father forgive them for they dont know what they are adoing. One morning early as i was going to milking with a soul full of sorrow that word come to me with some sweetnes these light afflictions[31] which are but for a moment shall work for us a far more exceeding & eternal wait of glory. But now scriptures seem only to help me while they are passing threw my mind. Lord i thank thee for here a little and there a little.

October 26 1769—i met with much sorrow from one that should be my best friend.

[*p.* 182]

O the sin which i continually hear i see great sins indulged great duties neglected. It makes my heart ake and my joints week. In the evening i went out a few rods from the house and tryed to pray but how shut up i felt. At nine [illeg.] i will try again. I know not that i can speak one word but i was scarse on my knees and i see my own & my families case and spread it all before my god. My very soul cryed. I wrastled hard in prayer to god. O lord have mercy or we perish. Now when i come in my husband was very angry because i stayed so long and he laid cruel things to my charge and said he believed it. I believe the devil stired up his mind by a story wee had just heard of a man in this place that cetched his wife away alone with another man. Now my husband spoke such dreadful words i dont love to write them and that night i might not sleep with him. The next morning he began to rage at me again. Then i up & told him that i was praying for him & for

[*p.* 183]

us all that as i spoke a shower of prayer come upon me. I fell to praying loud before him with a flood of tears for some time and that spirit in him was stild and i felt in some measure at his feet. I felt little then i asked him what that ment except you come as a little child you can not enter into the kingdom of heaven.[32] He said nothing but lookt smiling & pleasent and never said anymore about this matter. Ah my poor family are trying continually to beat me of from all my religion and to get me into hell tho they dont know it. Father forgive them but i am sometimes astonished at the nautines of my own heart that calls for such rods.

In november 1769 the day after thanksgiving i went without my dinner. I took my byble went into the barn & kept the afternoon in fasting and prayr (the weather was cold). I thought i felt my own & see my families wants. I had liberty in prayr part of the time. I thot i had light into the 34-35-36-37 of ezekiel. O lord come quickly. Fill thy promises.

[*p.* 184]

December 1. I went to meting it was the sabbath. The next morning i met with a sore trial. O methinks no sorrow is like my sorrow no sorrow like the hidings of gods face and being punished with sin for sin. Lord let me not dispair of thy mercy i fear i fear i shall. Methot i longed to find the grave where the wicked cease from troubling & the weary are at rest. I have roared by reason of the disquietnes of my heart—thine arrows stick fast in me & thy hand presseth me sore.[33] Ah methot ide fellowship with dear david in the 38 psalm but o i feel like one cast out and banisht from thy sight but my cries are to thee. O forsake me not utterly. One night i went into the cow house to pray. I had freedom to beg but when i come in my husband thot i staid twoo long & was very angry. I believe he knew as well as i did that i was praying & that raisd the evil spirit. He spoke many wicked hard words but i answered him not a word. Lord forgive him. Then he talked abusively against me before a neighbor that

[*p.* 185]

was by i believe 2 hours and i bore it tho at first it made me sweat. O my proud heart. It was in the evening but i slept quiet that night. Methot i see the hand of god in it all. Next morning i rose early took my byble read the sixteenth of john aloud but i could not read that soul comforting chapter without the tears running and i believe it took hold of my husbands mind for he went that morning & acknowledged his fault to the person & desired they would not tell of it and then acknowledged to me how sorry he was. I see god in it. Lord i thank thee for any regret in his heart for sin tho no reformation. I have ever been choise of my husbands carector. It was the reason i did not write what he said to me when i come from praying in the cow house. But now i must write one speech among the rest. As soon as i stept in he said i hoped you was dead and if it was so ide yoke up my team and sled you to hell. I said not a word ah dear man. O god do let death & hell appear in his vew naked without a covering. Convert him i beg then he will help me pray. But i must kiss the rod take warning saints o marry in & for the lord i beg.

[*p.* 186]

December 28-1769. Mr miner preacht in the place. My husband heard him twice & seemd convicted 2 or 3 days but then like the dog to his vomit & the sow that was washt to her wallowing in the mire. On new years week it was a trying week to me. Poor me of little faith i sunk down under my trials. I dispaired of gods power. It seemd to me as if my husband was so hardned in sin god could not convert him. Lord have mercy. I prayed hard for that place where the wicked cease from troubling &. I feel as if i was going down to the pit. O god dont turn a deaf ear to my groanings.

1770–1775

[*p. 186, cont.*]

Jeneware 10-1770. Mr nathaniel beech and mr elias bech apointed a fast. That word was upon my mind luke 24 & he made as tho he would have gone further but they constrained him & he went in & tarried with them. On this morning i got things ready to leave my family and unknown to them i privately slipt out at the back door and went to the house appointed and six of us spent the day the lord was gracious. O let us see the returns of this day. O help us to constrain thee to come into this neighborhood and tarry with us. Brother ebenezer baldwin at newark in the jerseys sent me a good letter in May 1770: o god build up his soul.

[*p. 187*]

Ah now my soul is vexed from day to day with the unlawful deeds of the wicked. Lord sin abounds. O make thy grace much more to abound as it has been wont to do. O save to the uttermost. Lord i come to thee for loaves altho tis midnight. O god look on my sufferings and my sorrow and pardon all my sins.

1770 fabruary 15. Mr beech preacht at serjant heatons from isaiah 64-1. I believe god helped him. There was a solemn attention of about 40 people. Gods providence looks smiling. O let me see lord thy word fulfilled on which i have hoped and rejoice in thy salvation. Altho the meetings on the sabbath are yet at wallingsford i can seldom go lord give patience to wait for thee.

1770 in the spring i kept a day of private fasting and prayr for myself & family. I read gods word & prayed had some sorrowfull groans to god for mercy. O god let me see the returns of seed sown in tears o let me reap in joy bringing my sheaves with me.

I was disgraced by the tongue of one that should be my best friend but i felt a still mind. Bless the lord o my soul but then i again meet with a tryal and i sunk under it. I thot i felt like

[p. 188]

cain my punishment was greater than i could bair. Methot i was like one driven out from the face of the earth & hid from gods face. I felt like a fugitive & a vagabond. I dispaired & fainted under the cross. Ah my sin in marrying. I dispised gods counsil and i fear the sword shall never depart from my house. Wo wo is me in this life here i am a stranger. I complain to none this 25 years. I have many times thot on that eclesiastis 4-1 behold the tears of such as were oppressed & they had no comforter & on the side of their oppressors there was power but they had no comforter. Now i want a lodging place in the wilderness. I dayly hear and see worse sins than there was in sodom. Here is sins against light and idlenes & fullnes of bread. But o i believe christ is mine still and when i am prepared for it he can give me such comforts in my soul that out bids all things below the sky. O give patience lord to wait for thee.

May 18 1770. At night this word seemd to strike my heart—i will heal your backslidings & love you freely.[1] It seemd to lead my mind to prayr and thanks to

[p. 189]

god that he had recorded such a promise in his word for me. O i hoped in his mercy. Next morning i wakt with this word—on my mind thou hast had five husbands & he whom thou now hast is not thy husband.[2] Now i see my legal spirit. I have been playing the harlot after other lovers. I have had idols in my hart and i see i did now hold some idol (ah methinks i see what it is) and so i dont injoy & live upon christ my spiritual husband.

1770 in june 2. A meere trifle at first discomposed my mind and i gave way & sind against god & wounded my soul. Also at this time earth did oppres whom heaven forsook. Nothing but greif i found for they that to my soul should look my soul did pierce & wound. O lord forgive them o lord help me look on the anguish of my soul i cant get back again. Anights i watered my couch with tears. For 3 days i was almost prest to death with sorrow. O my cruel sins against a lovely jesus. One morning i wakt with these lines—the children weep beneath the smart and move the pyty of their heart. It melted my soul. I could hold no longer. I went away pourd out my soul to god & he pytied & easd my soul & stayed the rough wind &. (Read 103 psalm). And there was a strange alteration in those that had afflicted me bless god o my soul o larn me to be more watchful i beg i beg.

[*p. 189a*][3]

August 12 1770. As it was in nehemiahs time in the history so it is now in the mistery. Christians are far one from another and there is much rubbish so they are not able to build. Ah how is love wanting the daughter of my people is become cruel &. Methinks i see the book of job a fulfilling at this day. Some christians i believe indure as much sorrow in their souls as job did in the nature of it and their christian friends mistake their case as jobs did his. O the hard reproofs i have heard. If the afflicted jobs at this day do tell their souls sorrows to others they will tell them it is all for sin sin &. O how it sinks the poor distressed jobs & they are ready to say as he you break me in peices with your words. But i believe god is trying his children now in the furnace apreparing them for great mercy and as for my part i tell not nor write my greatest tryals. I am the creature that sitteth alone & keepeth silence because i have born it upon me. Lord gather thy flock like sheaves into the floor make no tarrying.

August 16-1770. Mr beech preacht in north haven at his brothers from lamentations 3-24. I believe god was his helper. There was i believe about 40 people. Their minds seemd to be all sober and very attentive. I felt my mind helpt. He shewed what it was to have god for a portion and what it was the christles have for their portions both here and hereafter. Lord bless it.

[*p. 189b*]

Now i met with much opposition from my family. It sunk my mind much but that word come to me in ruth 2-7. Here i hoped i had been gleaning with ruth altho i have tarried longer than she in the house i mean lived out of the path of holines and obedience twoo twoo much. O lord dont let me rest till like industerus ruth i am subjected to the lord say amen.

August 30 1770. It was our meeting day but my soul was so sunk with tryals i was like them that could not go over the brook besor to battle with david they was so faint. But i toock my byble and went into the swamp and spent the time in reading & prayr lord help me.

September 7. This morning i had uncommon help while pleading for myself & husband. My very soul aggonized with them words over and over again my god my god why has thou forsaken me. As to shewing thy salvation in my house it seemd good to me to say my god lord give patience to waight thy time for thine own names sake o come quickly.

1770 september 9. Mr frothingham preacht at wallingsford & administred the sacrament to ten persons. His text was in 2 epistle of peter 3 chap 18 ver grow in grace &. It was good preaching to me. He shewed how christians might know when they did grow in grace.

[*p. 189c*]

Now i was enabled to give to twoo persons provision for their families such as they wanted. Lord supply their wants as to soul and body.

1770 the first week in october. I heard mr miner of lime preach 4 sermons. They all seemd lovely but the last was the best. He seemd to have a large portion of gods spirit. His text was in luke 23-31 for if they do these things in a green tree what shall be done to the dry. He shewed the sorrows of jesus in his life & the sufferings and torments he indured at his death. He was the green tree he said. Well might the earth tremble when the blood of the son of god was soaking into it. He also shewed that the christles was the dry tree by sin a fitting for the fire. He shewed at large that when gods justise toock hold on jesus for imputed sin what agonies he felt. Then what must the sinner feel when gods justise shall press him in hell. Now my soul traviled for sinners especially for my dear husband with whom i often plead to go to meeting with me & reform his life. And now all my family was at meeting. The preaching seemd to take some hold of my husbands mind. But alas he soon returned like the dog to his vomit his beloved sin quencht it all. O my soul was in distres. I feard he would never have another call. O my god my god do have mercy on him. I gave freely a little money to the dear imbasendor lord bless him. O my sufferings are great my sorrows greater than i can write or expres but i feel a still mind. O my sins are greater than all my sorrows.

[*p. 189d*]

I am not worthy of the least of gods mercies. Oh my god i bless thee that i am not in hell. O my god i bless thee that i believe. Shortly i shall be in heaven with thee where no sin ever shall interpose between thee & me forever. O i long my god give me patience.

Now i lit of a baptists book concerning baptism.[4] He brought much schripture to prove that they was right & that infant baptism was rong or contrary to the mind of god. I have been tryed about it before in years back but now it took such hold of my mind i was in sore distres. My mind was confused by studying but my soul cryed to god to know his mind & will in his word. And at the same time i was under other tryals that lay heavi upon me that i was under great disadvantage. My mind was sunk with discouragement. I felt bewildered. I had lost my way. But i hear mr mershah of winsor preach from psalms 46-10 be still and know that i am god. I will be exalted among the heathen i will be exalted in the earth. He shewed at large what it was to be still or resind and the soul that was there god would help them. They should know

[*p. 189e*]

that he was god & their god and so by shewing salvation to them exalt himself among the heathen and in the earth as he did with israel at the red sea and at jordan. O this was a good time to me. Methot i could see my path again that god had markt out for my feet. Be still & know that i am god was with me several days. I felt something of the power of them words. I felt a quiet soul and i hope & trust god will establish my mind about baptism in his own time. Now i gave something freely to this dear minister. Lord i thank thee for so far helping me. Now about this time methot i see the last chapter of zachariah a fulfilling at this day. I felt something of the power of that & asked my husband if he was willing to have no rain of righteousnes rained upon him for god said they that did not come up to jerusalem to worship and keep the feast of tabernacles on them there should be no rain. I cryed and talked to him to go to meetings with me and hear the dear ministers of christ. He heard me but i think he said nothing. Lord i beg he may yet feast upon christ a greater & more perfect tabernacle not made with hands—hebrews 9-11. Now i sent twoo letters to my friends one to my mother and brother at quoage on long iseland the other to my brothers &

[*p. 189f*]

sisters at newark in the jerseys. Lord i beg for thy blessing set home thy truths for jesus sake or all my pains will be intirely lost.

November 14-1770. The sorrow i met with this day & night is incredible for me to write. I shant be believed so i forbair. Now i was soon taken poorly in body but i felt a still mind & a prayrful mind by turns. Twoo days after my family was all gone i shut up the house spent part of the afternoon in prayr for myself & dear family and that night in my sleep methot i was with a company of christians and there come in a number of people and methot i begged the brethren to have a meeting but they refused. And now my heart was ready to break to miss of such an opportunity. Methot i felt cut of as to creatures and i spoke and said i will flee to my old promises. God has said i will gather jerusalem and stand in glory thare.

Nations shall bow before his name
& kings attend with fear
He sits a sovereign on his throne
with pity in his eyes
He hears the dying prisoners groan
& sees their sighs arise

[*p. 189g*]

He frees the souls condemd to death
and when his saints complain
It shant be said that praying breath
was ever spent in vain.

Now while i was speaking these verses i felt them. Methot i had faith in gods word. I see him acoming to build up zion. That word come into my mind—who are these that fly as a cloud and as the doves to their windows.[5] So i awaked. O my god help me by faith to keep fast hold of thy promises untill i have seen thy compleat salvation. Again i was helped with a free heart to give about 3 pounds of wool to poor folks that had not a sheep in the world. O lord pyty them & have mercy upon their souls & bodies. Now our meeting day is once a fortnit and it being near the devil and sin beset me not to go. The failings & imperfections of christians was held up before me but i kept crying to god to help me. I hoped in him and so i went altho sad and i found god did help me in that i feard. But when i come home o how did the devil rage in one poor sinner at me. Good things was broken out of enmity. I said but little then lord help me to be faithful with meekness

[*p. 189h*]

& humility when it is convenient. Oh god thou art stronger than the devil. O help me to resist him make him flee from me for jesus sake that dyed to destroy the works of the devil.

Jeneway 3-1771. Mr beech preacht in the evening from isaiah 9-6 to a room full of people. He seemd low in mind but was helped to hold up clear truths. The hearers was attentive lord set home thy word with power. Now my eyes sees my teachers but ah i want a thankful heart. O how is my soul prest down with hard tryals. Lord subject me to all thy humbling dispensations. O let me kiss the rod & be still and know that thou art. Methinks there is no sorrow on this side of hell like being punisht with sin for sin. I got up before day went to writing. It seemd good to me to say my god my god. Again i arose before day went to prayr then to writing. Methot i had freedom at the throne of grace while pleading with a flood of tears for my family and there i leave my petitions. Lord let my waiting soul & weary eyes yet see thy salvation. O dont say to me as thou didst to david the sword shall never depart from thy house because thou hast dispised me. Lord i dispised thy council when i married. O wash my soul and i shall be clean.

[*p. 190*]

March 17-1771. I am now 50 years old and o how much of this precious preacious time have i lost. Almost 20 years i lived in a state of nature and i have lost all the other 30 when i hant lived to god. O god be mercyfull to me a sinner. I have got almost to the end of my race. O god help me to redeem precious time if i live a few moments longer. Lord i thank thee that i am this day out of hell. I have been shut up this cold winter & endured much hardship. I have been to meeting i think but twice since winter on the sabbath. O the sinful practises that i have seen & heard in my family has worn much upon my body and mind but i fear to complain i have so many great mercies.[6] O how often am i drove to god in straights and have got help when i could hardly expect it. Ah i am one that must be hound to heaven. Lord pyty me. I fear when i write that i shall exceed on the right hand or on the left. I dare not put pen to paper without prayr.

Now i was helped with a free mind to give something to a poor person for clothing to help them to go to meeting. Lord bless them in soul & body.

[*p. 191*]

1771 march 28. This evening mr beech preacht in north haven to a room full of people from mark 10-46-47-48 verses concerning blind bertimeus. He shewed sinners what it was to sit by the highway side begging. It was to use all the means of grace they possibly could altho no promice belonged to them out of christ but there was incouragement. Ah who knows but they may have theyr spiritual sight restored altho born spiritually blind as all was by nature. I believe he had help from heaven. My soul was helped. Methot it was good to be there. The people seemd solemn & attentive. One man that come in laughing soon had a sober countenance. Lord i thank thee that my eyes sees my teachers according to thy word on which i have hoped. O help me to arise & wash myself in the jordan of thy blood. O make my leaprosie depart & let my flesh come again like the flesh of a little child. O lead me into edom. O bring me into the strong city.

Yesterday i was helped to plead with my husband for peace with tears & a hart affected with gods mercy. I told him it was

[*p. 192*]

free rich mercy that god had not writ us childless on his earth and it seemd to still him in some measure. Lord pyty us. O when the earth was curst for sin did not thy blood fall upon it and thy body was interred in it to take away the curse. O make application of this sin killing blood to our souls. O on thy cross thou did rectify all. There was death in the pot but thou threwest in meal and didst sweeten all.

Now i was enabled to give freely twice to them that was in want. Now it being lection week there come a woman and pleaded hard with me to sell her some wheat flour to make cake for a frolick. I told her i would not sell her none. I believed frolicking was a great sin. I would not have a hand in it i dare not. But poor woman pleaded as tho it was a harmles thing & so went her way lord pyty her.

On lection day i met with much sorrow. I concluded to spend some hours in the afternoon in prayr. Then i opened the byble that word was incouraging 1 of chronichels 28-20. Then i went into the barn the tears of repentance flowed. My heart confessed my own sins & the sins of my family. My heart cryed lord have mercy on us.

[*p. 193*]

Methot i could now use schripture arguments. Lord i am a bruised reed thou hast said thou will not brake me methinks the flax smoaks dont quench it[7] i beg i beg for jesus sake. O god be my guide in this day when the world lies in wickednes. O how many ways to ruin do i see how many nets do i see spread for me what a snare are even these outward comforts. How do they intice and draw away the heart o how many hundreds do i see beguiled some by one snare and some by another. O how do they forget god forget heaven & forget hell as if all this was but a trifle.

O how they can sing and dance & laugh as if all was well and one is helping the other down to hell. O lord wake them o lord raise the dead.

May 19 1771. Some gray headed live prayrles in their families & give themselves up to cursing and swearing & excessive drinking and will mock at reproof. I know one whose flesh and heart has often trembled for fear of gods righteous judgments. O are the righteous scarcly saved. Lord give me persevering grace to press through to glory say amen.

May 24-1771. This evening one of my neighbours was drownded while going to ealing. His life was extreemly wicked from a child & he was a mocker of the religion of christ and of gods children and as i was

[*p. 194*]

informed he workt all day and contrived his work for three years hence & dyed in about 3 hours after. He has left a sorrowful widow & 5 children. This death seemd to strike everybodies mind with a kind of horror. I see gods word fulfild proverbs 29-1. He was often reproved but hardened his neck & was suddenly distroyed. O how many times did that go through & through my mind in wattses psalm 73-8 commd. Methot i had a sence of the distinguishing mercy of god towards me and mine. It seemd to quicken me to pray for sinners. O god of mercy let not this awful loud call be lost i beg i beg.

1771 june 13. Mr beech preacht from luke 13 the 4 & 5 verce. He spoke of the multitude of sudden deaths that are at this day all around us and he insisted chiefly on them words except ye repent ye shall all likewise perish. He shewed what that repentance was that always flowed from a principle of grace wrought in the heart & he shewed the difference between that and a repentance flowing from fear of punishment. He closed the whole with a suitable improvement. I believe he had help from heaven. The hearers was sober & attentive a number was there that i little expected. Lord set home thy word for jesus sake.

[*p. 195*]

1771 in june. A young man one of my neighbours was brought out of darknes into a great nearnes to god. The family fearing he was going to be distracted got a doctor to him. The doctor said he was distracted & asked him how he felt. His answer was how do you think you shall feel if god should shine into your soul in the face of jesus christ brighter than ten thousand suns. And now the young man goes about to the neighbors and talks to them of the things of god. O this seemd to incourage my drooping soul. Lord i thank thee that thou hant yet forsook this earth. But now again i was overcome with sorrow. Many days i felt as if i should perish for fear of saul the sons of zurviah was twoo hard for me.[8] I reapt the fruit of my sin in my marrying. Take warning o christians. Wo is me that i sojourn in meshek and dwell in the tents of keder. Lord have mercy on me.

In july 28-1771 there was a hard thunder storm. Five persons was struck with the lightning in a sabbath day house that stood near north haven meeting house. Their flesh was blistered some holes burnt in their cloaths.

[p. 196]

They was much hurt but quick means being used we hope they will all recover. There was an oaak tree stood near this sabbath day house under which there was three horses and one colt killed dead. I had been to meeting at wallingsford & when i come by and see the dead horses lye streatcht out i see it was death by the immediate powr of god from heaven. I had some sence of the awful majesty and power of god. O lord blees this instance to them that was hurt and all this people & to me and mine. Lord i thank thee for thy begun mercy.

August 11-1771. The saparate meetings was set up here on the sabbath a third of the time. The wallings people would allow no more and was hardly willing to that. Ah methinks i see the ark brought back only as far as the house of obededom.[9] I mean we have our meetings but part of the time. Ah when david and his brethren brought the ark to jerusalem & put it in its place he danced before it and david dealt to everyone a loaf of bread a good peace of flesh & a flagon on wine. Ah he could then feed the

[p. 197]

people with that that holds forth christ to us. And that day david delivered a blessed psalm and it seems he got a blessing for his own house 1 of chronicles 15th &16th chapters. Ah methinks i dont wonder at the leannes of one that preaches for he ought to be the first to bring back the ark to its place. Ah here in north haven god brought out our brother benjamin to preach and god was with him in his improvement. Now since he moved the meetings away i have often thot of david that when he see that god answered him in the thrashing floor of ornan he sacrificed then and there.[10] Solomons temple was built i suppose there abraham offered up his isaac. Now this hold forth to me that where the rules of gods word are held up and where his presence is there wee ought to meet and there wee shall get the blessing and be built up.

One of my neighbours a church member altho a great opposer of gods work slandered and raged against his neighbors. Now it

[p. 198]

seemd in my mind i must write to him these lines—sir i pray you to read gods word and consider it without prejudis. The time is hastning when we must all be judged by it. Pray consider this he that privately slandereth his neighbour him will i cut off & cursed is he that privately slandereth his neighbor & thou shalt love thy neighbour as thyself.[11] O i fear some that are at the lords supper are eating and drinking their own damnation for by their fruits you may know them farewell Hannah Heaton. Now before i could send this to the man a cart wheel run over his head. He was much brused and hurt and lies now in danger of his life.

Last sabbath which was september 1-1771 mr beach preacht from revelations 1-7. He seemd to have some life & there seemd to be some consern upon some minds. Lord water the seed sown glory to god for meetings. September 11 was a day of sore sorrow to me by reason of the sin which i see & heard. I went away alone and spent some part of the afternoon

[p. 199]

in prayr. I read micah 7 it seemd to my very case. O me thinks this chapter was written for me. O me thinks i will bear the indignation of the lord because i have sind. O me thinks i will look unto the lord & wait for him. O me thinks he will turn again he will have compassion he will subdue eniquity & cast sin as into the depths of the sea—lord i thank thee that i am out of hell i thank thee that i am here where i may read thy blessed word i thank thee that i am here where i may pray to thee groan to thee weep to thee kneel on the ground to thee ring my hands to thee sigh to thee. O let me feel & see thy salvation in my own sanctifycation & in thy turning my family to thee. O build up thy zion appear in glory there. I thank thee for what thou art in thyself holy just and true.

October 6-1771. I had a heavi tryal on my mind. I got up before day and prayd with tears and secret groans to god to help me. I got some relief tho not delivered. O me thot i cryed with bertimeus[12] and will cry jesus thou son of david have mercy on me. Again i was helped to give three times to poor people that wanted o lord pyty their precious souls.

[*p. 200*]

Again i writ a religious letter to my mother and brother. O god bless it or all is in vain. I know dreams are not an infallable rule to be depended upon but i have sometimes seen them fulfild and i think i must now write one. I dreamd that a great mad bull come up to the door & was acoming fiercely at me into the house. I was much terrifyed but i thot i must go right to him and shut the door which i did and awaked. Me thot trouble is acoming. Then i dreamd that an old woman one of my neighbours told me that she was under some indisposition of body. I thot i told her that she was agoing to die. Me thot she cryed and said i dont know but i am. Now soon after this woman come to me and pretended to reprove me severely for a sertain thing that i had done which she said was rong but the thing that i had done was with an honest mind and a good consience & i had peace in it. Now in this womans discourse i vewed her to have a perty spirit. She had no sin to acknowledg but seemd cruel and bitter & provoking. I felt my nature rise but i never spoke one hard word to her but i felt a dreadful burden on my mind twoo heavy for me to bair. I could hardly sleep or work for some time but i followed god night and day to have mercy on me and help me for jesus sake.

[*p. 201*]

One afternoon i felt as if i should die with anguish of soul. Gods face was hid i thot i felt something like the damned. I went away & spent some hours in prayr and i felt afterwards releast and still when i had pourd out my soul before god but it soon come on me again. Now i was brought to see what it was for viz this woman that i have spoke of i believe is a christian but of a quick high temper & i have known for twenty years she has lived in some perticuler sins & i durst not reprove her but now i find i must or i must die for me thinks my flesh is wasting. Now i begged of god for an oppertunity. Wel soon after she come to see me but o unfaithful wreatch that i was i thot i could not begin & i let this oppertunity slip but o what regret did i feel afterwards. But then i impertuned the lord for another oppertunity. I think now of gideons fleece.[13] I prayd god to hold down her temper & to give me of his spirit. Now one morning i felt sorely distrest. Part of that verce struck my mind for a moment or twoo—christ our lord is conquerer still through all the wars that devils wage. It made the tears to flow. I believe satan had a great hand in my distres for i was much confused sometimes in my mind for it would look to me like quarreling to go and tell the woman

[*p. 202*]

of her sins & if i could then have got rid of my distres i should gladly have said nothing to her. But that word i did not get rid of—if thou has ought against thy brother tell it to him between thee & him alone[14] and i promised the lord i would do it when a oppertunity presented. But o i could not get rid of my burden. Now i awaked three mornings with this word in my mind in isaiah 33-21 the lord shall be to thee a place of broad rivers and streams wherein shall go no galley with oars neither shall gallant ships pass thereby. O i thot they was blessed words and hoped to see them fulfild that i shall yet be brought into the path of obedience & holiness. And methot i see that sin and devil was the galley with oars that shall not nor cannot walk in the path of holines and gallant ships shall not pass there by them. Gallant ships i thot was ante christ. Alas but i soon lost the sweetnes of this schripture. Now when i had waded about six weeks in this sea of distres one afternoon i took my byble & went into the barn to read & pray. Now when i had been there some time wrastling & groaning to god this woman just mentioned come to my house and

[*p. 203*]

not finding me there she come to the barn and knocked at the door. Now at that instant i was apraying for her. I was surprised but got up and went out to her and we come into the house. I felt in some measure a humbled soul & i soon began to talk to her. First i acknowledged my unfaithfulness & then told her of many things & how she had abused me with her tongue many times. And she never now shewed any resentment but acknowledged her fault & thanked me for talking to her and begged my prayrs & so went home. And now my soul was delivered and my burden gone then i thanked the lord. O christians our god is a prayr hearing god. O be faithful to reprove with meekness & humility acknowledging your own sins first. Ah me thinks i may say with hannah 1 of samuel 1-27. O how she pointed to her child and said for this child i prayed & the lord hath given me my petition which i asked of him. Lord make me faithful for thee for time to come and not let offences lie long between me and any. The lord raiseth up the poor out of the dust & lifteth up the begger from the dunghill. Bless the lord o my soul.

[*p. 204*]

1771 november 7. Mr beech preacht in the evening from revelations 1-7 behold he cometh with clouds & every eye shall see him &. He shewed what christ was acoming to judgment for viz to reward his saints & to punish sinners. Ah he said this jesus that is now like a lamb pleading with sinners to except of his mercy & love by & by will become like a lion full of fury his smiling face will be turned to anger—i believe god helped him. Now my family was all at meeting. O gracious god i beg i beg bless these warnings o water the seed. Now i felt my soul in some measure quickned. Bless god o my soul.

Again i was sorely afflicted with wicked company wo is me that i sojourn in meshek & dwell in the tents of keder.

The next sabbath morning i waked with a discouraged mind. It seemd as if it was in vain for me to pray any more. But that word come to my mind—it is not a vain thing but it is for your life.[15] Then i arose before the sun and got me away to prayr. I told the lord my sorrows with a mourning sorrowful soul and with god i must leave them. O god give me a strong heart to indure hardnes if thou see fit to lay it upon me.

[*p. 205*]

November 5 1771. I went to meeting exceding sad & cast down. O my family is a house of belial[16] for want of family government which is i believe one of the plagues of babylon. Wo wo to me i am daily reaping the fruit of my marrying contrary to the mind and will of god take warning o christians. Now as i have said i went to meeting on the sabbath at noon. I chance to see a letter with that schripture in it genesis 49-19 gad a troop shall overcome him but he will overcome at the last. This took hold of my heart i could hardly contain for weeping. O how often do i feel as if i was overcome by a troop of sins corruptions & temptations within & without. But o that melted my heart he shal overcome at the last. O methinks i see then it will be glory glory to god.

December 8 1771. Methot i went to god this morning with a sorrowful sence of the sin of my family. I pleaded hard that schripture (where sin abounds grace has much more abounded) my soul cryed lord sin abounds here. I begged for his mercy & grace. O god help me to give them to thee & leave them with thee. Lord where much is forgiven will thy not love thee much. By hard tryals god often prepares my heart to pray. O cause thyne ear to hear for jesus sake.

[*p. 206*]

December 1771. I being under gods rod still one night them words seemd to be whispered to me (enter into thy closet & shut twoo the door untill the indignation be over & past).[17] Now i felt my soul quickned to prayr. I arose as soon as it was light tho cold i got me away alone to pour out my complaint to my god. Now i had some raisd hopes to see gods salvation. Ah i find there is some comfort in gods quickning presence. Also god says when he has performed his whole work upon zion he will punish the stout heart of the king of asiriah & bring down his high looks. Lord hasten thy work destroy sin i beg i beg.

1772 fabruary 11. Mr beech preacht in the evening at thomas sanfords from revelations 6: 17. He shewed who would be able to stand in the great day of gods wrath and who would not be able. There seemd to be a sober sollemn assembly. O lord bless the seed sown. The next sabbath he preacht at his brothers from isaiah 26-20. He shewed who gods people was them that are invited to christ & bare fruit. He shewed what the fruit was and from this word he observed that altho it is a time of tryal & darknes to the church it would be short. Ah that night my spirit made deligent sarch. I see a necessity upon me to be subjected to god. Next day sin sat heavy and the cross sat light.

[*p. 207*]

Fabruary 28-1772. In this month i am 51 years old. Ah days and years hant made me holy. Ah i am yet a poor sinner. This day i can say as david psalms 119-53 horror hath taken hold upon me because of the wicked that forsake thy law my soul is afraid of gods righteous judgments. This day in the afternoon my family being all from home i read ezra 9 then i prayed. Methot i felt an honest free soul in confesing my own & my families sins. Methot i brought them up to god. Schripture pleadings flooed into my mind and i had some sence of the worth of precious precious souls if once lost forever lost. O lord i thank thee that there is yet yeildings in thy heart towards me or there would be none in my heart towards thee. Now the next day there seemd to be some reformation in the family some mildness and pleasentness. Lord continue it o send converting grace. From the ends of the earth have i cryed unto thee. Thou hast regarded the destitute and hast not dispised their prayr glory to god.

1772 the second week in march. Methot the devil was come down in great wrath. O how light does the wicked make of gods wrath forever more. O come lord jesus & deliver me or i shall die. My spirit fails before thee. I am like one going down to the grave. Lord help me (psalms 88).

[*p. 208*]

The last of march. I was at home alone there come a poor lame begger. I felt a willing mind & gave a bushel of rye. We talked of soul matters. Lord pyty the blind that know thee not. Make the poor rich in faith and hairs of thy kingdom. We are told in first of samuel 25-18-19 how abigail gave to david & she told not her husband nabal.

Aprel 9 1772. I spent alone in fasting and prayr and i writ some. I was low and dull all day. Next morning i arose before day. I had some groanings to god in prayr some lively sighs to him for myself and others. Lord i thank thee that i may pray to thee. I deserve to be in hell where there could be no yieldings in thy heart towards me nor none in my heart towards thee. There all sin is let lose and all restraints taken off. There they are pressed down with the wight of both thy hands but i & mine am here yet. Glory to god for his rich mercy & free grace.

Now again god in his providence has caused our inheritance to increase but i often see a worm at the root of the gourd.[18] O tempting devil o aluring world o wicked sinful flesh. Lord help me to honour thee more with my substance for all is thine to do it while i can for in a few moments i shall be gone till the heavens be no more.

[*p. 209*]

Aprel 19 1772. Mr beech preacht from psalms 107-8. O that men would praise the lord for his goodness & for his wonderful works to the children of men. He shewed in perticulers what gods children ought to praise god for not only for what he is in himself but for what he has done for them. He told sinners if they did not praise god here they never would in heaven and when he spoke to backsliders to invite them home to their fathers house again it melted me into tears. Methot he almost took zion by the hand. Lord hasten the day when the hunter shall bring thy children out of the holes of the rocks where they have been hid.

I have this hard winter met with trial upon trial twoo much for me to rite. That word in 1 of samuel 8-18 has been heavi on my mind—and ye shall cry out in that day because of your king which ye shall have chosen you and the lord will not hear you in that day. O my god give me persevering grace. O help me to win threw to glory. O when i get safe home will it not be sweeter to me than to any. O shall i not cry glory glory to god in the highest.

May 3 1772. Now my jonathan is fixing to go into the troop. It grieves me to see so much laid out to uphold pride while many poor suffer want and our publick newspapers tells us that in forreign parts of the world thousands and

[*p. 210*]

millions of late have dyed with famine & poor child he never as i know of gave one penny to support the ministers of christ or his dispised cause or to any poor person. O god i hope in thee for him yet. O grant that in him thy long suffering might be shown forth for a pattarn to them that shall hereafter believe 1 of timothy 1 chap-16 ver: jesus thou son of david have mercy on him lord help me to wait thy time.

Last week i was helpt with a free mind to give a little money to a poor sickly family. Lord pyty them for jesus sake. Now i just received a letter from long iseland. Lord i thank thee for the sweet truths in this letter. Bless my friends & relations there with the choicest of heavens blessings. O god bless again that ile in the sea.

Now i rit a letter to my 2 brothers & 2 sisters at newark in the jerseys. I was inabled to adres myself to each ones perticuler case that their souls was in as i thot. Lord bless this letter to their souls i beg for jesus sake.

May 21 1772. Now i went to new haven prison to see the poor indien under the sentence of death for murdering a white man. He stood at the window in irons with twoo books in his hands. I talked some to him but he seemd stupid & unconserned

[*p. 211*]

about his soul. Lord have mercy on him. O lord it is free grace that i am not in the like case this moment. O god i thank thee for restraining grace glory to god for it.

This week i was helped with a free mind to give some money to a poor stranger that had his house burnt. He seemd willing to hear council. In years back he was taken and abused by the indians so that he can work but little. O god be gracious to him & his family as to souls & bodies. Bless god o my soul that has preserved my habitation from fire wherever i have lived this 51 years. O god be merciful to me a sinner until and threw death then take the sting away for jesus sake. Let perfect love cast out fear for christs sake.

1772 july 5. Now we have the newspapers that tells of terrible things in righteousnes that god is doing threw the world. Now every mans sword seems to be against his fellow. There is great devisions in the perliment and among civil athority there seems to be universal devisions in churches & in families oppression by cruel laws pestilence and famine whereby millions have dyed & we hear of earthquakes in divers

[*p. 212*]

places. O methinks i see the scriptures fullfilling the end ahastning prepare to meet thy god o zion.

This day i met with a disappointment and my mind was discomposed. O lord subject my soul to all thy providences o that i may turn as thy providence turns. Lord when shall it once be. O let my will be swallowed up in thine i beg for his sake that was hung up between twoo thieves on the hill calvery.

1772 in july 1. I heard mr coal[19] a young man improve from hosea 10-12 and from romans 13-11. I believe god helped him his discourse was sweet and teaching to me now i thot with a free mind. I gave a little money to this dear young man but that night i was sorely worried in my mind about it because i did not inquire more about the mans worldy sircumstances for i ought to give understandingly. In the night i got up and read the 77 psalm & prayed and a good god took my burden all away (bless god my soul even unto death and write a song for every breath).

Again i was helped to give three times to them that wanted. Lord convert their souls let them be thine in that day when thou shalt make up thy jewels

[*p. 213*]

(psalms 18). This week the sorrows of hell compassed me about because of the sin i saw committed with deliberation. Terror & horror took hold of my mind but a good god supported my soul. Them words ran in my mind the god of abram will support his children lest they faint. But o sinners what will they do when the great god shall rend the sky & burn the sea and fling his wrath abroad.

In july 1-1772 the brethren that carry on in our meetings moved them to wallingsford five sabbaths going. Now i was so burdned in a days i had no peace and in the night my sleep went from me. At last i went & told them of it and i got help in my mind altho the cause of my burden was not removed. O thou who dost indulge sparrows & swallows pray give us meetings constantly fild with thy presence.

This week i heard and see much sin committed my soul was grieved. I was sore afraid of gods righteous judgments. I laboured with one for family peace. That ran in my mind—the mighty god will compass them with favour as a sheild.[20] I hope in his mercy.

[*p. 214*]

1772 september 2. Poor moses paul was hanged (the indien i mentioned before). I went to new haven and heard mr samson occom (an indien minister)[21] preach from romans 6-23 the wages of sin is death but the gift of god is eternal life through jesus christ our lord. He shewed something what the wages of sin was both here & in hell & something what the gift of god and eternal life is threw christ (people said the poor crimenal trembled)—Then i went to the gallows but when i see him stand upon his coffin with his hands lifted up praying just agoing to be turned off i turned about and went away. O god i thank thee for distinguishing grace and free mercy to me & mine. I deserve to be left of god who have left him so often. O god thy mercy is infinite or i should now be in hell lifting up my flaming eyes with divers.[22] This day i see this word to be truth the wages of sin is death[23] in my soul often ah the wages of sin prest jesus to death o sin sin o cruel sin. Now some thot there was 8 thousand people assembled at this time. O lord grant that what they heard and saw may be blest to them. I thank thee that by it thou hast quickned me.

[*p. 215*]

My husband labourd with for family peace o lord bless my poor endeavers. I find that sin drives me to god. Ah must i be hound to heaven.

October 4 1772. Mr beech preacht from psalms 84-11. He markt out the path of holines and shewed what it was to walk uprightly. It seemd good & teaching. I heard the wicked dispising good food. I asked them if they believed it was the price of blood. Jesus dyed to buy all our mercies (with tears i told them of the myrter woman just before she was burnt. They gave her some fish bone broth. She replyed what all this & christ twoo all this & christ twoo). O god do pyty sinners i beg i beg.

October 7-1772. The athority in this place went and took away three cattle from 3 of our saparate brethren to pay their rates to mr trumble.[24] Now when they drove mr beeches cow away his little gairl of [] years old ran crying after them begging for the cow. She told them they had stole the cow. They said to her they wanted to pay mr trumble. She told them her dade did not ow mr trumble anything but they would not hear the cries of mothers

[*p. 216*]

nor children. The babes father told me the child never heard anybody say a word about their steling the cow but had a mind to go after them of her own accord. Mr beeches wife talked to the men told them how rong they did & they must be accountable to god for it. One of the men cryed like a child but ah cruel antichristian laws of this nation bind even the concience while god & the king gives liberty of concience. I hear mr coal of the great swamp which i mentioned before has been in prison for refusing to pay his rate to the standing minister. And after the saparate meeting was set up at wallingsford the athority took 4 young women & one man and carried them by force & put them in new haven prison. The foundation cause was their meeting together to worship god according to their own conciences in the light of gods word &. Of that poor zopher tuttle that is now in the eternal world lay in new haven jayl about 2 months for refusing to pay his rate to the standing minister. Having much sorrow & hardship he was never well no more (as i heard) & dyed in peace saying the kingdom of heaven is come. O god do come quickly and destroy the whore of babylon o destroy ante christ by the spirit of thy mouth & the brightnes of thy coming. O god of abraham when shall it once be o give pyty to thy suffering saints.

[*p. 217*]

1772 october. I took my byble went to a private solitary place. I being under gods rod i spent about half the day in prayr for my own sanctifycation & for gods presence this winter when i am shut up and for my dear family. O lord let me see the returns o make thy power & grace come into my family so that we shall not be able to resist. O let us yet draw water out of the wells of salvation.

1772 in november. I sent one letter more to my dear aged mother and brother. O god bless it for the good of souls. O send another shower from heaven upon that isle in the sea that thou hast blest.

1772 in december my husband & i was taken ill. I had a cramp pain in my back so i could help myself but little sometimes & he had the argue in his head and was sick but god was merciful to us easd our sore pain. In about a fortnit we was comfortable in body but o god how basely have we served thee. Pray make us return and give glory to god as i have made it my practice almost thirty years to set apart some time every night for meditation reading & prayr

[p. 218]

unles hindred by something special and altho my house is not so with god yet o god make me now to double my deligence. Pray let me rejoice in this that thou hast made with me an everlasting covenant ordered in all things & sure.

But o what an unsubjected heart do i feel sometimes o what a coveteous heart. I remember in 1768 once i was ronged by a rich neighbor in worldly things. I talked to him but he would make me no amends and i was worried night and day by sin & devil twoo years about this. Sometimes my mind was helped a little while and then it would come harder again. Much time i spent in prayr to god and he helpt me when his own time was come. Now if i had been subjected to god as he is god over all blessed forever i should never have met with this sorrow. O god i thank thee that thou didst help me o let me through the cross be crucifyed to the world and the world be crucifyed to me.

Now i find i must hide my diery in a more private place than usial for a wicked person got it & hid it almost a month and said he intended to burn it. But at last he gave it me again. Once it was flung away outdoors another time twoo hands took hold of it to tare it to pieses but was prevented. I have no cause to be ancious god has & can preserve it.

[p. 219]

I do believe it to be my duty from gods word to keep a journal or spiritual history of my experiences tho the thing be not of commandment.[25] But i have thot how god kept a dieary of the creation of the world gen. 1 yea he keeps a book of remembrance for them that think on his name mal. 3 and shall not i write down his name works love. Moses wrote the goings out of the children of israel numb. 33-2 david paul & john & many others writ the dealings of god by his spirit & providences with them and their frames and carriage towards him. And sometimes in my darknes when i have read things in my diary that i had forgot it has seemd strengthning & quickning to me. I have many times written when weeak in body and mind amidst many distractions & trials of divers kinds which require them that read it to have the more favourable & charitable acceptance thereof. Sometimes when my mind was dark & unfruitful i have used the words of others to tell my mind

He left his shining throne above his bowels yearnd his heart was love
how loth he was to loose his bride to save my soul the bridegroom dyed
Twas the same hand that spread the feast that sweetly fourst me in
else I had still refused to taste and perisht in my sin.

My dear children.

I leave you here a little book for you to look upon
that you may see your mothers face when she is dead and gone.

[six missing pages]

[*p. 226*]

Gods hand is in it we had three hogs and a yearling coalt dyed wee suppose bit by the mad dogs. Now i believed god had a right to take away that which was his own & felt a quiet still mind. Lord i thank thee for thy mercyes towards me.

1773 may 28. The wife of joseph jacobs dyed suddenly. Was well for ought they knew one moment & dead the next. I went to her funeral o god of abraham do not let this warning be lost.

Now i sent another letter to a friend at long iseland o god bless it to saints and sinners. And now i got bunyans pilgrims progres.[26] It seemd pleasant but o that it was with me as in months past when gods candle shind upon me &. O lord give me patience until i come to the place of deliverance. O then my burden will fall of.

1773 in july. I met with a hard trial from the wicked. I was so overcome with grief that i got into a great fit of crying but at last my greif turned into prayr. This verce kept running threw my mind that day

My god in whom are all the springs
of boundles love and grace unknown
Hide me beneath thy spreading wings
till the dark cloud is overblown (psalm 57).[27]

[*p. 227*]

With this verce i prayed that day & a good god took off my burden. O lord i thank thee pray give me faith and patience to bare whatever thou sees fit to lay upon me. O god pyty my poor child of promis. This week i prayed that god would find a rebecca for my isaac. Lord have mercy on us all.

Yesterday i was helpt to give something to a poor woman. Lord help me always to pyty the poor.

O god when will thou come to my soul. O lord it is a dark day here with thy children the righteous sad the wicked mad. Why hidest thou thy face. O when shall the divisions of rubin be heald. Twenty years ago that word use to run in my mind judges 5-16 but i knew not then what was acoming. But i see now gods word is fulfiling altho i have been ready to say with gideon sometimes judges 6-13 when shall our officers be peace & exactors righteousnes &. But o this is gods word stil. Still i will gather them like the sheaves into the floor. O god come quickly.

The next sabbath mr beech preacht from hebrews 4-last & from psalms 116-18. He seemd to hold up dear truths. Soon after i got home i went away and pourd out my soul to god in prayr. O let my cries come up before thee.

[*p. 228*]

The next day there was a great croos laid upon me & my proud heart sunk under it. But one morning as soon as i wakt this verce in the psalms went threw my mind—bleess god my soul nor let his mercies lie forgotten in unthankfulnes & without praise.[28] Now i see altho i was under gods severe rod yet my mercies was many many and great & i ought to bless god with my whole soul and not forget his dear mercies. O how dear was they bought for me. Ah i am not in hell but hope for salvation through jesus christ. Now at first I felt a proud quarrelsome heart but this has stilld me. Bless god o my soul o thou that did preserve a lot in sodom from the infection of that beastly crew o preserve me keep me from the evil of this sodom world. O thou that did preserve elijah among an idolatrous nation and bid him stand up for & declare thy name o keep me o help me to percevere. O help me to honour thee in life and at death. O lord i deserve to be in hell but o thou that preserved jeremiah job abraham moses amongst pagans athists and infidels o keep me from the infection of the sin which i daily see & hear. O god I thank thee that i have been in some good measure preserved. O grant that i and mine may be thine for jesus sake o say amen. I have read that joseph was imprisoned thirteen years and then was governour forty years.[29]

[*p. 229*]

October 1773. Met with a close outward trial. I felt almost ready to sink & die for several days but i kept praying to god. This was on my mind (he knew not his children neither did he acknowledg his brethren[30]). Now i thot i must not know my own family with that natural affection so as to cover up or countenanc their sin. Now i prayed to god for an oppertunity that i might tell of it to a cartain good man one of my neighbours & soon after i went out to him as he was agoing by. I told him i had been praying to see him and altho it was a croos to me yet i must tell him that i was under gods rod & begged him to talk to them. I was melted into tears. He said that he and i had but a little while to live. I must pray hartily for them and that all was not so bad as hell or to that affect. I told him that i said to myself a hundred times when i see them so wicked this is not hell. There i could not pray. No said he there is no praying thare—there is nothing but cursing. Now he promised me to talk to my family. O god bless the rod to me o god have mercy on them o god bring me out of the furnace like gold tryed for jesus sake. Now my burden was lightened.

[*p. 230*]

June 28 1773 [*sic*]. Now my dear mother at long iseland sent me another afecting letter. She rit to me about the work of god there now among the dear indiens. She rit that a little indien boy about 7 years old preacht jesus & his dying love to poor sinners & a young squaw stood in a meeting & praised god so blessedly that she supposed a harp was given to her by the master of the feast and that the indiens have witnessed to gods ordaining one peter a malatto to be their pastor. Mother sent good counsil to me. She said she had almost finished her testimony & hoped to meet me where the wicked ceased from troubling & the weary are at rest. Lord jesus jesus say amen.

This morning as many times before i may sing of mercy & judgment. Last evening without accation i was threatned with much hardship but i committed my cause to god & was still & the lord turnd things so that i had great mercies (bless god my soul nor let his mercies lie forgotten in unthankfulnes & without praises die).

Ah i find it true where our lord says your chief foes shall be them of your own household & that the devil goes about like a roaring lion seeking whoom he may devour & sometimes the devil like a roaring lion tempts to sin

[*p. 231*]

i beleve on purpose to afflict me & hoping to devour them & sometimes when i am at prayr the devil comes craftily & are i am aware my thots are drawn away. And sometimes when i wake in the night o how many distressing thots are dasht into my mind ready to make me sweat. But then i fall to prayr and i have felt the devil sensibly to withdraw & i have felt peace. And o how many other ways has satan tryed to tempt me to sin and to distres whom i trust though christ he cant destroy. O what a strong enemy is satan how did he carry the sacred body of our dear lord threw the air to the pinacle of the temple & o what he indured when tempted by him. Saints are bid to rejoice when they fall into temptation for therein their are made pertakers of the sifferings of christ. In years back i was made to rejoyce at times in the love of christ but now my life is a warfare. Lord deliver me from the evil of temtation. O give me progressive and perservering grace for jesus sake. Now o god make me like them in nehemiah 4 chap to work with one hand & hold a weapon with the other hand. O god be merciful to me a sinner.

[*p. 232*]

A few nights ago our house was preserved by gods wonderful providence from being consumed by fire and wee in it. O lord thou seems loth to destroy us. Thou hast given us as it ware new lives. I thank thee o lord prepare us all to live thy praise.

Now i perceived that mr beech was a mind to have the meetings on the sabbath moved to wallingsford again because we was so few in number. Now this struck my mind with anguish. I was ready to sink under my burden but i went to talk with him. He seemd loath to talk but at last i got him to it. I dealt plainly with him and let him know i thot it was pride. He said that people had more of an hearing ear and their minds was more waid and he seemd to think i could go there if i would. I told him how weekly i was and how against it my husband was but we could not agree. But i come home with a calm mind having in some measure done my duty. But before we parted i told him his little children would be deprived of the meetings. He said they by going did more hurt than good they toock peoples minds. He had as leave have them stay at home as go. I said the promise is to twoo or three that meet in gods name & fear. I told

[*p. 233*]

him how it faird with him when he moved the meetings away to wallingsford in the year 1762 how he reapt the bitter fruit. I said to him you went to that meeting a year & half or 2 year & then you left it and went to our standing meeting about as long and then with sicknes confind in gods prison about a year or to that affect i said. But he told me my mind was rong. Now soon after was thanksgiving day & as i was informed he went to wallingsford & they concluded to have the meetings mooved.

Now it being the year 1773 november 28 on the sabbath his text was in john 6-66:67:68 ver. He seemd to insist on that—will ye also go away. He seemd to speak as if he thot he had more light and see beyond us all and must see for us all. He said them that was behind must act up to the light they had. I thot the import of it all was to make us follow him to wallingsford whither wee could or not. He never discovered to me that day the least motion of gods spirit in prayr or preaching. His soul seemed dry and barran. He seemd to have a scattering spirit but the spirit of jesus breaths nothing but love and unity. Now when meting was done he said the people at wallingsford

[p. 234]

have agreed that the meetings shall be moved there. It will be no more here for the present and it was a judgment upon this people that god had toock our meetings from us and that the body of this people would be accountable for it. Now one man spoke & said he thot it was not right to say god had done it. He has permitted it says he or to that affect. Now it seems plain that if towards fifty people here had constantly come to hear him as there does i believe at wallingsford i believe he would not have went away from us. Here was 12 heads of families that met constantly some children & a few did of other people come in sometimes. So mr beech ended now with saying i would not have any of you prejudist. I thot then he discovered gilt. Now I come home with a waied burdened mind. That night I slept but little but my cries & tears are to thee my god. Creatures fail o jesus lead me to thee the fountain that cant fail nor be exausted. O god help me to set creatures aside and be subjected to thy permissive will. Now i writ these lines to mr beech.

December 8 1773. Brother benjamin pray do read these schriptures carefully & prayrfully. Matthew 18: 6 and 7 verses—but whoso shall offend one of these little ones which believe in me it were better for him that a milstone ware hanged about his neck & that he war drowned in the depths of the sea. Wo to the world because offences

[p. 235]

wo to that man by whoom the offence cometh. I think in the nature of gods word you have killd the poor mans little ew lamb read 2 of samuel

12 chap therefore prepare to meet thy god & my god o brother benjamin because he did this thing & had no pity—Hannah Heaton.

Now after i had rit these lines i went to the fire to dry it & a spirit of prayr come upon me that god would bless it to him lord say amen.

December 25 1773. Stephen morris his negro man was drowned in the river. O god sanctify this sudden death to us all.

Now i sent three letters to christian friends at south hampton on long iseland. O lord bless them for the good of souls or why should thou cause them to go over the ocean. It seemd as if my heart was inlarged to write o get glory to thy name thereby.

1774 genewary 28. Poor lynus sanford a young man in his prime was killd in a moment by a tree that was lodged. It fell on him and crushed him to death in an instant. O lord do make sinners to hear thy voice. They wont mind it if thy spirit dont set it home.

Fabruary 6-1774. This lords day morning i retired & wept sore before the lord. Methot i longed to serve the lord in the assemblies of the saints. My soul cryed to the lord of the harvest hast thou not said thou wilt hear the cries of the destitute & wilt not dispise their prayr. Lord hear let my cry come up to thee.

[*p. 236*]

December 15-1773 [*sic*]. Mr beach writ these lines to me. Dear sister. Christ is speaking of some in matthew 18 that was of such a humble child-like temper that they was not ashamed to own and confes him before men and not such as was ashamed to confes him before men of whoom he says he will be ashamed of before his father. Hence christ teaches the danger of offending one of these little ones that is humble and subjected to own him in his laws and ordinances before men. It is them we had better cut of a hand or foot than offend. That is self and flesh pleasing ways rather than to offend one of these little ones that are subjected to follow christ and own him in his laws and ordinences. Therefore if i should offend one of them i should be a transgressor. So if i should take away from them that i have covenanted with to eat and drink at one table and give to travellors that is to those that are without that are not under no covenant obligation to be nowhere to spare myself and to gratify myself i should be without pyty to christs poor lambs and should make myself an offender—benjamin beech.

Fabruary 6-1774. Now i writ these lines to him. My dear dear brother in the lord. Them schriptures i wrote to you are not only a rule for you

[*p. 237*]

to practis in towards those you have covenanted with but for all that are in the covenant of grace for they all belong to the church of christ altho some by reason of hinderences are not brought into the bond of the covenant. Nay i see but one rule for all flesh the lord has told us to give none offence to neither jews nor gentiles nor to the church of god. O how tender was st paul lest he should offend a week brother in christ. I believe god never laid that to his charge in jeremiah 23-first and second ver. Dear brother the spirit of christ breaths nothing but love and unity but i think you in your letter call all that have not outwardly covenanted with you travellers and those that are without. But if wee was humble and had a childlike temper of mind we should be as willing to meet with twoo or three of gods children in gods name and fear on the sabbath as to meet with a greater number of people. Our meek and lowly jesus was not ashamed to promis to come to such a meeting himself. Dear brother i think you have put me into prison but i hope to spend an eternity with you in heaven in loving & praising our jesus. O there will be a meeting that will never be broken up to eternity glory to god—Hannah Heaton

[*p. 238*]

Last year a christian neighbour one that use to pray in our meetings defrauded us as i thot in a worly affair. Now it sized my mind so that i could not hardly think of anything else. My sleep went from me altho i knew christs rule was to go and tell him his fault. Yet i let it lie about ten months hoping and crying to god that it might vanish and die. And now in this time when i use to go to meeting i did not want to have this man pray but at last i had a oppertunity and i had in some measure a meek spirit and i told him what i had against him and he seemd humble. He said he did not think i had anything against him. He said he was willing to pay all the damage but i found the damage was paid for. I was delivered o god i thank thee. Here i must set up my ebenezer[31] the lord has helped me. O that i may not stay long no more in the place of the breaking forth of children. O how lovely are the rules of christ. Lord bring me to thy supper take away all things that hinders. O that i may publickly give myself to god.

Now when i had staid at home from meting about 3 months mourning to god for a subjected soul to him mr miner come in fabruary 1774

[p. 239]

and preacht 2 sermons at mr nathaniel beaches one from luke 10-21. He shewed who the wise & prudent was and how the mysteries of christ was hid from them. Also he shewed who them babes was to whoom these things was revealed. The other was isaiah 66-2 he shewed who them poor contrite ones was that trembled at gods word and how they trembled and when he had done he said he had great hopes of north haven or to that affect for he was made welcom to the throne. There was a considerable number of people they gave good attention. There was some melting among them but as for me i felt a hard heart most of the time. But when mr miner was telling the temper of those babes to whoom jesus reveals himself to (the child says he believes what the father says to it as much as if there was no such thing as an untruth) o i felt it. Me thot i knew i had believed what god had said to me. Bless the lord o my soul for this oppertunity. O lord water the seed sown cause it to take deep root downward & bair fruit upward for thine honour sake amen. I was enabled to give something freely to this dear minister.

March 6 1774. Now of late i have felt much of the power of sin in my heart much drynes and barrennes of soul except at times but i take it as my croos. O how many

[p. 240]

arrants i now have to the throne my mind seems fixed on god i seek the lord tho sorrowing. This morning it being sabbath i hoped from chronicles 2 book the twentieth chap. O i hope yet to stand still & see the salvation of the lord altho there is a great multitude come against me for within and beyond the sea of sin in my own heart nay from as far as hell. Yet i hope to gather the spoil of my enemies and see more than i can carry away a large portion for zion. Lord jesus do come quickly.

But o how has my heart aked this winter to see how the wicked fear not god nor regard man. It has caused i believe this great weeknes in my body and mind so that it seems sometimes as if i was near to the grave [crossed out]. Them was when christ was upon earth and [crossed out] his [crossed out] to this and the other [crossed out]. O thou that did cast many devils out of mary magdelene[32] hast not thou made me to pyty him and pray for him thirty years and sometimes with groanings that cant be uttered and through thy rich mercy i have in some measure sat a good example before him. O jesus cast out the evil spirit and take up thy abode in his heart. O jesus thou son of david have mercy on him tho made many slips.

[*p. 241*]

1774 March 13. Yesterday morning this word ran in my mind—thou sayd to me seek ye my face my heart answered thy face o god will i seek.[33] Now i took this for an incouragement to prayr. O god help me to seek thee from my heart root and never let thee go till i have got the blessing. O this morning methinks i was made welcom to the throne of grace for my husband. I cried o make thy word like a hammer to break & like a fire to melt his rockey heart. O lord hear let my cry come up unto thee.

Now there has lately dyed near us five that sprung from three families all next neighbours to each other. Three of them was only daughters grown up. O lord be gracious to the mourners. O bring them all into union with thyself by these dispensations towards them. O bless it to our youths. O take away my ingratitude and make me truly thankful that we are spard yet while others are summoned away by death. O turn us to thee blessed jesus we & shall be turned.

This week the lord heard my prayr and took of a trouble that had lain long on my mind. O lord make me truly thankful. But o how can it be when i have lost my first love. Pray do away my unbelief. O give me faith to believe thy word. Now i find it is easier to believe anything then to believe thy promises. O look on the anguish of my soul. O heal my backslidings. O love me freely. Thou hast said i will but lord when when shall it once be. My soul flutters like noahs dove & can find no rest here below. I have no nehemiah to go before me none to take me by the hand. My meetings are gone. Lord give me thy love and i am well.

[*p. 242*]

April 1-1774. This morning i awaked with this word in my mind psalms 37-37 mark the parfect man & behold the upright for the end of that man is peace. That day & the next day it ran in my mind. I see altho i was a great sinner by nature yet in christ i was perfect & upright and so my end would be peace. It seemd teaching but i wanted jacobs ladder[34] to bring the comfort down. Yet it made me pray.

April 13-1774 was the publick fast. I retired to a solitary place alone and spent part of the day in prayr had some groans & tears to god. Towards night i read & rit a little. When o lord shall i love thee when shall i love thee above everything when shall i be swallowed up in love to god. O give me faith & patience to wait for thee. O let me see the returns of this day. O let my groans come up before thee.

Now my husband ran greatly in det for land which caused us much carnal company. This was a trouble to me besides i met with a disapointment and i was now seisd with much weaknes & pain of body. But this was the worst of all these things took my mind now i felt at an awful distance from god. I could not scruple my interest but i felt darknes that i could not pray except now & then. O i felt now no love to god nor to his word. I see i am afraid to trust god to deal out to me as he pleases. I am weary of staying at home on the sabbaths tho i endeaver to watch & keep the lords days as well as i can. O methinks my case is almost like jobs. O that my greif was waied would it not be heavier than the sand

[*p.* 243]

of the sea but i find i mourn more now for my misery than i do for my sin as it is against god. O god is thy mercy clean gone forever. Doth thy faithfulnes fail forevermore. Ah i may say as habakkuck 1 2-3-4 verses. Last sabbath my heart was affected in reading that book the souls preparation for christ.[35] I hope in gods mercy & truth o thou that delivered joseph job & jonah when he cried out of the belly of hell o bring me to live to thee and honour thee and make my rod like arons to bud blossom & yeild the peaceable fruit of righteousnes o god.

May 16 1774. This lords day morning met with hard things from the wicked but lord thy strokes are fewer than my crimes and lighter than my gilt. I went to god twice this morning with a heart affected in confessing sin & pleading the promises. O lord pyty my dear family. O pyty my weekly body. O sanctify my sinful soul. O thy mercies towards me are more than i can number. Lord help me to recover my first love to thee o jesus. Yesterday morning this word affected my heart in luke 21-19—in patience posses ye your souls. Methot i see the love of christ in these words. They seemd good to me while they passed threw my mind. O lord pray give me a double portion of patience as my days are so let my strength be. I hant been to meeting on the sabbath in about six months

[*p.* 244]

and i hant been complained of as i know of yet. I hope babylon is fallen dayly. When o lord shall i worship thee with the assembly of thy chosen. Lord help me to wait for thee.

May 21-1774. Samuel smith in this place had a child drowned in the river about 3 or 4 years old. Lord pyty the parents and turn them to god. A few weeks ago the widow bassett of this town was found one morning burnt to death. It was supposed she dyed [crossed out]. O lord sanctify this to all especially to drunkerds.

Now of late this word has been upon my mind commit thy ways to the lord and he shall direct thy paths.[36] Lord help me to do it in faith also that in hebrews 12 chap 11-12-13 ver seemd to come with a melting power. O i hope yet for the peacible fruit of righteousnes. O i hope yet to lift up the hands of faith that hang down and confirm the feeble knees. Lord help me to make straight paths for my feet. O bring me into obedience to thee dont let that which is lame be turned still further out of the way.

June 12-1774. This week past i had troublesome company for several days. One of the men was a drunkerd. Ah this seemd no burden to my husband & sons but when he was going away i felt in some measure a meek spirit & i told him his sin that it would sertainly shut him out of heaven if he persisted in it and death was

[p. 245]

acoming. I talked with him alone. I told him what dreadful thing it was to have no intrest in christ. I pleaded with him to pray in his family and to be sure to forbid young people frolicking at his house or i feard his children would be ondone soul and body. The man thanked me for my advice and said he should think of what i said to him and went his way. Then i went away and pourd out my hert to god. Wo is me that i sojourn in meshek & dwell in the tents of zeder o god pity sinners that dont pyty themselves. Lord be merciful unto me. O subject me to live to thee in the midst of a croked & perverce generation.

Ah alas yesterday a small temptation come upon me and i felt vext and some rash words broke from my lips and i felt immediately sorry then prayd to god and acknowledged my sin with sorrow. Then i acknowledged it to my husband. I told him i was sorry i had done rong. Now my husband looks smiling but o lord against thee i have sinned and done this evil in thy sight. I cant forgive myself. O that i that am under gods rod should sin against him. Dont lord say to me as thou did to israel i have chastised thee and thou was not chastised. I will chastise thee no more till i cause my fury to come upon thee. O help me to thank thee for thy rod. It is not so hard to bare as thy fury is. O god give me a new pardon. O give me the grace of watchfulnes. O keep thou the door of my lips.

[*p. 246*]

July 3-1774. By reason of the indulged sin i have of late been quickned to pray. I spent part of this day in prayr for my poor family that are all christles. I had some freedom to plead some groans to god. I can say as job 16-20 my friends scorn me but my eyes poureth out tears unto god. But o i hope yet in gods mercy o let my cry come up unto thee o lord hear.
This week i was sore afraid of gods righteous judgments upon the family. I feard utter destruction was at hand by reason of great sins. O how did it drive me to god. I labourd with them with a flood of tears and warned them to forbair and flee to christ with an aking heart. I intreated them with many arguments now they said but little and seemd to hear better than usial. Then i begged for a blessing o lord grant that the consumption decreed may overflow with righteousnes. O lord humble me greatly that i should want such a rod. I thank thee for thy patience but o it will be at an end. O fit us for it o let me lay my head in thy bosom thou that sits on the waves. O let me be still and know it is god. This week my soul has been vexed with evil communication of sinners. O lord how do they run upon the bosses of thy buckler and sin as with a cart rope. I labourd with them i pleaded for them o god send help from heaven or they must drink down thy wrath eternally in flames.

[*p. 247*]

Last sabbath was a good day to me tho at home alone yet not alone. My mind was helpt by reading a letter of cousen john cooks. That word seemd supporting and strengthning—i will send the comforter that shall bring all things to your remembrance whatsoever i have spoken to you.[37] Methot i rested on that promis. I had a sweet peace threw the day.

I slept well but arose this lords day morning with a dejected soul. It sent me to god with tears & groans. Something easd but not delivered but i went on in the duties of the day. The lord is daily exercyseing of me. O god thou dost speak terrible things to me in righteousnes and when god speaks where are the lips that will not quiver at his voice. Into whose bones will not rottennes enter. And i believe only they are happy that now tremble in themselves that they may rest in the day of trouble—habbakkuck 3: 16.

August 28-1774. Had carnal company several days & i found it took my mind but i kept pleading. This morning this verse come upon me with power. It seemd to lead me to god.

How can i sink with such a prop
as the eternal god
Who holds the earths huge pillers up
and spreads the seas abroad.

[*p. 248*]

O how good was them words how can i—as tho god had said to me—it is impossible for you to be lost or perish or sink altho in the midst of so many sins & tryals for i am your god of almighty power & truth. O this morning methinks i was made welcom to the throne of grace. Schripture arguments seemd to flow into my mind for myself & dear family. Some raised hopes to see gods salvation on my soul [some] seemd to make matters of discouragement matters of faith and boldness. I was loth to leave praying this morning it was so delightsome. O let my cries come up into thine ears o god of sabaoth.

But the cares of the world pulld me & the sins of my heart drove me and ere twas night i lost this sweet sence and now i feel a dejected soul indeed and a weekly body. Several worly losses and many crosses come on me but mortifyed flesh feels no pain and if i was weand from the world and subjected to god i should be happy come what would. Now the last day of this august was a publick fast threw this colany on acount of old england army that was come and lay at boston.[38] I spent part of the day alone in prayr but was dark dry and barren. Once my heart moved while complaining to god. Behold lord how i need mercy and the more needy i am the more need there is of thy shewing mercy. O i hope lord thou hast some good design towards me that thou art trying of me so uncommonly. I beg i beg i beg for patience.

[*p. 249*]

1774 september 14. Upon hearing the great commotion there is in new england on account of the old england army at boston the king has sent to change our constitution and how our land rulers and people are set against it.[39] Last night we hear there was a mob at new haven and they abused some that had spoke for the kings conduct and they have raired up liberty poles as they call them in the towns around us and all the trading to england is agoing to be stopt.[40] Now these things sent me to god to plead for our english nation. O god thou knowest what the end of these things will be. Ah we deserve destruction but this was supporting to me that god rains & governs and i hope the glorious day is near. O blessed lord when will thou do as thou hast promist to make our officers peace and exactors righteousnes. O when shall violence be heard no more in the land wasting nor destruction within our borders. O come quickly let us call our walls salvation & our gates praise.

Last saterday night there come in a stranger to stay with us till monday. At first he seemd to be a great opposer of the saparates but when my husband told him i was one he seemd to turn and up & told his own experiences and they seemd wonderful but i soon see he had an extreem appetite for sider and that broke

[*p. 250*]

up my fellowship with him. Now when he was agoing away i went out to him & told him it lookt likely wee should see each other no more till wee met in eternity and i should not have a good conscience if i did not tell him my mind and he seemd willing to hear. Then i told him he had twoo great an apppetite to strong drink which was rong. I told him we was told to let our moderation be known to all men & whatever we did to do all to the glory of god and o what a bad example. I charged him to think of it and not to rest till he knew christ was his. I think he wisht me well & went his way. O lord have mercy on him. O how many are there that live upon past experiences and no fruit to god no sanctification. Now i had peace in my talking to him.

October 5-1774 [crossed out] & anger with god [crossed out]. O i thot how dear was that food bought no less price then the blood of christ jesus [crossed out] now this [crossed out] till good part of the [crossed out] was spent the food [crossed out] what will such [crossed out]. O god thou art shewing me hard things. O let me yet reap the peacible fruit. In my first book i have written a little how i was used for not going to the standing meetings. One of them women that

[*p. 251*]

went to be an evidence against me has lost the use of her limbs and has lain a long time bedrid full of pain and always like to lie so while she lives. Of late seven have dyed that sprung from her family. The other woman has gone to the grave with three of her children. Lord open their eyes have mercy on them.

October 1774. Mr beech preacht a lecture at his brothers from leviticus 26-25. He shewed how christians now like israel had on their part broken covenant with god and in many respects god had brought the same judgments on us as he did on them and no reformation. The last was their quarleling and fighting among themselves that proved their final overthrow and now our nation is threatned with the same destruction. He said he believed our nation would be destroyed without a reformation but without gods spirit what is all preaching. Lord i thank thee that i could hear him willingly. O that the consumtion decreed against our nation may overflow with righteousnes and mercy in christ jesus.

November 10 1774. Again mr beech preacht at serjant heatons from psalms 89-15 blessed are the people that know the joyful sound. He shewed how & who them was that had seen themselves by gods law condemned to eternal misery and how they was made to hear the joyful sound of the gospel or salvation by jesus christ. Now i knew it was truth but i wanted faith. Now i went out sad and returned sad o the darknes of my mind of late. Lord bless thy word dont let it fall to the ground.

[*p. 252*]

November 23 1774. I met with a croos. I was much worried in my mind i could not take it up and submit. The next day was our thanksgivingday. I wanted for several hours before i did to go and pour out my soul to god. At last an oppertunity presented and i retyred away. But o how did my heart go out to god with thankfulnes for the great priveledg of prayr. O if i was in hell i could not pray but glory to god i am here another thanksgiving day where i may tell thee all my wants all my sorrows and all my sins hoping and expecting help and mercy for the sake of him that had nowhere to lay his head. I had heard mr miner was coming to preach. My hart went out to god that the god of abraham isaac and jacob may come with him and cause his preaching and prayrs to do good to souls. O touch his hart and tongue as with a coal from thine alter. O make him a hunter to bring thine out from the holes of the rocks where they are hid and o lord i thank thee for this help. O subject me to thy whole will.

A few days ago there come a poor man from hartford that had lost a great estate he told me at sea and just buried his wife and twoo children. Now i examined him how it was with his soul. He said he knew he should go to hell if he dyed as he was for he was not nor could not be conserned about his soul for his mind was overcome and drownded in his worldly sorrows. If he attempted to pray that would be dasht into his mind and be uppermost. Now i laboured with him about the worth of his precious soul. I shewed him some kindnes and gave him a little money. O god pyty him pray give him an intrest in thy love and favour. Convert his soul for jesus sake.

[*p. 253*]

December 4. It is above a year since i have been to meeting on the sabbath except once i went to new haven and heard mr mather.[41] I thot death was on the minister and people tho he spoke truth. O god give me contentment in my prison for i have lived at a distance from god above twenty years tho i have a little reviving sometimes in my bondage. But o i feel a carnal worldly mind and an opposition against reading the word but i read constantly and find it teaching tho almost lost the conforter. But i hant lost yet a hope in gods mercy and promises. Come lord jesus and deliver me from this body of death & o here is hardness of heart and blindnes of mind [crossed out] my proud heart shakes [crossed out] my stroke is heavier than my groaning. I have lost my faith it is carried captive. I cant see nor long after jesus but i hope i hope yet in gods mercy. Lord look on my sorrow and my pain and pardon all my sins.

December 5-1774. Mr miner as he was on his journey to the nine pardners come here and preacht twoo sermons one from luke 15 concerning the prodigal the other from matthew 20 concerning the labourers being hired into the vinyard. Now in these twoo sermons how did he labour with souls to come home to their fathers house & to come into christs vinyard to work. He insisted on their being converted and shewed what it was and the fruit. The lord helpt him. When meeting was done i went to him and took hold of his hand. I told him i had prayed for him that god would come with him and said he is come. Now i think he said it is by the prayrs of the saints wee get the blessing. There was a full meeting. I gave him a little money. My husband

[*p. 254*]

and twoo children was at meeting but i fear untoucht but through free grace i felt a watchful mind afterwards a desire after holines. O lord bless this thy servant with the choisest to heavens blessings prosper him in his journey for the good of souls. O make thy word here to run and be glorifyed. Lord jesus come quickly.

December 1774. Poor jason cooper dyed in a fit. O god sanctify this sudden death to all especially to the widow & fatherless children. O be the widows god & the orphans father. O when wilt thou thaw this frozen world.

December 13. Methinks i had freedom to the throne of grace this morning for myself my family for our nation and land for zion that she may arise and shine and with god i must leave all. How long holy just and true are thou. Wilt come terrors are upon our nation and this land. Waatts says terrors attend thy wondrous way that brings thy mercy down psalms 77-7. O i hope yet in gods promises. I hope against hope (tho times look dark ill tidings spread that fills the peoples hearts with dread). My heart is armd against this fear. I hope the god of grace is near [page torn].

[*p.* 255]

O when shall i get away from them that are not my people to live with them that are my people to say farewell meshek and keders tents. O my god make my soul bless and praise thee for distinguishing grace and free mercy to me the chief of sinners while others are left holden with the cords of their sins.

December 30-1774. Read in hosea prayed went to bed but could not sleep. I could say as habbakkuk 1-2 & 3 vers. Also i prayed for our nation. O god spare spare this threatned north amarica. About midnight i arose & drest me and pleaded for mercy. O why wilt thou not save since mercy is thy delight. O pour down thy lovely spirit and that will heal devisions. Towards day i lay down and slept a little. O god help me in faith to trust in thee come what will o give subjectednes. [lines crossed out, page torn].

[two pages missing]

[*p.* 258]

The next day after the fast fabruary 2 about one of the clock there was a terrible noise heard in the northeast towards boston supposed to be a commet that broke. Many see the fire and heard the loud noise. Ah god is calling to us to prepare methinks for his righteous judgments. O god turn us to thyself. There was such a thing seen and heard before the last war in new england. O that we may be reformed. O god do stir up thy strength and come and save us. O pyty souls.

Fabruary 19-1775. This lords day morning had some meltings in prayr. In confessing sin i was affected in reading some passages in my diery. O god bring me and all thy children to live to thee in this day of calamity. O how busy are people in hearing and telling news about this war but who says alas what have i done what shall i do to be saved. Pray father turn us and wee shall be turned to thee.

Fabruary 27-1775. This saterday night was broken of my rest as i often am by reason of indulged sin. O what will aged sinners do when that god that thunders and makes the commets blaze shall rend the sky and burn the sea and fling his wrath abroad. The next day was sabbath. I was poor and week i was not pacient anuf. O god let me not be offended with the stone that hits me without regarding the hand that sends it. My husband laboured with for family peace. He seemd to hear and said but little. O lord give me to speak right words that are forcible. O give me charity that suffers long and is kind but o horror hath taken hold upon me because of the wicked that forsake thy law. O my faith is gone jesus is gone and cant lament after him as israel after the ark but i hope yet. O christians take warning marry in and for the lord.

[*p. 259*]

Went to the funeral of an aged man 84 years old. A wonder that i and mine are alive when others are dead. Lord thy patience is infinite. O draw us that we may run after thee after thee.

March 7-1775. An angry contention arose in the family now i had a great sence of the sin. I went crying to one and to the other pleading for peace with tears. I begged them for christ sake to leave of. I told them to forgive each other. I said o how much worse have we done by god and christ than any creature can do by us here and he bares it o forgive forgive. Now at first they seemd offended at me but i kept on crying to them to make peace. At last they began to come twoo and preasently made up the matter went about their work pleasantly. O god let me not forget thy goodnes and thy great mercy to me and mine. O bring us near to thee. O give us always a christ like temper of mind. O make us all to live with thee eternally o lord and join the glorious throng crying holy holy lord god halalujah.

The week past i was afflicted with vain company that was entertained. Ah my unequal yoke galls me sore but now i must indure it. On the lords day morning my heart melted in prayr before the lord. O god shew me thy salvation. O let my cries come up unto thee.

Again for several days had carnal company. I felt twoo much of a carnal mind but had anguish with it. Had some meltings in prayr but o lord help me help me to recover my first love. O make my heart faithfull to thee o god faithful to souls. O i hope yet in thy mercy o convert me again.

[*p. 260*]

Aprel 8-1775. Now after much prayer & consideration i believe it to be my duty in the sight of god and we have taken one of my husbands brothers daughters to live with us about thirteen years old. And tho my house is a house of beliel for want of family government o lord grant that all parents may first govern themselves and then their children. I believe god can keep her from the infection of the family. I feel consernd for her soul as for my own. O my god grant that i may daily preach christ to my family by my practis. O let me commend the ways of holines to all by my daily walk. I beg i beg dont let me never again bring up an evil report of the goodly land.

Aprel 19-1775. Was our publick fast day threw the colony. I arose week and low in soul and body was obbliged to take my bed. This was occasioned i believe by my last nights tryal. O to see sinners go on daring god to his face to strike them down to hell almost breaks my heart but ah i marryed a christles man. O my sins has procured this rod my sins are ever before me. Every night i have to mourn for the sins of that day. O what a fountain of sin do i feel and see in my heart more than i can count or recon up. O lord surely thy patience is infinite or wee should have been destroyed long ago. I plead it lord sin abounds here. O let thy grace much more abound.

Aprel 20 1775. I heard mr cammel[42] of new milford preach from mark 10-first and the people resort unto him again as he was wont he taught them again. He invited christians not only to resort to christ again to be taught but examined them if they got this teaching. Ah how he spoke to backsliders and warned sinners. He shewed how the woman of samaria

[*p. 261*]

was taught that she was a sinner and about worship twoo and how he shewed her that he was christ and he is now wont to teach his children. I had some meltings of soul while i heard the gospel preacht. When i come home there was carnal company intertaind that night and i was sorely distrest with sin and devil. My sleep disturbed so i almost lost my melting but i plead yet o god be merciful to me a sinner. O pyty my dear husband that would not go with me to this meeting pray bless thy word.

Aprel 22-1775 there was a great alarrom threw all this country on hearing that genneral gages army and the boston men was fighting and great numbers was killed on both sides.[43] Now great numbers of our men are gone and going to help them. O god have mercy on our nation. O bring them in to nearnes to thee and to one another. O the sin of this land in refusing and trampling on an offered christ imprisoning christs ministers that he has sent forth to preach. I beg lord give our rulers to consider while they are fighting for liberty what cruel bonds they have laid upon thy saints.

Alas it is a trying day with new england. I know if god doos by us as we deserve wee shall all be distroyed in this affair. My soul has been in anguish for this land but last thursday morning i awaked early with them words—the angel of the lord incampeth about them that fear him and delivereth them.[44] O i hope and believe god will take care of his children let what will come. Also on sater morning i awaked with them words—the wildernes and solitary places shall be glad for them.[45]

[*p. 262*]

This word seemd to raise my hope. It kept running threw my mind on the lords day. Hope the daughter of faith seems to be in exercise in my soul. I hope the glorious day is near altho saints have lost their first love and hath sind with the wicked. But the affairs of zion are all yet in good hands. Lord our robes are defiled o when shall they be washed white in the blood of the lamb.

Again there was a croos laid upon me and i felt an unsubjected heart. Much disturbed in my sleep that night but next day i felt still about it and was so several days. And now in a way at first unexpected restitution was made to me bless the lord o my soul. The next lords day morning my soul went out to god in prayr when i was retired into a lovely secret place. My heart melted while i pled them promises that i believe god gave me many years ago to see his salvation in my family in this place and gods glory to be seen threw the world. I cryed but behold lord now we are threatned of having all our priviledges both temporal and spiritual taken away. I have looked for peace but behold trouble. We are alooking every day for the kings ships of war to come into new haven harbour. Now here is a trial of faith but i hope yet in god. Now all new england is alarromed and preparing for a hot war. I must expect my husband and twoo sons must be in it. O lord pyty their souls. O let me lay my head in thy bosom who sits on the waves.

[*p. 263*]

Things look as if god was about to redeem zion with judgments. I find these threatning desolations are a means of quickning to me. It makes me to look up my former promises and carry them to god. O i hope in gods word when moses cryed when samuel prayed god gave his people rest. Our nation i believe are the wickedest nation on earth because wee have abused our great privelidges. But o are not gods mercies greater than our crimson sins. O has king george forfited his coronation oath. O does lord north govern king and perliment and must he govern america twoo by papists laws.[46] O lord we have no might against this great company. Thou o god has stiled thyself the lord of hosts. The lord is a man of war o help for thine honour sake. O lord thy judgments are upon us make us larn righteousnes. Shower down of thy spirit o turn us to thee for jesus sake.

O when lord wilt thou take away my stony hard heart. I see and feel myself to be a great sinner this day. O my sins my sins that has caused god to hide his face. The treasures and glories of gods word are almost hid from me. I fear a total dissertation. The book is seald and i can scarcely cry. John wept much when no man in heaven nor

[*p. 264*]

earth neither under the earth was able to open the book neither to look thereon. But ah how was his holy soul pleased when one of the elders told him the lion of the tribe of juda had prevailed to open the book and loose the seven seals thereof.[47] And o how he felt when the lamb as it had been slain came and took the book out of the right hand of him that sat upon the throne—read revelations 5. O god thou canst why wilt thou not give me the same faith and grace that john had in kine if not in degree. But o for a large portion that i may honour thee much before my head is under the clods.

May 30-1775. Arose this morning took my byble. I lit on the 24th of genesis. Was at a stand whether to read it or not being i knew it all so well but i read it as well as i could. But o how was i put to it to get along my hart melted under the word. O to see the faith of blessed abraham verce 7 how he kept hold of gods promis and o dutifull eleazer like christs ministers obedient to their master and lord. Eleazer had rather do his masters arrant and court a bride for his masters son than to eat when he was weary and hungry just like christs embassendors. O says faith the lord before whom i walk will send his angel and prosper thy way &. Ah what follows faith and obedience prosperus blessing upon blessing. O lord do give me like precious faith and obedience. O help me now to bles thy name. O give me as much santification as i have light.

[*p.* 265]

O what a creature i am. But a few days after gods word was so precious to me i was much provoked and spoke some fretful words in a rong spirit. Immediately my husband was in a rage at me and spoke bad language. Quick my hart smote me i went mourning and praying. O did i not only sin myself but caused him to sin twoo. Lord keep thou the door of my lips for time to come. O help me to bair with patience every wheel of gods providence that rools over me. O god keep me o let me know i am forgiven.

June 5-1775. I went to new haven saw a woman that told me that my oldest brother at quoage on long iseland was dead but i could not raily realize the thing but it seemd to put me in mind of my own death. O lord make me actually ready for it whenever thou shalt call me away. O take the sting away let perfect love cast out fear.

June 20-1775. This sabbath morning my hart melted before the lord. I pleaded hard for a blessing for myself and family had freedom to confes sin to beg

[*p.* 266]

for a labourer here after gods own heart. The lord has bid me pray for it and will he not hear. O to serve the lord on his day with the assembly of his saints how precious is it. O god give me a double portion of patience to wait for thee and tho i tarry at home let me divide the spoil. O did i leave my lovely long iseland where means in plenty i injoyed liberty of concience and freedom[48] and marryed here contrary to gods mind & will. O my rod o my bonds they often make me groan to god. But ah my day is gone i have done it but i hope in gods mercy yet.

[page cut off]

[*p.* 267]

That word that come to me long ago because he could not do it by day he did it by night—judges 6-27. This is i think sin in the same nature of that alter of baal that gideon destroyed. Ah how many times have i got relief by my doing thus privately or methinks i should not be here now o lord do make him give up his idol. O thou that can do everything that pleaseth thee and o bless bless the rod to me i beg i beg.

1775 July 3. Yester we had the

[page cut off]

[*p. 268*]

enmity he has in his hart against the religion of jesus christ. O [crossed out]. This morning being the lords day i was reading in mr john flavels large book of 2 voloms[49] the 31 sermons from john 19-27—then saith he to the disciple behold thy mother. My hart melted. I wept & red and prayed for a blessing [crossed out]. Calvin come and sat down by me lord i thank thee. This sermon shewed how jesus took care of his mother upon the cross. It shewed the dreadful sin of disobedient children and what judgments had befell such as was so [crossed out]. Glory to god i have a schripture hope for his soul. I hope against hope lord have mercy on me and mine. O pyty our nation. O rain down a rain of righteousnes. O for a reformation i beg i beg o that the work of god that wee hear is at rhode iseland may spread from sea to sea and from the rivers to the ends of the earth.[50] O god keep me keep me & fit me for what is coming. O be merciful to me a sinner i beg i beg i beg.

[*p. 269*]

I have such a weaknes in my breast and head it hurts me much to write. If god is not pleasd to give me new strength i must leave off. Now one morning i awaked with this in my mind—the 6 of romans and the 7 verce. I got up and read it. The words are these—for he that is dead is freed from sin. Here i was taught that i had got a legal spirit that i was not dead to the law and so not freed from the condemning power of sin. Help me o lord by faith to look to and take fast hold of jesus who has fulfilled the law to a tittle paid his father the full sum. I have cause to believe for my soul. O how long how long are it shall once be with power upon my soul.

One day this week just at nighth my calvin come to me and told me to get his best cloaths ready for he was going to hire one of the neighbours horses to ride away and off he went in a hurry. O how my hart acked. I laboured with him but to no purpose. O the sin of froliking is a great sin. O precious time. Now when he was gone i went to prayr begged of god to stop him to disappoint him and i believe god heard my cry for he come home again. Bless god o my soul.

[*p. 270*]

But ah poor child [crossed out] poor children they take liberty from one that should teach them gods fear. One of my children is now 31 years old the other 20 almost and i cannot perceive they ever pray alone. This day the 20tieth of july is a publick fast i exhorted to prayr. I felt so [crossed out]. I said i will pray while i can breath but o god when shall sin [crossed out] cease [crossed out]. O it sinks me and as job i am broken in pieces with their words. Ah mr flavel says there is a vile generation in our days that instead of calling upon gods name by prayr do call upon it prophanely rending and tearing that great terrible name with the language of hell. O saith he let god rather strike me perfectly dead while i live than afflict my soul through my ears with these dreadful dialects of the damned. But o my soul is almost continually afflicted through my ears. But o god i thank thee that i am not hearing such language in hell. But o my spirit is broken my heart wounded i fear destruction. How can i endure the destruction of my [crossed out]. O it makes me pray. Yesterday my soul got near to the throne. I cryed for patience to hold out. I cryed why art thou so long acoming o come lord jesus

[*p. 271*]

with salvation. O how many years have i been praying for them and they growing worse and worse. Lord help me still to fall down before the lord as at the first like moses not regarding the bitter reflections and anger at me for spending so much time in praying and reading. But o i hant prayed half anuf. Lord help me to yet get a blessing.

A few mornings ago i believe god moved a good neighbour of ours to come early on purpose to talk to my husband. I believe he fell before him for that day i see an alteration no bad language. O now i begged for him but it was soon over as the dog to his vomit. O lord have mercy.

But o i may sing of mercy and judgment. The child that i have taken to live with me is a great comfort to me. Ah she is the answer of many prayrs when i suffered for want of help. I use to plead this with god that rebecca had a deborah[51] and sarah a hager and if it was for gods glory that he would find one for me altho i have been afraid to take a creature that had a soul these many years into the family because of sin [crossed out] but now my necessity has drove me to it.

[*p. 272*]

I love the child and her hart seems towards me o lord dont let her larn to sin. Pray make me an instrument of good to her soul. I pray reward her obedience to me with an intrest in thy love and favour. Lord i thank thee for this answer of prayer. O keep me keep me dont let me sin against the child.

Now a contencious spirit arose [crossed out]. I laboured and strove for peace. My soul almost trembled with fear but i prayed to god and found that he was above him for that evil spirit was still. Lord i thank thee.

July 30-1775. Had a praying mind this lords day morning. Had peace threw the duties of the day. Lord help me to give thee the glory that is due to thy name. O make me willing to suffer anything thou shalt see fit. O make me willing to wait thy time for delivrance.

August 13. Had fredom this sabbath morning in confessing sin. I thanked god for his word that has told me that he has forgiven great sinners or how could i ever have had a thot that such sinners as manassah magdalene and saul could ever have been forgiven. O the sin i saw this sweet morning distresst my soul. O sin abounds here. He aynt willing i should read or pray or talk of religion. O make thy grace much more abound here.

[*p. 273*]

1772 [*sic*]. I think i must give a few hints of my tryals about baptism. I have been tried about it by turns i believe almost twenty years. Sometimes so distrest i was almost overcome and ready to die. My husband use to tell me that i distrest myself about things that i had no need to but i could not see threw it and it was fastned to my mind. But this i believed all the time that dear abraham was under the same covenant in the nature of it that believers are now. For above twenty years ago that word was powerful upon my mind—abraham rejoyced to see my day he saw it and was glad.[52] Ah i thot when i vewed christs comeing into his kingdom i felt just like abraham when he see christs day. Now when i read those books that pled for infant baptism i did not know but they was right and when i read the baptists books i did not know but they was right and so i was confused in my mind. At last i concluded to fling by these books and i serched the schriptures looking to god for mercy. I think this was the first schripture that took hold of my mind l of peter 3 chap 20-21 verses and romans 4 galations 3-29. These and many more has seemd to establish my mind. I believe that we gentile believers are under the same covenant now that abraham was tho under twoo

[*p. 274*]

different administrations the latter much clearer than the former. The jewes and their infant seed injoyed the priviledges of the covenant but they were broken off. They and their seed we gentiles was ingrafted. I say them that are believers are abrahams seed as raily now as the jews was [crossed out] and have a right to the same covenant priveliges now as they and their infant seed had [crossed out]. Pray do read carefully romans 11-17. I hant time to to write what good we have been taught by gods word about these things only a few hints. Glory to god for his mercy towards me a great sinner.

The word is plain that baptism is in the room of circumsition but before i was tried about it when my children was young i dare not have my minister to baptise them nor join to his church as i told him because i believed he was not nor his church in gods rule. But ah i thot infant baptism was right because i was taught so by tradition and never see it by gods word till now. Lord take away my ignorance.

September 6-1775. O what a stupidity do i feel and see it so all around me altho surrounded with judgments the noise of wars and our heaven as iron and earth as brass. O how is the earth burnt with a long drouth. Lord thy judgments are in the earth pray pray make us to larn righteousnes.

[*p.* 275]

Altho i have lost the comforter the faith of evidence yet the faith of adherance or relyance remains. I feel some trust to the truth and faithfulnes of a god in christ and i feel a tendernes as to sin in myself and others. Ah when the candle of the lord shined upon my head i doted upon my graces and caused him to withdraw.[53] O now his heavi rod is upon me. I fear i shall sink. He has hedged up my way with thorns—hosiah 2-6. He has put a clog upon me. O i cant thirst cry nor long after jesus. O this week my hart has aked and my eyes flowed with tears to see a sinner go on so boldly in sin and excuse & justify themselves in it and say it is all because of the new light. O but it drives me to god with groans sobs and siths night and day. Mr. flavel says god uses afflictions as we do sope to clense out spots and take away filth. O refine me lord in this furnace. O i fear when i think how good joseph larnt to swear by the life of the pharoah. God keep me keep me dont let me with the froward larn frowardnes. O give me a christ like temper of mind. O help me to overcome evil with good for jesus sake.

[*p.* 276]

A good god has sent rain on the perched erth. O send down a rain of righteousnes upon dead souls dead in trespasses and sins. Thou canst make them live.

September 1775. I was reading and [crossed out] spoke the truth but i fear with a rong spirit. I was sorry afterwards. Lord help me to do so no more. Some think my sorrows are great but o they are not so bad as hell. Now a few days after i sat reading in mr flavels book [crossed out] but i did not speak one word but i hope committed my cause to him that judgeth righteously. O lord i thank thee for this act of self denial o keep thou the door of my lips.

Last week i was dealt so hard with and heard such [crossed out] words that on saturday evening after my family was all in bed i took my bottle of beer and some food my testament and my diery and ran away into the woods. Like hager i took my lodging where i could. I spent the sabbath in reading and prayr had liberty to pour out my distrest soul to god. With him i leave my petitions hoping to see the returns. The second night i laid me down and slept sweetly. Thou lord made me to dwell in safety. On monday morning i come home & found the evil spirit was lowerd for some days after. Lord i thank thee that word was upon my mind—when they arise he stilleth them.[54] O god i pray bring me forth in thy own time like gold tryed seven times more fit for thy use. O purge out the dross and take away the tin. O jesus interceede for me to thy father.

[*p. 277*]

Went to hear a sermon preached at a tavern to the man of the house who tis thot is nigh unto death with the dropsie. He has been a great drunkerd and a prophane swarrer. He semd to have a secure stupified mind. He told the minister that he did not know whither it was worth while to pray for his life only to pray that he might have an easy passage into the other world. And i heard him say that he did not care if he dyed before sundown and not a word to evidence a change of hart. O what an awful sight to see a great sinner going down to hell easy with a benumed soul. This minister said to him a person had need to be well assured of his good estate to be willing to die. O unfaithful to say no more. His text was ecclesiastes 7-14—in the day of adversity consider. He preacht some truths in a general way. Ah he ought to have took the precious from the vile and have given to each one his portion but o how unfaithfull to god and to this poor dying man. O lord pray hasten the time thou hast spoken of in ezekiel 34—when thou wilt deliver thy people from false shepherds and give them faithful ones after thine own heart. Send out thy ministers o lord that preach alone for thee—that use no notes nor take no hire. Send help and make us free.

[*p. 278*]

November 12-1775. I saw a sinner raging fearing not god nor regarding man. My soul was in pain. I cryed to him to forbair. I pleaded with him but to no purpose. Wel i told him i will go and pray for you and with a hart full i pourd it out to god for him. O to see god dishonoured and his soul sinking into hell. O how awfull o lord do cast out the devil and reign king in his soul. Thou can make him holy with thee i leave him o lord. I thank thee that i believe and hope through christ for a country whare the wicked will ceace from troubling and my soul be at rest. O give me patience.

Soon after i was [crossed out] called all to nought by a wicked person before a vile young man that was by but god was gracious and gave me some measure of patience & meeknes. I never spoke one word and afterwards had a still mind. O god forgive him that has tried to make thy religion and children vile. O let me never be angry at the stone that hits me without regarding the hand that sends it. O let thy hand humble me at thy feet and keep me there.

1775 november. Our yearly thanksgiving day. When i could leave the house i retired alone pleaded hard for myself and

[*p. 279*]

family o pyty our nation and land. I had i hope in some measure a thankfull hart for gods dear mercies threw my life. O god i thank thee for jesus christ. I thank thee for thy written word. O let me for time to come live thy praise.

November 1775. Elias beech buried twoo of his children one [blank space] the other 4 years old. He told me the first was taken away because he was not faithful anuf to his neighbours altho he had been to some of his neighbours and warned them to get an intrest in christ. But he got unbelieving and wanted a sign and one day he was away alone and god who is a spirit spoke thus to his spirit—i am acoming into your house with my judgments. And he said he cryed mercy mercy mercy lord but that word seemd spoke to him—be still and know that i am god. And then his hart seemd to surrender. This he takes to be his sign that he must press forward to warn sinners. He spoke at the graves of both his children and warned the christles and now he goes from house to house pleading with souls. A few evenings ago i believe god sent him to my house to talk to my husband. As soon as he began he fell before truth. He prayed twice and warned pleaded and laboured with him to get an intrest in christ.

[*p. 280*]

He talked to the children and to me. It seemd to stir up my drowsi mind but my husband soon shook it all of. Ah the man told him that this night will be remembred in eternity but he soon got to his old way of sinning and abused & reviled the man for coming to talk to him. O lord if it is not twoo late pray have mercy on him for jesus sake.

December 8-1775. Awaked this morning with a great melting in my soul accasoned by the last nights tryal. I cryed o how long have i been looking up for gods salvation but no deliverance. O methinks i feel like david in sauls court.[55] Ah i wantted to get me away from keders tent. O if my hope was only in this life i am of all women the most miserable. Now this same morning that word come into my mind—the upright shall have the dominion over them in the morning psalms 49-14. And while i was milking it kept running in my mind o methot i see something what a difference there would be between now and the resurrection morning. Now the wicked has the dominion over the godly but then the godly shall have the dominion over the wicked for the saints shall judg the world. O god i thank thee for this o my god give me patience to wait for thee lord help help.

[*p. 281*]

December 15-1775. This night went to bed slept a little awaked with a pain in my breast. Got up took my byble lit on the seventy fourth psalm. My hart was much melted. While i red me thot it pointed right to this day not only in temporals being under the calamity of a terrible war but it held forth the dark state of zion at this day which now lies as with her face as on the ground. Ah the psalmist says there is not any among us that knoweth how long. He begs god to pluck his right hand out of his bosom and then tells god what mighty things he has done for his people his great kingly power and he makes this an argument to plead with god that his name was reproacht and that he would not forget his forever. He begs for the dark places of the earth remember thy covenant and for the oppressed for the poor and needy that god would plead his own cause. Pray and read it carefully o all gods people go thou and do likewise.

Now there come one and told me that my calvin was going at night to a frolick at the tavern to keep chrismus. Hearing this struck my mind. I fell to weeping to think that this was their way to spend that

[*p. 282*]

night on account of a saviours being born. I talked to him and would have had him staid at home. Then he begd me to lend him money but i told him i will not i dare not do it. But what i said was to no purpose for his father indulged him in it and gave him money to spend. O god pyty this land for now thy rod is lifted up they will not see o pyty the risen and rising generation or wee are ondone.

Now there happened a sad contention between us and a neighbour of ours on worldly accounts. I was much tryed in my mind and carried the case many times to god begging that the matter might be made up. Now when it had lain long one morning i went down to his house begging mercy of the lord. Now when i come there he seemd lofty and high. He spoke bad words and seemd very angry but i laboured to get down his spirit and begged him to come to our house. At last he said he would come in an hour. Then i sat away home and in about an hour he come. At first my husband was much enraged and the man twoo but i pleaded and laboured with them not to be angry. Now after a long while their spirits come down and they made up and made friends. Lord i thank thee that thou art a god hearing prayr. O make me at all times a peace maker and o that the peace makers blessing may truly rest upon my hart lips and life.

1776–1779

[*p. 283*]

January 7-1776. On the lords day morning i retired and pourd out my complaints before the lord. O hear thy poor creature and send salvation. O give me thy appointed means of grace on the sabbath. O convert my poor husband and children. O how long holy and true are it shall be. O how dear are their souls to me at any time when they go from home my prayrs goes with them. Lord keep them from evil have mercy on them bless them in their lawful undertakings return them safely. But o their precious precious souls are above all lord pity them.

January 17-1776. A fast on account of the war threw the colony appointed by our rulers. I was ill in body and low in mind but i attended on the duties of the day by reading gods word and secret prayr and teaching the little garl that lives with me. Lord bless the child and o let this be such a fast as thou has chosen. O prepare thy childrens harts to pray and cause thine ear to hear. O give our rulers now to consider that while they are contending for liberty in their temporal priveledges how they have made cruel laws to take away the sacred privelidges of the people called saparates and will not alow them the liberty of conciance. O lord pyty our nation and land. O let it be a land cared for of the lord from one end of the year unto the other. God be mercifull to me a sinner.

[*p. 283 sic*]

February 1-1776. I had a sad provocation. I said but little but felt my corruptions raised like an ocian ready to overflow and drownd me. I could hardly pray at all but i tried to beg for mercy. It lay like a mountain upon me for several hours but a kind god took it all off. O lord let me live thy praise. O what a hell would it be to me if god should let loose the corruptions of my heart. O god keep me keep me keep me. O take away my adamentine sins that wont be refined. O jesus when shall i rejoice in thy love again.

February 11-1776. This lords day morning weept and prayed before the lord. I told him that he hears the ravens when they cry and feeds the sparrows and young lions[1] and has bid me pray to him the lord of the harvest to send labourers in to his harvest. O give us here thy appointed means of grace upon the sabbath. O give me patience to wait tho at home alone. I say to wait thine own appointed time but ah i have lost my faith what shall i do. O must i never feel thy love again in this world as in years back. Is my lot here nothing but tryals and distresses but o this stills me many times and i see it to be a great means to bear my punishment here for sin and not have it reserved for me to bair in hell. O god i hope yet in thy mercy lord help me.

[*p. 284*]

March 10-1776. The week past i saw so much sin in my own hart and in the practice of my family that i was almost ready to sink and die under it but it made me pray to god. One day i being prest down retired away alone. I felt such freedom to pour out my soul to god. I told him with a travil upon my soul that he had told me in the sight of that sun that he would have mercy upon my jonathan and make him an isaac. O make good make good thy word. I have waited 24 years and he is in his sins still [crossed out]. O help me to hold fast the word of thy patience. Lord give me patience to wait thy time. Now after i had pourd out my soul i felt in some measure relieved and quiet. Lord have mercy help me to endure to the end.

This winter a wicked man got my dyery out of enmity and kept it from me almost a month. I prayed god to bless it to him if he red it and to preserve it from harm and if it was his will to restore it to me again. I believe the lord heard for it was brought to me again in an unexpected way. Lord i thank thee o that i may write to thy honour and glory. Now i lit of the book of the second spira.[2] O how did he prize the love and favour of god. Said he oh that i was to lie broyling in that fire a hundred thousand years to purchase the favour of god and be reconciled to him. O god take away my ingratitude o quicken me according to thy word lord have mercy.

[*p. 285*]

1776. O the sin that my sin has been punisht with this winter [crossed out]. Here my body poor week and low my faith gone my spirit wounded. Jesus gone i cant thirst after him but i mourn i mourn i mourn but this often stills me—i am here i am not in hell where there is no hope no praying. Here i may pray and hope to be heard [crossed out] but there i should be filled with nothing elce but cursing and blasphemies to eternity. I say this stills me. Ah my mercies are great mercies i hope yet in gods mercy and promises. God is just and righteous in these tryals that he daily lays upon me for i marryed contrary to the teaching of gods word. Also i was warned against this sin of being unequally yoaked with an unbeliever just before i married by a faithful minister of christ but rebel that i was i made light of all. I think that night i married it made way for the curse of god upon me and my family in this world. 1 samuel 3-13. Also as david i fear the sword is never to depart from my house but my heavi hart looks up to him that can bring good out of evil. Take warning by me christians.

[*p. 286*]

March 24. Had carnal company all last week and i gave way to their talk in some measure but o what guilt i felt and stupidity afterwards. Lord come near to my soul and help me for time to come to do better. O make me faithfull to souls for christ sake or i shall dishonour thee and wound my own soul.

March 25. Mr cammel of new milford preached in the evening at mr nathaniel beeches. His text was in ecclesiastes 8-11—because sentance against an evil work is not executed spedily therefore the heart of the sons of men is fully set in them to do evil. He seemed to have nearnes to god in prayr and o how did he labour with christles souls. He seemd to think gods salvation was coming to this people. The preaching seemd good to me while i heard it but alas i am where i was before. O god when shall the hunters bring thy children out of the holes of the rocks where they are hid. O send help to zion that sinners may be converted unto thee.

March 31-1776. Now my twoo children are gone from me. Jonathan i know not where he is he being offended with his fathers froward treatment towards him has we hear been to boston and calvin gone into the army down to new york. O god do do do pray father come quickly and send that spirit that elijah had and turn the hart of the fathers to the children and the hart of the children

[*p. 287*]

to the fathers lest the earth or such families are smitten with a curse—malachi 4-5. O for jesus sake pyty the souls that thou hast made. O lord keep my children from sin & evil. O convert them return them safely. If not if i must not injoy them no more in this world o lord grant that they and i may meet in heaven at thy right hand never to part. Threw gods mercy i felt in some measure a still mind at parting with my twoo children. One went nine days before the other. I never shed one tear at parting with them and i have felt since in some measure subjected to the cross. Lord make me wholly subjected to thee in all thy dispensations of humbling.

Aprel 14. Now jonathan is come home lord turn him to thee. But calvin wee hear is sick at new york with the plurisy.[3] Now his father cries & groans in distress for him but not for sin. Lord grant that this may be the time for thee to work. O turn his iron heart to thee and o pyty the poor childs soul. O convert him prepare him for death if he must go now but pray spare his life if it is for thy honour and glory. Thou knowest i know not what is best for me o subject me to thee. I feel in some measure a still mind lord i thank thee. Now i dare not pray that he may live and come home only so that if it may be for gods honour and glory. Now my husband was so distrest he could not eat. His appetite was almost gone he could sleep but little he would say

[*p. 288*]

my all is gone calvin i believe will die and come home no more. Now his haughty spirit fell. He would let me talk to him. I told him if he had given his children to god he would feel still that i had done it and i felt still. My portion is not here god is my portion i said. To see him live such a wicked life and sin against god as he had done i thot was worse to me than if both my children was dead. Now i believe my husbands conscience roared upon him like sampsons lion for his bitter carriage to his children. Ah i told him how dear and choise gods mercies was and how little wee have prised them and what examples of holiness parents should be to their children. Now my husband heard that jonathan was come from boston and stayed at the neighbours unbeknown to me. He sent for him and was pleased when he see him come home and sent for a man to write and he leased land to him about 300 pounds worth in land and he has took a bed and his things and gone intending to live in his own house. O god how great are thy wonderfull works. O christians i know god is a god hearing prayer. O pray pray dont be discouraged tho it tarry wait for it. And o god as thou hast begun to shew mercy and has turnd their harts to each other pray go on turn their harts to thee. O is not jonathan a child of promis shew mercy lord o lord forgive.

[*p. 289*]

Now after we had heard calvin had been exceeding bad with the plurisy god was mercifull to me till he come home. I felt still except once thinking of his death and not knowing every day but wee should have the news. It seemd to strike my hart for a moment with terror and went of my cry now for a resined will. But o the justice of god that i see if he should deprive us of that child by reason of sin. But ah wonderful mercy a mercy that wee could scarcely expect for before the child had been sick a fortnight he come home comfortable tho weak. I looked upon him to see how his flesh was gone how poor and hollow eyed he looked. Methot i see the wonderful works of god glory to his name. O god except of our private and publick thanksgiving i cant dissern any affect it has had on the child's mind only he reads and keeps at home. One poor man dyed at new york that went with him of the same sicknes he had. He told me when he heard he was dead it struck his mind. O lord turn us all to thee. O that we may have harts to live thy praise. Glory to god my eyes beholds his eyes.

May 7-1776. I dreamd a wicked man angry at me had a whip in his hand trying to strike me and me thot i had wings and i mounted and flew away from him and received no harm. Now i soon saw it fulfilled [crossed out] parson. O the

[*p. 290*]

sin that i see and heard for several days anuf to make ones heart to ake and flesh tremble. Now i was poor week in body and sunk in mind. God is punishing my sin with sin but my prayr was to god for mercy. This verce are i was aware come to me and kept running threw my mind

When desolations like a flood
ore the proud sinner rools
Saints find a refuge in their god
for hees redeemd their souls.

O methot i had a glimps of the vast difference there would be in a little while between the saint and sinner the one must sink down to hell and the other mount up to their eternal refuge who has god for their refuge here.

May 19. This lords day morning my soul was in anguish till i could retire from the family. I poured out my hart freely to god. I had great freedom to pray for jonathans conversion. My soul held up that word on which i have hoped for his souls salvation that he should be an isaac.[4] Lord i have been looking for it this 24 years. How long how long lord when shall it once be pray give me paytience to wait for thee.

But o my hart akes my body is week lord forgive sinners for they know not what they do. O make them pattarns of thy long suffering and patience to them that shall hereafter believe.

[*p. 291*]

June 9-1776. This sabbath i felt all day a body of sin and death except in one prayr some desires. O when lord wilt thou come to my soul. 2 years and a half i hant been to meeting on the sabbath but one day. Lord i have no nehemiah to go before me. Thou lord of the harvest has bid me pray to thee for labourers and wilt thou deny me. O when shall i see the fair feet of them that bring the news of gladnes. O lord pray give me patience i beg i beg to wait for thee.

Now again one morning i awaked with that verce which i have just written when desolation like a flood over the proud sinner rools &. I know not but i must soon see a sudden death. O lord fit me for what is coming. Last night i felt the power of sin so strong in my hart i was in sore distres. I had some rest in the morning. I retired and poured out my confessions and complaints to god with tears got some relief but not delivered. O when lord shall i be delivered from this body of sin and death. Oh for patience to endure. My heart is heavi my spirit wounded none but god can cure me. My body is weak and poor i can write no more.

[*p. 292*]

One day as i was mourning i thot thus i fear i shall never have a meting to go to on the sabbath i fear i must stay at home always. This word passed threw my mind with a melting power—why are ye troubled why does thoughts arise in your harts.[5] O such loving words made the tears to flow. The lord reigns i hope in his mercy.

June 31-1776. This lords day morning my hart was affected while reading chronicles 1 book 21 chapter when david had sinned. He offered up an exceptable sacrifice to god and the plague was stayed. I have sinned o that i might offer up myself in faith both soul and body to god in and threw jesus christ. O that the plague of sin may be stayed and destroyed.

July 7-1776. This lords day morning jonathan sat away to go down to new york in the troop to join our new england army they say 30:000 men to withstand the old england fleet that is there. O thou that garded cyrus to the fight tho he knew thee not[6] o cover the heads of our army if they must come to a day of battle. Lord have mercy upon my poor child. This morning i prayed for him as a child of promis. Lord if it is for thy honour pray let him live and come home. O turn him to thyself pray make him afraid to sin.

July 14. A few days ago i talked to a young woman about her soul but soon upon it she did something that provoked me and i spoke to her with a rong spirit tho no bad language. But o what a guilty distressed soul i felt several days. Ah o how soon may wee distroy with our practice the truth we hold up in words. O lord have mercy on me o give me a meekness & patience for time to come.

[*p. 293*]

July 21-1776. While jonathan was gone to york there was a battle with the enemy some killed but he spared. Lord i thank thee. While he was gone i begged god for him to keep him from sin and all evil and if it was for his honour and glory that he might come home again tho i knew not every day but that i must hear of his death but i felt in some measure still begging for submission. Now in thirteen days he come home sick with the fever and purging. Bless the lord o my soul that he is yet alive when others are dead. O turn him to thee. Now the next day after he come home he was from his own house with us and a traveller broke into his house and stole about 7 or 8 pounds of his best cloaths and went off. Now jonathan soon missed his things and tho weak and poor rid to new haven and sent out men after the thief which found him and all his cloaths and he was put in prison then whipt and then sent off in a priveateer vessel. Now when wee first understood his things was stole it seemd next to impossible for him to ever get them again for wee knew not which way the thief was gone and our land now fild with soldiers and travellers. Well i knew the lord knew the poor child wanted them cloaths. I retired and prayed that if it was gods will he might get his cloaths again. Methot i see it all in the hands of god and i kept praying that day and for the poor thief that he might be converted and when i heard next day the cloaths was found i was astonished and wondred at gods providence. This i have writ to show that

[*p. 294*]

god is a prayr hearing god. O christians let us go to god with our small things as well as with our greater matters our jesus got all by prayr. This day my hart was affected when reading mr shepards sound believer.[7] O lord sanctify me by thy truth jesus has prayed for it.

July 4-1776 [*sic*]. O the sin i saw and heard last week the very language of hell. How often was gods holy name took in vain ah how my hart aked. On the lords day morning part of the 35 psalm in wattses ver kept running threw my mind. Lord help me to make this psalm my pattarn. Hast thou not by it incouraged me to pray for this poor man. I hope in thy mercy lord i beg bless the rod to me.

August 1776. Benjamin perdee of n h was driving his team and fell that the cart wheel run over his head and masht it and he dyed instantly. We hear also at new york there has been a terrible battle that a thousand of our men are killed and taken. My mind is wayed it makes me pray. O lord spare new england wee as a people have dispised an offered christ to us belongs shame and confusion as it is this day but to thee belongs mercy and forgivenes. O get glory to thy name by mercy and forgivenes. O make thy will my will. I awaked one morning mourned and wept to think of gods anger against us that such a vast number of our dear young men are dead and o the rest are not turned to god. That word with some power ran in my mind—scatter thou the people that delight in war.[8] I thot how did i know but david then prayed by the holy ghost against this old england army

[*p. 295*]

that seems to delight to shed innocent blood and i prayed so two. Lord scatter them cause them to return ashamed the same way that they come. Ah lord we have sinned but i hope in thy mercy.

September 18-1776. A day of publick fasting and prayr threw this colony on account of the war. In the morning i retired into the woods had freedom in confessing sin. O that this may be such a fast as god has chosen to undo heavi burdens and let the oppressed go free and that every yoke be broken that binds the conciences of gods people. O lord pray hear o lord open the eyes of our rulers to consider what they have done o grant a true reformation.

October 6-1776. I may say as jeremiah did abroad the sword bereaveth at home there is as death. The camp distemper is spreading threw the country it seems. Some our neighbours are dead and some lie sick and o the stupidity of saints and sinners. I cant mourn as i ought for mine. O lord send down a rain of righteousness upon our nation and land or wee are ondone. I am poor and weak in body and soul lord forsake me not. Hast thou not promised to thine the conditions of the covenant as well as to convey the good of the covenant jeremiah 31-31 32 33 34 hebrews 12-12. But o my soul is afraid that some that i live with must perish eternally while they are a scurge & rod in gods hand to punish me for my sin. O god bless the rod. O help me to except of the punishment of my sin o that by it i may be crucifyed to the world and the world crucifyed to me i beg i beg. A few days ago cousen james heaton and his wife and son was carried to the grave. I awaked one morning wept and mourned in some measure affected with gods distinguishing goodness that we are alive while our neighbours are dead. O god open our eyes to see and our ears to hear thy voice that our souls may be turned to thee and live

[*p. 296*]

October 20-1776. Afflicted several days last week with troublesome visatents. O the wicked that wont bare reproof now o how will they bare to be streached upon the rack in hell & never let down to eternity. Last night had a heavi croos my soul sunk down under it o how will heaven dareing sinners sink by and by. I sat up late after my family was in bed to mourn for sin. I mourned and prayed till i was week and weary went to bed god was merciful i slept comfortably. In the morning it being the lords day the last nights trial seemd to drive me to god. O lord did not twoo sons once rise upon the lords day morning the son of god and the son of nature. O make me feel the power of thy resurrection on this day. O give me grace give a strong hart to endure with patience whatsoever thou shall see fit to lay upon me. I am in the field.

November 26. Again as it is commonly practiced by a wicked person [crossed out] o the [crossed out] language that i heard went threw my soul [crossed out] anuf to make the wickedest man in the world tremble. Now i felt no anger at him my soul often pyties him and i can threw gods mercy constantly do as well for him as if he was never so good to me. My children talks to me for it and say it makes him worse and they think i am rong in it

[*p. 297*]

but poor harts they dont know that wee must take jesus christ for our pattarn o lord that i always may. Ah when he was reviled reviled not again when he was threatned committed his cause to him that judgeth righteously. Now the next day after this tryal i was poor and weak and went to bed. O what will them sinners do in a little while that have sold themselves to do wickedly. Lord have mercy on them o christians take warning by me. O marry in and for the lord o that my sin may be a warning to others lord have mercy.

December 8. This morning i see a man take away the property of another unbeknown to him which i knew would make a great quarrel. I talked to the man with an aking hart but to no purpose. Then i prayed to god that he would touch his hart and make him relent and restore that he had taken away. So i begged and hoped in gods mercy and behold the next morning this man sent back to its place that which he had taken away before the owner knew it was gone. That word come fresh into my mind—blessed are the peacemakers they shall be called the children of god.[9] I spoke out and said with tears o how much better is it to be called a child of god than to be the child of a king. Glory to god for his great goodness and mercy not only in preventing sin but he has shewed himself a prayr hearing god lord i thank thee. Why should the wonders god has wrought be lost in silence and forgot—Hannah Heaton

[*p. 298*]

December 29-1776. Felt low and dark lost almost all love to god and to his word. Ah i live and have lived this twenty years at a low rate but god is gracious to me for i feel my corruptions more and more subdued. O my god dont let them rise up again. O how fraid i feel to pray deliver me from self confidence. I say went to bed low and dark awaked with these words

Corruption earth & worms shall but refine this flesh
till my triumphant spirit comes to put it on afresh
Arayed in glorious grace i shall this vile body shine
and every limb & every face look heavenly & divine.

I sung it and felt my spirit raisd. Lord i thank thee i hope the time is coming when all my sins will be done away & when jesus does appear i shall appear with him in glory.

January 12-1777. This lords day felt a hard hart. I could not mourn for sin nor thank god for dear mercies. O god help me to wait upon thee still has not thou promised to take away the hart of stone and give an hart of flesh. O how has sin raged in me this winter. Lord have mercy on me in my prison. O that he that come to preach deliverance to captives and to open prison doors to those that are bound would have mercy on me.

Jenuary 14-1777. Calvin sat away to new york to join our army. Now i felt a still mind. O god help me to give him up to thee and leave him with thee wheather in life or death. God be gracious to thee my son. Fareweel.

[*p. 299*]

Genuary 19. Now i had not been to meeting on the sabbath but one day in above four years but hearing that mr miner was to preach at wallingsford on the sabbath god being gracious my husband helpt me to go tho as pharaoh who thrust out the israelites in haste. Now mr miners text was in luke 13-7 about the barran fig tree. He shewed how god was cutting us down by the sword pestilence and our men in captivity by famine starving to death in prisons and god still threatning to go on with his judgments. But he told gods children they had no more need to be concerned now tho it lookt so dark than the disciples had when jesus was in the grave. He shewed that soon glory followed. He said he expected greater glory than in the apostles days. Afterwards i felt my soul quickned and encouraged to seek god. Lord i thank thee.

Now this same week i heard mr miner preach twoo sermons more at mr nathaniel beeches one from romans the other from matthew 24-25. Behold i have told you before. He shewed how god always shewed his children beforehand when he was about to bring any great thing to pass. How gods word was fulfilling in his providences ante christ was agoing to be destroyed. He opened several places in revelations that was fulfilled and fulfilling. He shewed the signs of christ coming in to his kingdom and then lift up your heads for your redemption draweth nigh. O god i thank thee that thou hast not shut the mouths of my ministers at this day they are still calling to saints and sinners. Now freely i gave him a little money.

[*p. 300*]

Fabruary 9-1777. Now when calvin was gone down to the army i felt in some measure patient. I kept on praying for him that god would keep him from sin and all evil and if it was for his honour and glory he would spare his life and cause him to return again to us. And in about 3 weeks he come home again alive and well while others that was in the same fights that he was in some was killed and some wounded. O lord i thank thee but alas i hant made thee such returns as the mercy calls for lord help me.

A few days ago i kept a fast alone for myself and family and for new england. I felt dry and barran but i kept on pleading lord thou has said thou wilt not dispise the prayr of the destitute o pyty pyty the souls thou hast made.

Now i dreamed i was talking to jonathan and calvin. I thot i told them to pray to god that they might live threw this deluge of war for when it was over i believed there would be greater glory seen in new england than ever was before. And while i was speaking i thot my soul was raised with a vew of the glory and so i awaked. O lord i hope to see thy word fulfilled on which i have hoped. Pray come quickly and send down a rain of righteousnes.

[*p. 301*]

March 2-1777. This lords day i felt in some measure comfortable threw the day. In the morning i cryed my soul cryed jesus thou son of david have mercy on me. O was i born first from beneath and then born from above. Am i a child of man and god. Oh rich and endless love.

March 23-1777. On the sabbath i went to wallingsford and heard mr cammel preach from the first epistle of peter first chap 3 ver. He shewed what it was to be begotten unto a lively hope and who had it and who had it not. His afternoon text was first of romans 16. He shewed what the gospel was and what and who them was that was not ashamed of it. Ah said he pauls whipt back did not make him ashamed of it. Now i heard him again at mr elias beechs from first of john 5 chap 12 ver. He shewed who them was that had the son and had life and what that life was. Ah sweet gospel preaching i believe god helpt him to preach. Now i was enabled freely to give him. Glory to god. I felt my soul quickned afterwards but oh the [crossed out] i heard soon after for several days together struck my soul down again. O wo is me that i sojourn in meseck and dwell in the tents of keder. O my god my god have mercy and o god dont let me never forget to thank thee for a reformation in my jonathan. Now he lives at his own house pray carry it on to a sound conversion i beg i beg i beg for jesus sake.

[*p. 302*]

Aprel 23-1777. My brother jonathan cook from long iseland come to see me and brought me 65 pounds in york money a legacy of my deceased brothers estate. O dear i found at first it took my proud worldly mind but o how did i beg for mercy lord help me. Ah i thot the rod must come now if i was not turnd to god. Now i intend to take out some of it first (before it is laid out) and keep it by me in store to give to uphold the religion of christ. But now this word run in my mind—wine and new wine take away the hart.[10] I thot that wine held forth the riches & prophets of this world and new wine the pleasures of the world and these when confided in takes away the hart from god. But o god of truth and faithfulness i trust has helpt me. O wash me from all sin and make my gilty concience clean.

Aprel 23-1777. My jonathan was married [crossed out]. O lord make them like isaac and rebekah like zacharias & elisabeth.[11] O make them helpmeets to each other in the way to heaven. O make their house a house of prayr for jesus sake i beg i beg lord i hope in thy mercy.
Now a few days after calvin went down to fairfield with our army to face the enemy some thousands we hear are landed. They have burnt houses stores a great deal

[*p. 303*]

of provisions and carried of much. Our men had several skurmishes with them they suppose about 60 of our men was killed and double of the enemy. This was done before calvin was got to them. He see a number lie on the ground dead roold in blood. Glory to god that spared him and brought him home again. O lord turn us to thyself o pyty poor new england say it is anuf. A few days ago we had news that 400 of our enemies was killed in the last fight.

May 11-1777. I heard mr miner preach 2 sermons from romans 8-24 25 26 27 verses. He insisted much on gods children praying and pleading with god for the accomplishment of the promises of christs comeing into his kingdom. He said the war was just at a close the glorious day is near. He seemed to labour to comfort gods children. O it was lovely gospel preaching. I was poor and ill in body that day but i felt my soul quickned afterwards for several days. I was helpt with a willing mind to give him more than ever i did at once before. Glory to god for his goodness. My constant practice always is to pray for my ministers before i hear them & i always get something when i hear them. If i dont feel comforted in the lord jesus i then see what a unbelieveing wicked heart i have and that is a great mercy to see ourselves. Lord jesus come quickly have mercy on me.

[*p. 304*]

June 7-1777. My husband being gone to meeting i prayed with calvin and the child that lives with me but o when i come to plead for new england my heart went out to god with freedom for mercy. O that we may be turned to god that we may see thy salvation. Lord let my prays come up before thee this sabbath. This week calvin come home from cheashire sick but the lord blessed means used that he soon grew better.

O lord wake him up to consider that thou art often calling upon him by illness of body to turn to thee. I thank thee for thy goodness and mercy. O make us live thy praise.

June 14-1777. Last week there come a young woman mourning to my house and told me that her brother that went down into the army was dead. Now it seemd to strike my mind for i knew he had been a wicked young man. Now it seemd as if my mouth was opened & my tongue loosed. I told her with a melting heart that god called her to get an intrest in christ. Life was unsertain o dreadful to die out of christ &. She seemed willing to hear. O lord set home truth o that her brothers death may bee a means of the life of her soul. Ah lord how art thou cutting down our young men by the sword and by sicknes while i and mine are spared. O turn us turn us to thyself and we shall be turned.

[*p. 305*]

June 29-1777. Yesterday morning while praying this verce seemd to affect my heart viz—let those that sow in sadnes wait till the fair harvest comes.[12] They shall confess their sheaves are great & shout the blessing home. And this sabbath morning when i desired and prayed to god methot with my whole heart that i might serve the lord with my whole heart and be devoted to him here not for the reward but methot i see a beuty in holiness. It is gods likenes i ow him my all. O when shall i give him my all and o lord thou hast told us to pray to thee for labourers & will thou not hear my cries to thee for one here to labour in word and doctrine. O wilt thou not do that which thou hast bid me to pray for & thou hast said thou wilt hear the prayr of the destitute and wilt not dispise their prayr. Ah i am destitute on many accounts but i hope in thy mercy. O that i may come again with rejoysing bringing my sheaves with me.

July 6-1777. I hear mr marshal preach twoo sermons this sabbath from proverbs 6-9 10 11 verses. He shewed how sinners was asleep in their sins and loved to sleep altho gods judgments are so heavi upon us old england against us and they have hired a forreign nation to come against us blood running cannons roaring and yet assleep. But ah he told them what their poverty & want would be in a little while. He seemed to long for the conversion of souls. He said he believed angels was in the assembly ready to carry the news. He believed some souls was intangled in the gospel net. Come saints help me to draw the net ashore. There seemd to be some shaking among the dry bones. The same week i heard him again from matthew 25-43—i was a stranger & ye took me not in.

He shewed how jesus was the stranger that sinners would not take into their harts. Bless the lord o my soul for these oppertunities.

[p. 306]

O lord have mercy upon that poor wicked man that come up to mr beeches window and spoke out reproachfully to this dear minister of christ while he was apreaching and others laughed. I hope the lord is near for the devil is busy. Now i was enabled to give to this dear minister freely. I felt stired up to pray for sinners. O lord cloath thy ministers with salvation and let thy saints shout aloud with joy.

But alas a few evenings after i being provoked felt angry and spoke with a rong spirit. But o how did my heart ake quick upon it. Now i thot i must sleep none tonight but i went to bed god being gracious i slept comfortably. A verce run in my mind which i have forgot. Arose in the morning and retired into the barn. There i powred out the tears of repentance before the lord. I prayed hard for patience meeknes and humility with strong desires and a good god took off my burden and i hope has forgiven me. O god i thank thee with jabaz.[13] I beg o let thine hand be with me. Keep me from evil that it may not grieve me. O that like him i may have my request for jesus sake.

August 17-1777. Awaked this lords day morning very sad and discouraged. Ah thot i i fear i must live and die just as i am at a great distance from god. But in a few moments this melting word come to me—i will restore to you the years that the caterpiller the canker worm and the palmer worm hath eaten.[14] It seemd to affect my hart glory to god i hope yet in his mercy. O methinks that will be a lovely word. Behold the brides groom cometh go ye forth to meet him. Was sick now poor and week fild with pain for several days but god had mercy on me and raisd me again. O make me live thy praise.

[p. 307]

Awaked with that word where samson when ready to die pleaded for water.[15] The lord gave it to him and his spirit come again & he revived. O lord pray give me of the water of life then i shall be revived and my soul shall live o come quickly.

Last sabbath heard mr cammel preach from psalms 50-15 and from luke 14 concerning the great supper. Methinks this was lovely gospel preaching. Afterwards i felt my mind stired up to pray. Jesus hast thou not prayed for my sanctification and shall it not be done for me.

September 3-1777. Was a day of publick fasting and prayr on account of the war ordered by the congress. I retired away alone had some nearness to the throne. O god hear the cries of our poor distressed amaricans this day. O that it may be such a fast as god has chosen to ondo heavi burdens and let the oppressed go free. O turn us to thyself i beg i beg.

October 18-1777. Last saturday about sundown a man did an injury to another and went off. Now the injured man was in an extreem rage and spoke many dreadful wicked words and sat away to go and fight with him. Now it struck my mind exceedingly. I pleaded with him for the lords sake not to go. I told him he would be sorry afterwards if he did not mind me. I laboured but it seemd to no effect. Away he went but a good god stopt him before he got far & caused him to come back again. Glory belongs to thy name o lord if i must never do nothing else for thee. O make me an instrument to prevent sin. O pyty sinners that dont pyty their own souls. O send a rain of righteousness i beg i beg for jesus sake.

[*p. 308*]

November 23-1777. Mr frothingham preached at wallingsford in the fournoon. He expounded from ephesians 9 and in the afternoon he preached from the twoo last verses. It being sacrament day he shewed what a church of christ was and how christians must be qualifyed to be meet members and he invited sinners many times to come to christ and warned them and several of the brethren arose up and exhorted. I believe the lord was there and that he was worshipt by some in spirit and in truth. I went out low but with something of a spirit of prayr all the way and i returned low but had something of a prayrful mind. Now i was enabled to give something freely to this minister of christ. I have of late had my mind wayed about coming to the lords supper. O lord remove everything out of my way. O let me feel a necessity upon me to honour and wait upon thee. O sanctify me by thy truth and o thou lord of the harvest thou hast bid me pray for labourers. O send forth a labouror here to us one to labour in word and doctrine. O fit us for such a blessing i beg i beg.

Now a good god by his providence has cast twoo good books into my hands. One is mr stoddards safety of appearing in the righteousness of christ.[16] The other is mr hildersams that has an hundred and eight lectures in it preacht above a hundred years ago.[17] O lord i thank thee for such helps in my way to heaven.

November 20-1777. Was our publick thanksgiving. The next day i retired away alone spent a few hours in prayr and reading gods word felt low and barran but had some sighs and groans to god. O that when grace dont pray that nature may to my latest breath. God hears the ravens when they cry.

Now i gave something to a poor woman that was sick. O lord have mercy upon her soul o convert her i beg.

November 23-1777. This lords day heard mr beech preach from solomons song 1-6—they made me the keeper of the vinyards but mine own vineyard have i not kept. He shewed how that at this day when wee are under such a sore calamity of war how apt wee all are to say it is for sin but we are laying of it upon others upon this sex and the other sex

[*p. 309*]

and so secretly excusing ourselves when we all ought to be serching for our own sins that has provoked god and repent before him with true reformation and he doubted not but then god would take away his judgments from us. O that wee may not no longer keep the vinyards of others to the neglect of our own vineyard. Mr elias beech i believe prayed and exhorted in the spirit of christ. Bless god o my soul that my eyes sees my teachers.

December 18-1777. Was a publick thanksgiving threw our land on account of bourgoine[18] and his army being taken prisoners by our men (alas i lament it my hart was low all day i live like bunyan with a burden on my back or rather on my soul.) Ah i see the wicked growing worse and worse [crossed out]. He is angry at me if i read or spent time to pray and i may not reprove sin in the family. He scarcely ever goes to meeting himself and i must go but seldom and when i do he pharioh like i am thrust out in haste. He is offended at me for writing what god has done for my soul. Ah no family prayr poor children taught to sin [crossed out] but tho it is all with contention and i cant take the comfort of gods mercies. Yet he is a good provider. God makes a raven to feed me as great a maricle methinks as was done for elijah. O christians take warning by me be sure to marry in and for the lord as you would live to the honour & glory of god and the peace of your own souls. But o what cause have i to admire and wonder at the mercy faithfullnes and power of a god that has maintaind and upheld his own grace and fear in my heart so long and in the midst of so many sins and oppositions both from within and without. Ah methinks it is as great a maricle for a christian to live in the world and be kept unspoiled by the world as it was for the three noble jews to be in the

[*p. 310*]

fiery furnace and not be consumed by it glory to god.[19] O our god is a god of truth and faithfulness. One night last week i had raised hopes for my husbands conversion by my having some vew of gods omnipotency what he has done for his people of old. He is the same god still. Is anything twoo hard for the lord jehovah. O lord help me to give him up to thee in faith although he is dead in trespasses and sins.

Genewary 11-1778. This night i waked with this word—you only have i known of all the families of the earth therefore will i punish you for your iniquities.[20] O methot i see the perticuler sins that god was correcting me for with his heavi rod. Ah he is punishing my sin with sin. But o i may sing of mercy and judgment lord i thank thee if i may have my punishment here and not have it reserved for me to bear in hell. O help me to except of the punishment of my sins to be still and know that thou art god.

Genuary 18-1778. Now when i had waded threw several months of sorrow more than common heart ready to brake [crossed out] which i heard from day to day reproof scorned and sometimes i could not pray my soul seemd prest down under my burden. But i followed god as well as i could in some measure and glory to his name he hath heard the voice of my groaning and has now for several days reformed the tongue & softned the spirit of the wicked. O my god let me not sin away this great mercy. O keep me keep me help me to watch keep thou the door of my lips and o lord grant true repentance in his heart accompanied with a real reformation for thine is the kingdom the power and the glory forever amen.

Genuary 22-1778. Mr miner preacht from john 21-22— follow thou me. He shewed that when persons especially young ones when they

[*p. 311*]

begin to be concerned about their souls how loath they are to take up the croos and forsake their merry companions to have their names cast out as evil. But how did he plead with souls to walk up to the light they had received and to follow on to know the lord. He told sinners the dreadful state they was in while out of christ and he charged christians to improve their gifts. Elias beech got up and warned sinners to flee to christ but before he had done one wicked opposer got up and in a high spirit said you have no right to judge. O i hope this is a good sign for i believe satan is afraid of loosing his captives. Lord bless thy word o that thou would pour down of thy spirit here. O turn many to thyself i beg i pray. Now i was helped freely to give something to this minister of christ. I felt afterwards a sober mind and more of the fear of god before my eyes.

February 11-1778. This day my jonathan had a son born glory to god that has suffered such a sinner as i have been to live to see my seeds seed. O that i may see peace upon israel salvation come to zion. But alas alas with an aking heart i must write for jonathans wife had been brought abed but four days when a purging took her and a feever set in which caused a sore sickness which put an end to her life. She dyed march 11 1778 much lamented by all that knew her. I think she was the pleasentest tempered person that ever i was acquainted with. She was a neat prudent wife and a loving daughter. She said once in her sickness as she had been lying with her eyes shut that she had been to heaven and had seen god and christ and that god bid her begone for she was a sinner.

[*p. 312*]

but she said christ bid her come and take of the water of life freely and she said several times that she was willing to die and she has appeared to me to be concerned about her soul when she was well. So i must leave her the judge of all the earth will do right. Now a few words more she spoke in her sickness comes to my mind. She warned her nurse not to put off to seek first the kingdom of heaven and the righteousness thereof and all other things shall be added.[21] She told mr trumble that she had put off for a more convenient season. She desired him and others to pray that she might have an intrest in christ. Once i asked her if she felt any heart to pray. She said yes and she said she knew if she dyed she should go to heaven. I asked her if she loved jesus christ. She said yes i have a good many times and seemd to think if she should get well again she would try to live better than she had done. She seemd when well to subject to gods providences and when sick a pattarn of patience. The lord is holy and just in all his ways but it is wee that have sinned. O lord do bless the rod. O lord do bless the poor motherless babe with the choicest of heavens blessings. O pyty o have mercy on the poor mourning bereaved child of mine that mourns to excess. O that his mourning may be turnd into the right channel. O make him mourn for sin truly dont let him recruit again with creatures. O god be his portion in time & to eternity. Ah i feel my mind overcome and drownded in sorrows and distresses. My cries are to god for help and i hope in his mercy and if it please him to deliver me i intend to write this trial of god will.

[p. 313]

Now i heard mr wright[22] preach from these words—if you believe not that i am he you shall die in your sins.[23] O how did he warn sinners to flee to christ and several of the brethren exhorted. This was a good meeting the truth held up and i believe gods presence was there but my sunk soul could not arise and praise god. O lord have mercy on me a great sinner. O remember that thou has said i will heal your backslidings and i will love you freely. O come quickly lord.

Heard mr beech preach from revelations 7-14. He preacht weel several of the brethren exhorted. I believe it was a good meeting to many but i felt a sorrowfull soul that day. O lord let me feel the power of thy resurrection.

May 24-1778. Methinks i must wright a few lines about my distrest mind since my daughter in law dyed. Now a few nights after she was buried i was in bed & it was dasht like a sword threw my mind that i was deceived about her illnes and that she had not right things applied for her help nor the doctor was not sent for soon anuf. This i was followed with day and night. My mind was drownded in it my appetite almost gone. I could sleep but little & when i first awaked it was cast into my mind constantly. It was the worst anights. I was afraid to go to bed & use to sit up while near or quite midnight that i might sleep the better and as soon as i awaked i must get up quick. My body was weak a great distress seized my breast and sometimes i could not pray but i kept trying. I sometimes pourd out my distrest soul to god and felt some release. A little while i thot it was from the devil but i did not know it was. O how did i beg when i could that satan might be bound thrown into the bottomles pit shut up & a seal set upon him. And now

[p. 314]

this added to my sorrow. And strange it was for about the same time and in the night when jonathan was in bed the very same things come upon him & he soon come mourning and told me of it. O it struck me with new anguish but neither he nor i could exercise our rational understanding. Now my god was withdrawn from me no help nor comfort could i get for about twoo months. O i was perfectly miserable on earth. But o with a heavi hart i went to meeting at wallingsford but o the darts did fly threw me while there. But my hearing the gospel preached seemd to relieve my mind and these distresses seem to be decaying ever since. Glory to god for his mercy to me a sinner. O god help me to give away my all to thee. Now in may i hear mr cammel preach five sermons. Ah it was good to me. I believe god was with him in every sermon. I felt my soul comforted while he preacht from these words—blessed are they that mourn for they shall be comforted.[24] I with a free heart gave this minister something. O lord cause him to see thy salvation. Oh let saints shout aloud for joy.

June 1-1778. Now soon after jonathans wife died he seemd concernd about his own soul. Behold he prayeth and he reads much. There appears a great alteration in him and people sees it. Truth seems to be shining into his mind. He has been to the saparate meeting twice but o how does the devil now stir up his instruments against him. O the false stories that is spread abroad about him. Report it say they and wee will report it. Now my cries are to god for him. Thou who art the almighty jehovah can carry him threw. O i hope the time of the promis draws hear. Lord prepare me to give glory to my name.

Now i have been to wallingsford to meeting often of late and i believe god is with that people doing a work there. I believe god

[*p. 315*]

has a vine there of his own planting. Now it is upon my mind to offer myself to joyn with that church altho i feel myself as it ware in chains. Yet christ command is the same still—do this in remembrance of me.[25] O what a tried mind i feel i fear i fear i am afraid of many things. A few mornings ago this word come to me with some power—take my yoke upon you larn of me for i am meek and lowly &.[26] It seemd to me i was called to put my neck under christs yoke and one day i was thinking that it was fear made me go forward to join more than it was love to god. And this word seemd to speak to my mind—noah being warned of god threw fear prepared an ark.[27] O lord help me to go forward in thy true fear. O give me thy true love. O let me come in by the door. Jesus dont let me climb up some other way.

This week went to wallingsford to a conference meeting. My nature opposed hard but i was enabled to come forward and tell what god had done for my soul. After i was asked some questions. Twoo of the brethren spoke and incouraged me to come forward. What the rest thot i know not but i had peace in it.

Now the time drew near when i was to have my tryal. It seemd a resemblance of coming to gods bar. I felt terrifyed with fear. I believe satan had a great hand in it and my sinful nature opposed my duty but at times i was in some measure willing to be bound and thot it a priveledg to be bound to serve the lord. Now i prayed almost continually to god for help and strength and that i might come in by the door into the church or be shut out. O that i may not climb up some other way. One day i retired away into a solitary place a few hours

[*p. 316*]

as i could to pray and when i come back my husband seemd displeased & said i had been gone five hours. Poor man o that he might pray now before it is eternally twoo late. Now i held this up many times before the lord that i thot he had called me to come forward to put my neck under his yoke & would he not now own his ordinance. But now my mind was tried for fear i should soon brake covenant if i joind to the church. But this word helped me—he sends none a warfare upon his own charges.[28] Now i see it to be duty to ask my husband if he was willing tho i was afraid to. I see it in numbers 30. I asked him he then seemd pleasent but his answer was—i shall not tell you. I begged again that he would tell me but his answer was the same again and in as much as he did not forbid me i went forward.

July 31-1778. Was a fast before the sacrament at wallingsford. I told my experiences again & was asked some questions and the brethren received me. Now when i was coming home i felt sweet peace in my soul and a prayrfull mind. O thou that said father keep them from the evil of the world o keep me keep me help me to perform the vows i have made or i shall soon drop and o when shall zion arise and shine. Ah i have heard lovely preaching of late but o my barren soul. Ah alas at my first adventure to the lords table i felt low in mind. Lord lay me at thy feet and if it shall please thee to let me live to come to another. O that i may feed upon thee by faith to my spiritual nourishment and growth in grace. O that i may by faith eat the flesh and drink the blood of the son of god. Lord have mercy.

[*p. 317*]

September 1778. Felt a tryed distrest mind one morning. This word come to me as soon as i awaked—here wee have no continuing city but wee seek one to come[29] even a heavenly whose builder and maker is god. Ah methot i see that i had but a little while to be here. No matter then if i have hard fare it will be soon over. O what a sight will it be to be with jesus in heaven to behold him where he is not as a man of sorrows but in his exalted glory.

September 7-1778. Poor john barns while attempting to waid threw or across a creek was drowned. O lord grant this sudden death may be a warning to all. O how often does god send death like a thief in the night while they are (i mean the wicked) asaying peace and safety then sudden destruction cometh upon them. Take warning o ye christles sinners.

One thing i must wright to shew the mercy and kindnes of the lord to me. O the lord is a prayr hearing god i know it. Now after our poor daughter dyed the dear babe was left destitute. We tried but could get no woman to come and suckle it and keep jonathans house. But o now i went to my god with tears and groans. I told him he heard the ravens when they cry he provides for the sparrows and swallows. I begged him to find out a way for this poor babe to be brought up now and how soon after there was a woman in the neibourhood a christian i believe that offered to take it. This was unexpected for this woman lost her own child when it was first born and had brought up a motherless child to be a year old and had weaned it and then she told me her milk come afresh into her breasts again. Our child is now about eight months old. It grows and is forward. O lord let it not steal our hearts and provoke thee to take it away. Lord have mercy upon its soul. O make it an heir of thy heavenly kingdom. O that i may live to thee devoted to thee.

[*p. 318*]

Now in october 1778 i workt twoo hard overdid myself and then got a sudden cold which settled in my back. I was in distressing pain both night and day. It is now above twoo months but god is gracious to me for my pain is not so hard as it was the first month. O god give me submission if i am never well no more. I have indured much hardness [crossed out] from one that should be my best friend. They have rewarded me evil for good when they ware sick. But god has been good to me he has enabled me to justify him in his dealings towards me and to in some measure except of the punishment of my sin. Ah what a mercy to be corrected for sin here and not have it reserved for me to feel in hell. One morning that word seemd good to me are i was aware it come to my mind—lift up now thine eyes from the place where thou art and look towards the north and towards the south towards the east and towards the west.[30] Ah methot altho i was layed low yet i might look from the place where i was and vew my large inheritance in christ jesus. O my god let me never fear to dive threw jordan that i may go into possess the goodly land. And now methot i see why it was that the lord brought this illnes upon me because i was twoo coveteous and ancious to get my work done. I was not carefull anuf of my body which is the lords. I have had some mournings for sin. Lord give me a greater measure accompened with real reformation and o sin abounds in my family o i beg make thy grace much more abound as thou has been wont to do.

[*p. 319*]

Jenuary 6-1779. This evening went to my sons nearby to hear a sarmon preacht by mr john cornell[31] of new haven who i trust the lord has lately called forth to preach the gospel. His text was in matthew 5-25—agree with thine adversary quickly whiles thou art in the way with him. He told sinners in many perticulers how they might know when the lord was in the way with them and the vast importance of their case and the reason why this agreement must be made quickly and what the agreement was. He seemd to have light and assistance by gods spirit. There was a large number of people who seemd sober and attentive. This sermon seemd good and helpful to my soul. I gave him a little money freely. Bless thee lord o my soul that has caused jonathan to intertain the gospel in his house. Pray open his heart that he may receive jesus christ into his soul. Lord bless thy word for jesus sake.

Fabruary 25-1779. This evening mr beech preacht at my son jonathans from daniel 7-10—the judgment was set and the books ware opend. He shewed what a dreadful day was coming upon sinners how all their sins will be laid open and they judged according to law and their refusing and trampling upon the gospel that speaks nothing but love. O how cutting that will be. There was a considerable number of people they all seemd sober and attentive. O lord bless thy word i beg. Now some of the brethren with jonathans consent have set up meetings at his house sabbath evenings to pray and talk of the things of god. O lord harken and hear it. O write a book of remembrance for them that fear thee and think upon thy name.

March 28-1779. Now so well as to ride to meting this sabbath. Heard good preaching had thanks returned to god for his goodnes to me that when weak low and much pain for about six months has so far restord me that i am allowed to meet again with his children. O let me live to thee that has had mercy on me.

Now soon after there was a great croos laid upon me and o my hart sunk under it o lord lead me not into temptation but deliver me from evil for thine is the kingdom the power and glory forever.

[*p. 320*]

1779 july. I awaked with these words one morning—it is for my praise that i cut thee not off.[32] That day it kept running in my mind. Now soon after there come news that the brittish troops was landed at new haven on purpose to burn the town.[33] But a good god did not suffer it. Only they burnt a few stores abused women plundered houses carried off what they pleasd. At the same time some hundreds landed at east haven burnt 9 houses and many buildings disstroyed and carried off much. Some carried captive about 36 of our men killed and many more of the enemy. It is supposed they stayed twoo days but o what a sorrowfull soul i felt to hear the cannon and guns roaring. I had some sence of the justice of god but i kept pleading for mercy not knowing how quick they would be in my house. But glory to god that said hitherto shalt thou come and no further. O i hope i hope that he that once asked a draft of water of the woman of samaria is about coming to set up his glorious kingdom in amarica. Come lord jesus come quickly. Glory to god for sparing mercy. O turn us turn me to thyself for jesus sake. Now a few days after the enemy was gone i kept a day of fasting and prayr to him that has the key of death and hell that he would hasten the time that satan may be bound thrown into the bottomless pit shut up and a seal set upon him. Mark satan is to be bound with that great chain that was in the angels hand. The love and promises of god as a chain are all linkt in christ who is to destroy sin by the spirit of his mouth and the brightness of his coming come lord jesus come quickly.

The first day of last february in 1779. I toock home my poor little motherles grandchild to keep before it was a year old and since i have but very little time to write. O my god pyty its soul.

[*p. 321*]

October 11-1779. Job beech dyed with a cancer in his face he was about 17 years old. Mr cornweel preacht his funeral sermon from the 2 of samuel 12 ch 22-23 verses. He had wonderfull assistance in teaching the parents subjectedness to the will of god. O how he warned the aged middle aged and youths to flee to jesus christ. Now while god was calling so loud to us there was a large number of people and a considerable melting among them. O lord bless this dispensation to all. O bless thy word for the good of precious souls.

1779 in december. Calvin sat away alone for boston. Now early that morning these words was powerfull upon my mind—the lord of hosts be thy defence till wee do meet again. And my hart went to god with them words that day. Now the next day there was a terrible cold snow storm and wee had news that there was a man murdered on boston road the week before calvin sat away and it was extreem cold and stormy almost all the while he was gone. But o i felt my soul supported the lord of hosts was his defence till we did meet again. Glory to god for undeserved mercy. He was gone about 12 days o lord do convert him o gather him into the number of thy sanctifyed ones for jesus sake.

1780–1789

[*p. 321, cont.*]

Genuary 1780. One night in my sleep methot i had a soul ravishing vew of the truth and faithfullness of god to my soul. O how joyful i felt and so awakt glory to god but a tryal soon followed a cross was laid upon me and my corruptions got the upper hand. The devil i believe was very busy to distress my mind but i kept praying to god for help. On the lords day i spent chief of the day alone at the barn tho very cold had some hope of mercy. Now above a week i waded in sorrow before my will bowed. One morning i arose with this word

[*p. 322*]

in my mind (salvation) and soon after i felt in some measure a subjected mind the distress was gone from my mind. Ah i thot ide rather loose worldly things than to have god dishonoured. Glory to god for free grace to me a loathsom sinner.

March 1780. Having been confined this long tegious hard winter i pyned much after meetings. But this word met me and kept running threw my mind—go to the ant thou sluggard.[1] Consider her ways and be wise which have no guide overseer or ruler provideth her meat in the summer and gathereth her food in the harvest. Ah thot i how unsubjected i am that in the service of god i am so sluggish and am wanting a guide when the little ants hant none and yet lay up meat and food in harvest. Lord make thy word quicken me to lay up in christ jesus.

One day as i was going into the celler this word come sweetly into my mind—he that overcometh will i grant to sit down with me in my kingdom even as i overcome and am set down with my father in his kingdom.[2] O how precious was these words while they rolled over and over in my mind. O i hoped threw jesus to overcome all oppositions all temtations sorrows sins. Ah lovely to sit down with jesus in his kingdom of glory. Glory to god the father son & holy ghost.

Aprel 2-1780. Having some diffeculty with a christian woman one of my neighbors she discovered anger and spoke hard words to me & went home. I felt greived but kept praying to god for a forgiving spirit. And now it was but a few days before this woman come and asked me to go alone with her. She said that she had an arrant she believed from god and acknowledged her sin begged forgivenes and prayr for her. Now i felt my hart move when she spoke and we made all up. O jesus how sweet are thy rules glory to my god for this salvation. O help me to live to thee & upon thee i beg.

[*p. 323*]

Aprel 16-1780. Went to meeting once more heard truth but christians seemd low and when i was coming home i had a fall from my horse but without hurt. That word come into my mind—he keepeth all his bones that none of them is broken.[3] Glory to god for his salvation.

But o this week i have been sorely distrest with sin and devil. I was so worried that i felt almost ready to die with anguish of soul but i kept praying to god to bind satan and on the lords day morning when i awaked this word come to me—(come praise the lord and let us exalt his name)[4] adored be free grace in christ jesus. I felt my soul delivered had a comfortable day. Lord keep me keep me lead me not into temtation but deliver me from evil for thine is the kingdom power and glory forever amen.

On the lords day hindred from going to meeting sorrow overcome my weak mind. I went into the barn stayed there till past noon wept out my sorrows before the lord. Read mr stoddards book of the safety of appearing in the righteousness of christ. Had my mind stild and quieted bless the lord o my soul. Ah christians take warning by me poor me o marry in and for the lord. How can twoo walk together except they are agreed their souls agreed.

May 19-1780. Between 11 & 12 aclock there was a great darkness over the land. It appeard very awfull so dark that some people lit their candels to find things in their houses. It held about one hour then went off leaving a bright yallow coler upon everything. It is thot to be no eclips but a forerunner of some great thing coming on the world—the sun shall be darkned and the moon turned into blood.[5] O god make this a means to awake saints and sinners come lord jesus come quickly lord have mercy.

[*p. 324*]

July 2-1780. I dreamd that a fine dove come into my house and flew round its wings seemd flowered as with yallow gold. I cried out so loud in my sleep for my calvin to cetch it that he heard me up in the chamber and called to me. Methot he hastned to cetch it but it flew threw the roof of the house and was gone so i awaked. This made me renew my requests for poor calvin. O lord send the heavenly dove down into his soul. O how conserned i have been for him of late. O lord have mercy on him.

1780 in september. My jonathan was taken sick with a slow fever. Now i persieved all summer that he had a carnal careless mind and i talked to him several times. I told him his company was a snare to him and would have him avoid it and that i expected gods rod for he needed it or to this affect. Now in his sicknes he seemd to have a stupid scatterd mind. I was tried about his dying if he never manifested anything that he was converted. And now i see that if the lord held me in sight of that promis about his being an isaac as he has held me above twenty years i should have comfort in his death and if not i shall let it go and be miserable unless i am subjected to his justis. O how often do i pray thy will be done on earth as it is in heaven. But o i find now my deceitfull heart cant say so as to jonathans death. I went down again to see him he seemd

[*p. 325*]

to grow weaker. It was the lords day morning. I asked him if he desired prayrs. He said yes. I told him now he wanted a jesus and then he would be fit to live or fit to die. I see the tears burst forth. He seemd as if he could bair no more. I kept pleading for him and that i might be subjected and a good god being gracious in a month or more he rode to meeting and he and i offered up our thank offerings to god. O lord bless this rod sanctify this mercy. Thou has given him a new life. O give us harts to live thy praise i beg for jesus sake.

1780 in october. One sabbath morning i read the second of samuel ninth chap. It affected my hart greatly especially the 7 & 8 verses. Me thot i see david was a type of christ and he had sworn to jonathan to shew kindness to his house forever. Methot i see jesus was made a preist with an oath to shew kindness forever to them that are his. They shall eat bread at his heavenly table forever. I spent chief of the day in the barn. O that i may see gods salvation. These lines seemd good to me—up to the hills where christ is gone to plead for all his saints. And on our publick thanksgiving day my hart was drawn forth in prayr and praise to god for his great mercies to me a grevious sinner and for his mercyes to new england that wee are not yet given into the hands of our cruel enemies. O god do send salvation o pour down thy spirit o bind satan destroy the man of sin by the spirit of thy mouth and the brightness of thy coming. This day i was melted into tears reading samuel mathers book the self justiciary convicted & condemned[6] page 67 come lord jesus.

[*p. 326*]

January 7-1781. Now about twoo months i have been sorely distrest again with a pain in my back but little ease onely when warm in my bed and one that should be my best friend my daily groans deride. God is just just it is i that have sinned in my marrying against the light of gods word against the warnings of a faithfull minister of christ mr davenport. O take warning dear christians by my fall to the braking of my bones this 36 years o [crossed out] how has my soul been tormented threw my ears. I have suffered much for want of womans help but i have been enabled to justify the lord and condem myself. O i have thot my torment was like the torment of a scorpion when he striketh a man—revelation 9-5 but o i am out of hell where i deserve to be and i am surrounded with mercies a hope of salvation threw jesus christ glory to god. Then shall i see his face and never never sin [illeg.] (while from the rivers of his grace) drink endless pleasure in.

January 1781. Have been kept awake great part of many nights i beleive by the temptations of the devil for i have been filled with dispairing distressing distracted thots and everything apeared to be rong. But many times god has helpt me to resist him by prayr sometimes by singing himns or psalms softly so as not to disturb the family and i have gone to sleep comfortable. Ah i have cause to be ashamed to say it but true it is i have been afraid many nights to go to bed for fear of the devil that he would distress me. But how often have i found that he was bound and my heavenly father has got hold of the end of the chain. And i have had my mind of late much stirred up to pray for satans being totally bound according to revelations 20 the first three verces. O lord bless the rod to me subject me to thyself i beg for jesus sake amen.

[*p. 327*]

February 8. I retired to pray these words was good to me—they that sow in tears shall reap in joy.[7] He that goeth forth and weepeth baring preacious seed shall come again with rejoycing bringing his sheaves with him. Now my soul cryed lord let my poor husband and children be amongst my sheaves. Also this word affected my hart—light is sown for the righteous and gladness for the upright in hart.[8] Methot i see that if i was righteous and upright threw the imputed righteousness of jesus christ then light and gladness was sown for me in him. O i hope in the lord that i shall yet praise him.

February 18-1781. This lords day morning i retired into the cow house powered out my heart before the lord in confessing sin. O my sins are ever before me o teach me to know thee more and myself more o wreached one that i am because of this body of death o my hard heart o my ingratitude towards jesus the lord of glory. O let me feel thy love o allure my soul to love thee o create in me a clean heart and renew a right spirit within me.[9]

Now my jonathan and twoo of my neighbours had a bad contention arose between them. Now this man and his wife and i myelf are in covenant with each other and now i heard of threatning words by this man of going to the law to have the case decided and there appeared anger [crossed out] for there are such things in our church already anuf to make ones heart ake and i see go and tell him was my duty tho my nature strove hard against it. O what a creature i see myself to be but i kept pleading with god for several days for help and strength that i might go as abigal did to david. At las i went but o what a sweet spirit i see in this man and his wife. They both said they was willing to leave it to men

[*p. 328*]

and so i come home in peace glory to god o perfect thy begun goodness.

March 27-1781. This winter past it has seemd as if i have heard the sound of death acoming and o how many times have i begged of god to fit me for what was acoming. And as i have written in page 320 i took home my poor little grandchild. It was agrowing well likely child and it went out of the door but a few moments before to a fence about ten rods from the house we suppos to climb over and pulled twoo rails down upon its neck and breast which killd it we suppose in a moment. He was three years old and a month wanting 3 days. Thus the lord was pleased to take away the desire of our eyes with a stroke. He dyed march 8-1781. At first it seemd so aufull sudden and surprising my soul seemd to tremble but the lord made me in some measure like aaron when his sons was slain[10] to hold my peace and i have been supported since and felt a still mind. This word was with power on my mind—behold i stand at the door and knock if any man hear my voice and open to me i will come in and sup with him and he with me.[11] I told my husband of it that it seemd to me as if jesus was knocking at our door for entrance. O that wee mant shut him out. Now jonathan seems to have considerable of a still mind but my husband mourns to excess and will not be comforted. He goes almost without food and pines away. O lord turn his mourning into the right channel o make him mourn for sin o make it a means of his convertion o that wee all may be humbled down at gods foot & kept there o jesus do come in to my house and dwell here o let us not idolize dust any more.

[*p. 329*]

Aprel 8-1781. It is a month since the dear babe dyed. The lord is good to me i feel his supporting mercy. That part of a verce in wattses psalms 91 second part (ill bear their joyfull souls above destruction and the sword)—this has been sweet to me glory to god o rich grace o free mercy that god should help me to bair it with some patience while my husband sinks beneath the stroake. He pines away daily and is looking at second causes. O lord do shew him his sins o make death and hell appear naked in his vew without a covering o make him fly to jesus christ for mercy—a good god hant ronged us. I know sin is the cause i see a multitude of sins to mourn for in my heart and life. My son give me thy heart but i have given twoo much of my heart to this poor babe and so was unfit to have it o lord sanctify this stroak o bless the rod. Wigglesworths verses has helpt my mind.[12] Bless the lord o my soul god be mercyfull to me a sinner.

Aprel 29-1781. My husband still mourns to excess his appetite to food almost gone his body seems wasting away many fear his sorrow will work his death. And now i believe the lord has sent his servant mershal here and he with many others has taken great pains both by prayer and exhortation to turn his mind of from worldly sorrow that works death and that he may mourn for sin the cause of all sorrow. Now he seems to love mr beech and mr mershal. He has heard mr mershal preach four sermons and with much persuasion is now gone with him to guilford to hear him preach there. He lodged with us twoo nights and preacht a funeral sermon from the first book of kings 14-12-13 vers. I believe god was with him in every sarmon. Today this word was in my mind—from the uttermost parts of the earth have we heard songs even glory to the righteous.[13]

[*p. 330*]

May 3. Since my husband come home from guilford he seems more composed but i fear he has no sence of sin. O lord have mercy on him. Poor man goes groaning about with worldly sorrow. Lord make him groan for sin grant he may reform his wicked practices. He hant been to meeting i think on the sabbath but once in above three years but now he has been to wallingsford with me twice lord i thank thee. A few days ago i see that if it was not for electing love and grace there never would be a sinner converted nor saved. The father was first in electing and the holy ghost proceeding from the father and the son compells them to come in glory to god for this mistery of salvation by a blessed jesus.

A few days ago in this place there was a child scalded with lie and dyed the next day and at this lection a great company of negroes was froliking about twoo miles from us and one stabbed another with his knife who dyed instantly. Lord let this be a warning to all not to suffer merry meetings in their houses. O lord grant a reformation or we are ondone o pour down thy spirit for jesus sake. We hear a child at wallingford a few days ago was scalded and dyed soon after.

This lord's day morning are i was awake it come into my mind that i must pray for the downpourings of gods spirit. I felt some meltings of heart. O lord have mercy upon the church at wallingsford. Thy spirit is withdrawn and we are full of divisions and contentions. Pray come quickly lord jesus. Methinks i see this lords day from the 3 of matthew that the lord is apreparing his way for his coming. It seems he is saying to us repent for the kingdom of heaven is at hand. His fan seems to be in his hand he says prepare ye the way of the lord make his paths straight. Alas we have

[*p. 331*]

made many crooked paths but the path of holyness is very straight and it seems god is calling his children into it in a special manner at this day. Oh the dreadfuls that are upon us both in church and state there seems to be turning & overturning. O that he whose right it is to reign would come quickly lord jesus. O what a distrest soul have i felt this 3 months sin rageing the devil throwing in his fiery darts. Lord my cries continually are to thee o help me to stand and having done all to stand. One day my calvin went to cetch pigions as wee supposed on the other side of the river where he had been before but before night there come a man and told his father he had been to his stand and he want there. His father hasted home and told me they supposed he was drowned atrying to croos the river. O how i trembled and cryed lord have mercy on me subject me to thyself. I began to think where the linnins was to lay him out but behold in the midst of my sorrow he come home alive and well. Glory to god for his mercy o let me yet live to thee o pyty poor calvins poor christless soul o that my weary eyes and soul may see thy salvation before i lay my head in the grave.

October 1781. As soon as i awaked these was good words—thrones are prepared for all his friends that humbly love him here—and mr stodderd says it is much for the glory of god to have multitudes in heaven to be spectators and admirers of his glory john 17-24. And that was good to me revelations 7-9—and after this i beheld a great multitude which no man could number & cloathed with white robes and palms in their hands. O if they was without number in johns day what will they be at the resurrection.

[*p. 332*]

Last sabbath mr beech preacht from psalms 58-3-4-5 verses the word was good to me methot i fed upon it. But o now upon hearing that coronal walles[14] and his army is taken o what shooting frolicking and heathenish rejoysing is there in our towns instead of giving glory to god for this smiling providence. O god do pour down thy spirit o give us true repentance. Pray dont let us provoke thee to give us up yet in to our enemies hands. Ah wee deserv it—but o the righteous are sad the wicked mad while thou withholdest thy hand o pluck it out of thy bosom[15] lord jesus come quickly.

November 10-1781. I having a sence of expected trouble where there is so much sin as i went to fetch a pail of water this word come to my mind—let us indure hardness as good soldiers.[16] Now i thot that a good soldier will not turn his back for fear of a few bullets but will press forward hoping to gain the victory. O my god give me persevering grace i beg i beg o to wair that crown of victory that jesus will give in that day to all that love his appearing.

I have been of late distrest for calvin. My soul cryed lord have mercy on him if he goes to hell but one or twoo of thy attrebuts (justise & soveranty) will be honoured but if thou will save him thou wilt honour them all o pyty his poor stupid soul. And o god i thank thee for a reformation in jonathan ever since his wife dyed he goes constantly to the separate meeting o sanctify him by thy truth o make him devoted to thee and if it is for thy honour & glory find a rebecka for my isaac who lives solitary alone.

November 13-1781. This day was a publick thanksgiving day for the mercies of the year and for wallis and his army being taken by our men. This morning this word was with some power on my mind—humble yourselves therefore under the mighty hand of god that he may exalt you in due time.[17] Lord humble me i hope yet in thy mercy.

[*p. 333*]

Ah christians find it easy to trust in god in a time of prosperity when their souls are fild with his love but oh it is a great thing to trust in a hiding god a correcting a killing god and say in faith tho he slay me yet will i trust in him. And now dolefull things are breaking forth in the church at wallingsford. It appears as if it must be broken up. We hant had a sacrament in []. Now while i was praying this word come with power in the 5 of judges 16—why abodest thou among the sheep folds to hear the bleatings of the flocks for the divisions of reubin there were great serchings of heart. Now it seemd as if god calld me to come in to obedience to him and not to stay hearing the flock bleat as sheep when something is the matter with us. Wee have got away from god and have got a perty spirit—one is for this perty and the other for that and o the talk how the flock bleats. O god dont let us stay here pour down thy spirit o draw us and wee shall run after thee.

January 20-1782. Having warning of trouble coming i feard i should be a prey to sin and devil. I was cast down and sad but that word come to me are i was aware and helpt me—settle it therefore in your hearts not to meditate before what ye shall answer.[18] O i do now hope in gods mercy that i shall win threw safe. O lord help me to honour thee let what will come.

February 3 1782. This sabbath at home in much pain. This verce affected my heart (he is all love he is my love o do not him abuse—do not again put him to pain dear christians turn not jews). And now my husband seems almost stilld about the death of the child but seems to have no sence of sin as it is against god. O when i vew him stupid stark blind as to seeing himself wedded to his sins and will not leave him let what will come i say when i vew him it strikes me down before the lord. O jehovah nothing is twoo hard for thee. O reveal thine arm and the work will be done.

[*p. 334*]

A few nights ago my calvin went to a slaying frolick with his vain company. When he was getting ready to go i examined him about it and he said he was not agoing to slaying but soon after he sat away news come that they was gone to the merry meeting. But o how distrest i felt for him that he should sin against god. I thot i should sleep none that night so i did not pull of my cloaths but lay down with them on in the night. I could not sleep. I got up and prayed for him o lord convert his soul. Next morning my heart melted & my eyes slowed with tears to vew the dreadful state of the christles and christians asleep. O lord do come down by thy spirit and thaw this frozen world o lift up thy feet and march in haste aloud our ruin calls.

This lords day morning my heart was affected while reading the first of kings 17-4 & 5 verses. O methot i see something of the love and care the lord takes of his obedient children that live to him o sanctify my soul that i may live to thee yet in the land of the living.

February 28-1782. This day my jonathan was married again i hope to a rebekah. O lord i beg for a blessing from heaven upon them. O let his house be a house of prayr and every soul that shall dwell in it a tent for the holy ghost to dweel in. O god thou hast as it ware cut him down twise by taking away by death a pleasant wife of his youth & afterwards an onely child in a sudden and aufull manner and he was left alone drowned in sorrow. But i hope o lord thou hast ment it for good to his soul. O spare their lives i beg i beg and o that they may live devoted to thee. I have many times begged of thee that if it was for thy honour and glory he might marry again and if not that he never might and now thou hast ordered it to be so. O lord make them to be helpmeats to each other in the path of holiness which is the road to glory o grant it for his sake that dyed without the gates of jerusalem.

[*p. 335*]

Now when i had staid at home from meeting about 5 or 6 months by reason of pain and weekness in my back in march my husband and i went to the church to hear a funeral sermon mr beech preacht from job 16-22—when a few years are come then i shall go the way whence i shall not return. He spoke well to the people and to the mourners. There was a sober attentive assembly. O lord sanctify this death bless thy word i beg.

This lords day went to meeting felt a dark mind and see death all round me but rode homeward 5 or 6 miles with a christian woman i trust. We talked all the way of the things of god and his dealings with our souls. This was better to me than all i heard this day. Glory to god i felt comforted and strengthened and i felt helpt next day. O what am i o lord that thou should ever have a thot of mercy towards me.

May 1782. Mr cornwell preacht at my son jonathans from malachi 3 and 2—who may abide the day of his coming and who shall stand when he appeareth &. He shewed first who they was that would not stand and who they was that would stand in the dreadful day. He preacht charming well. People was sober attentive. O lord bless thy word i beg. Now before i went to meeting i prayed that god would bless his word to all and to me that it might be a time of comfort to me or if it was most for his honour and glory that i might see myself. And now while i heard the word it seemd good to me and after i come home but the next day i felt sin rage in me o i felt like a devil almost all the day tho none knew of it but god. But glory to the lord he toock the burden of o lord do take away my adamentine sins that wont be refined o santify my soul that i may love and praise thee in the land of the living i beg i beg.

[*p. 336*]

August 3-1782. About a year and nine months i have been sorely afflicted with a hard pain in my back scarcely easy a moment onely when i am warm in my bed then i am easy. Lord i thank thee. Shall i that am so vile receive good at the hands of the lord and not evil. Ah it often puts me in mind of death and it looks likely i shall die with this illness. Lord if it is thy will subject me to it. Here i have but little or no pity lord pity me a poor sinner o pyty my poor husband and poor calvin. Pray convert their souls altho sinners wax worse and worse. Lord is anything twoo hard for thee. O let me see thy salvation in my family before i die. Sometimes i am so discouraged about praying for them but i keep on and sometimes am encouraged. O father turn and look this way aloud our ruin calls.

And lord i thank thee for thy mercy to jonathan when he was left alone. Thou hast lent him a pleasant kind companion pray affect our hearts with a sence of thy kindness. O bless them with the choisest of heavens blessings.

This day was hindred from going to meeting it was from the wicked but o what a unsubjected heart i felt tho i was inabled to keep it to myself as to their knowing of it but o it made me pray. Lord i see something of my own wretchedness and i believe if i saw it to the full as it is i should sink. O help me to see a lovely jesus that knew no sin if thou canst do anything lord help me.

August 18-1782. I heard mr spenser preach from this word—quench not the spirit.[19] He shewed in many perticulers how the spirit was

[p. 337]

quenched and sweetly exorted all to cherish its motions in the afternoon. He preacht from jeremiah 18-4 o it was good preaching i believe god helpt him and he helpt me. And i was comfortable next day a part of it but o i was obliged to be with vain carnal company all the week. O what a torment it was to me reproof was not regarded o i thot what a hell it would be to me to be shut up forever with such company if there was no wrath of god nor no devils to torment. O lord sanctify me fit me and let me go to my own company. Ah heavenly glory is twoo great to come in to us we must go in to that. On the sabbath morning i awaked with these words—god works wondrusly[20]—and when i was amilking they was on my mind with a melting power. O i do hope god is acoming to work wonders come lord jesus come quickly or we are ondone.

On lords day heard mr cornweel his mind was much confused i went out sad and returned sad. Ah at this day wee must i think pass from the watchmen to find our beloved. One morning last week i had a raised soul in prayer. O how i wanted to have sin done away and my harp tuned to joyn with the hundred forty and four thousand in giving glory to god. O how reasonable a thing it is for redeemed ones to love christ above everything. Musculus says in heaven there are angels and archangels but they do not make heaven christ is the most sparkling diamond in all the ring of glory and i think so twoo. Glory to god for jesus christ it is heaven and happyness enough to see christ and be forever with christ.

October 2-1782. This sabbath heard mr miner preach from matthew 18-4—whosoever therefore shall humble himself as this little child the same is greatest in the kingdom of heaven. He shewed in many perticulers what it was to

[p. 338]

have that sweet child-like christ-like temper of mind. I felt the witness in my own soul to the truth of it. The sacrament was administered which has been neglected a great while by reason of diffeculties in the church. There was some that did not pertake and i among the rest by reason of diffeculties. O lord when will thou make good that word on which i have hoped above twenty years that thou will gather them like the sheaves into the floor. Come lord jesus come quickly. And now this lords evening when i came home poor in body and weary i suffered much from a wicked person but god helped me to hold my peace for i never answered a word nor pleaded my own cause for i felt as if i could bair it with sweet patience. I thot god bairs it why maynt i. O it is just just that i should be punished with sin for my sin o i did not marry in and for the lord. O how do i long for his and my childrens salvation. Lord i believe thou art able to save them o compell them to come in. I want to have them give glory to god here and to bow before thy throne in glory.

On friday evening i sat up late after my family was in bed to read in the history of the myrters in queen maryes day and to pray but o how angry was one at me. I was called all to nought for it. Oh the wicked words and actions father forgive him ah dear man. The next night i read in the same book again but it was toock out of my hand in anger and flung away. O the enmity i see the same spirit now that there was in bishop bonner[21] and the baptsts. O lord pyty poor blind sinners that dont pyty their own souls. I hartily desire their salvation o thou that wept over jerusalem.

November 21-1782. This morning my daughter sent me three pyes but o how it affected my hart with tears and thankfulness that god was so kind to me a poor sinner. I wondred he should take so much notice of me. Lord bless her o make her like rachel and like leah which twoo did build the house of israel.[22]

[*p. 339*]

I having hinted before of diffeculties in the church that i could not with a good concience pertake with twoo of them a brother and sister. I thot they had talked and done rong. I went five times to talk with them but to no affect the wound was rather worse. Now i see a brother offended is harder to be won than a strong city and now it is not possible for me to write what i underwent for above a year and a half. The devil i believe had a great hand in my distress. I felt a rong mind towards my brother and sister & god says he that hates his brother is a murderer that is he has the same spirit in a degree that murderers have. O dreadful o the hundreds of arrants i had to the throne of grace that it might be made up that wee all might have a christlike sweet forgiveing spirit. And sometimes my mind sunk i thot it never could be made up and then i was ready to wish the church to break up for it loookt likely to that i might get my neck out of the yoak. I was sorry i ever joyned to this church. I was tormented in my mind day and night above a year. I let it lye and then i went the sixth time to talk with them again and for above a week before i went i was so fraid of them i did not know what to do. Now i do suppose it was from satan he was loth i should do my duty but i felt willing to acknowledg my rong mind and went to see them. When i come there my fear was gone. I felt a calm mind i found the stone was rolled away. I acknowledged my rong mind towards them and they was very pleasant and mild. The woman was freer in acknoledging than the man was but however my burden was taken off. O my god dont let it come again unless it be for thy honour and glory. It is now about three weeks since i talked with them and god is still good to me. Here i must set up my ebenezer—the lord has helped me. I can bair witnes for god that he is a prayr hearing god glory glory to his name. O lord if all the world does rong dont let me be holden with the cord of my sins—proverbs 5-22.

[*p. 340*]

Now a woman at wallingsford sent me a book a very old book the date was gone there was left about six hundred pages sat out by thomas brooks. The title was heaven on earth.[23] I have cause to bless god for it. It was the meanes to shew me where i was and what i wanted. O it made me pray glory to god for such men of god glory to god that has said lo i am with you to the end of the world.

1782. This summer past there has been a extream drouth for many months the heavens were like brass and the earth like iron wells and brooks dryed up. Flocks and hards ready to starve for grass our latter harvest almost cut of and o at this time the righteous sad the wicked mad whilst thou withholdest thy spirit. O thou jehovah pray reveal thy arm pour down thy spirit and wee shall be turned.

January 7-1783 this day my jonathan had another son born about noon and it dyed about sundown a large lovely babe now glory to god for sparing the mother when in such eminant danger for many hours. Lord heal her restore her to health if it be for thy honour and glory and o bless the rod to us all for jesus sake.

1783 in jenuary. Mr miner come and preacht at seth heatons from ephesians 4-15—where is then the blessedness you spake of. He spoke cheifly to christians who generally have at this day all departed from god and have got twoo to much after the world. O says he has the beauty of absalom[24] bewitcht you that you are gone after him and by & by there was no place on the earth for absalom to stand on and his followers slain and fled. O says he where is the blessedness you spake of meaning in years back when gods preasence was with us. He said if gabrel was set to tell the blessedness of the saints he would even blush to undertake it. O how he calld us to come home again for the lord was still saying i am bone of your bone and flesh of your flesh. There was some meltings in the assembly o jesus come quickly.

[*p. 341*]

January 7 1783. Jonathan had another son born it lived about 4 hours and dyed a lovely babe and his wife was taken sick about twoo months. Some of the time her life was but little expected she was so low. Now i prayed hard that if it was for the glory of god that she might live and o that she might live to god for i have a hope that she is a christian but has lived in the spirit of the world many times before she lay in. I reproved her for her light spirit and carnal mind but on she went. Now there was an impression on my mind of terrible things in righteousness that was coming on that family. Jonathan also was very sick when his wife was and i think in about a fortnit he began to amend and now goes about his work. She is gaining. Glory to god. Now she laments her past life and seems to have some vews of the love of christ from these words—he has prepared for them a cyty that has foundations whose builder and maker is god wherefore he is not ashamed to be their god.[25] She told me how cutting it was to think how she had lived. O lord bring her into obedience to thee. Dont let her go back again like orpah to her gods.[26]

March 30-1783. A few mornings ago i awaked with these words on my mind—they come where there was twelve wells of water and palm trees. They seemd great words. I got up and serched the schriptures and at last found the place in numbers 33-9—and they removed from marah and come to elim and in elim were twelve fountains of water and three score and ten palm trees and they pitched there. O i do hope that the church is agoing to march and will draw water with joy out of the wells of salvation and that they will flourish as the palm tree and grow like the ceders in labanon. O my god make no tarrying do read revelations 21-12-14-21 bow thy heavens o lord and come down touch the mountains and they shall smoke.

Aprel 11-1783. I have had a pained weekly body this winter. I am now 62 years old but a step from the grave but i feel my anchor fastned above not below. But o the sin that i have seen and heard this winter has fild my soul with anguish but i in a constant way am enabled to subject to gods justice in it because i did not marry in and for the lord. Pray christians take warning by me i beg i beg act for god in your marrying.

[*p. 342*]

Aprel 18-1783. A sorrowfull day waded threw [crossed out]. What will willful sinners do when god shall rend the cave of their hearts and wound the hairy schalp of those that go on still in their trespasses. O almighty father reach down thine arm these words in job 16-18 was on my mind—o earth cover thou not my blood and let my cry have no place. Altho jobes friends did not shed his blood yet their cruelty to him in his affliction resembled it and it seems he wanted to have his innocency appear to the world as it did afterwards. And o lord let my cry have no place to rest in here below o let them come up into the ears of the god of sabaoth in the night. That in the 18 of luke run in my mind about the widows coming to the unjust judge to avenge her of her adversary. Now i find my greatest adversary is my sins that brings on me all my sorrows and tho he put her off a while yet for her importunity he said i will avenge her and will not my heavenly father do better than that wicked judge. O when his time is come he will help speedily this and see i think with a clearer light than ever i did before. O how great is his goodness to me a reabel sinner. Lord give me patience help me to hold on praying and waiting altho it is the eleventh hour with my poor husband. Yet i do hope in thy mercy for him and the children. O i do hope to see that head that was once crownd with thorns & have time to thank him to his face.

1783 in aprel. Peace being made my calvin has been over to long iseland and brought me news that my dear aged mother and my twoo brothers are alive and weel. O lord i thank thee pray convert my brothers bless my mother with the choicest of heavens blessings. O that her few moments may be devoted to god pray fill her soul with thy love.

In may 1783 i was taken sick with a fever and was in much pain. A docter was sent for but i continued very poor. One morning i was taken in a fit. I was in great pain my flesh twitched my breath and strength seemd

[*p. 343*]

almost gone. I loocked many times to see if the blood did not begin to settle under my nails. I told my husband i believed i was adying he had better go and call jonathans wife but it loockt likely i should be gone before he got back. But they soon come in and i told her where to get things to lay me out. But o how i wanted to be gone i told them i had lived long enough. O methot i could sweetly bid farewell to all things below the sky. I gave all up. This world lookt like a hell to me. I felt as naked as adam in the garden. My all was the truth and faithfullnes of god and there i must venter. I told them when i was converted and that i believed christ had bought me with his own blood. O that word they cry—holy holy holy lord god almighty which was and is and is to come.[27] I see god was from eternity and now is the same and will be the same to eternity (and is to come) sunk the deepest. O the joys and happyness of saints shall last as long as god lives but ah poor me in an hour or twoo i grew better. Then i tried to cry but could not. Many times i said to my husband what shall i do to get threw jordan. I was sick about a month and i never prayed for my life unless it was for the glory of god. And now methot i had got almost threw jordan and now i must come back into this world of sin and temptations. O my god keep me keep me from the accursed thing. O let me never be afraid again of dying. O subject me to thyself. Now i see that word fulfilled—god works wondrously glory to his name.

In may 28-1783 calvin had a son born a lovely babe o let it live if it be for thy honour and glory. O god grant that the parents and child may be brought into the number of thy sanctifyed ones. O pyty them for jesus sake o make them to know christ and the power of his resurrection. O bless my seed while sun and moon shall indure for jesus sake o send salvation to zion.

[*p. 344*]

September 1-1783. This lords day morning my heart sent to god for the downpowering of his spirit. O lord jesus come quickly or wee are all ondone. O come lord jesus and thaw this frozen world. A few days ago dyed the aged woman 86 years old that was an evidence against me when deavenport had me up for not going to meeting. She has been confined ten years could help herself but little. Her hands and knees was drawn up she was a terror to herself and to them about her. She was so impasient pevish unthankfull and discontented she departed without being desired by her own children. We do believe they all are glad she is dead for their pacience was worn out with her. I have written in my first book about my being had up for my not going to meeting in page 120.

October 1783. This morning my heart was affected while reading a letter sent to me from the nine pardners. The lord bless that woman with his presence forever glory to god for any help in my journey when i deserve none.

October 19-1783. This lords day at home had strong desires of soul for my husbands conversion. O thou that prepares hearts to pray wilt thou not cause thine ear to hear. O pyty his stupid soul o is anything twoo hard for thee no no. I do hope yet in thy mercy pray come send salvation.

Now i was greatly boarn down with sorrow again to hear the prophane language of the wicked. These words come fresh into my mind—the archers have sorely greived him and shot at him and hated him but his bow abode in strength.[28] O it helpt me glory to god. My bow does abide in strength. I can say from my heart root god does right in punishing me with sin for my sins. O i have often thot that when i come to heaven i could tell john that when he was on earth he saw a great wonder in heaven but behold john here is another the greatest sinner that ever was is come to heaven.

[*p. 345*]

Twas the same love that spread the feast that sweetly forst me in:
else i had still refusd to taste and perisht in my sin.

1783 in november. Mr mershel preacht at my son jonathans from these words—take not thy holy spirit from me.[29] It was good preaching but poor sinners made light of it and few days after mr frothingham come and preacht at mr seth heatons from revelations 3-16. Methot i could witnes to the truth of it. Lord i thank thee that in this trying day thou hant shut up their mouths. O when wilt thou send salvation to zion. Do come quickly for here the righteous are sad the wicked mad while thou withholdest thine hand. Brethren fall out by the way here. Thy purchast crown seems tumbling down this lion seeks no prey: confused hurld is this mad world. Sweet jesus hast away but glory to god that has not forsook his world for wee hear that in many places up the country and at long iseland god is pouring down his spirit in a wonderfull manner. O let that stone that was cut out without hands become a great mountain and fill the whole world. I hope it is begun.

Jenuary 1-1784. O how was my sins punisht with sin. One of the hardest of all tryals to see and hear god sind against willfully. My soul was in anguish last night and this day. I thot of the burning bush that was not consumed and them words come into my mind—fight the good fight of faith and lay hold of eternal life.[30] I see i must have opposition. The christians life is a warfare and this verce helpt me—shall simon bair thy croos alone: and other saints be free: each saint of thine shall find his own & there is one for me. Oh happy christians be not loth to have a courser fare. Saints that have had no tablecloth had christ to dinner there. Now i thot of manasah when he fild jerusalem with innocent blood taught them to sacrifice their children to devils and he was a wizzerd and i have red how he caused the dear prophet isaias to be sawn asunder and o lord he was not twoo bad for thee for thou did turn his iron heart to thyself.

[*p. 346*]

Lord pyty them that dont pyty their own souls. Nothing is twoo hard for thee and now these words was in my mind—we wrastle not with flesh and blood but with principalities and powers and spiritual wickednesses in high places.[31] Lord help me in the field for thou has overcome and art sat down in thy fathers kingdom. God be mercifull to me a sinner. Thy stroakes are fewer than my crimes and lighter than my gilt. O that saints would marry in and for the lord. Alas how can twoo walk together except they are agreed.

January 1784. By turns i have been sorely assaulted with unbelief. It would come upon me thus how will you do to bair the aggonies of death and it may be you will be lost between death and heaven. How can you get threw such a great journey heaven is so far above earth. But one morning i think between sleeping and waking this word come to me—he will send his angel to help you in all your way. It was good to me but i did not certainly know that it was schripture i searcht but could not find it. Then i thot i would ask some christians when i see them if there was not such a place of schripture but before i see them i found this word in exodus 23-20—behold i send an angel before thee to keep thee in the way and to bring thee into the place which i have prepared. This angel is jesus christ and in that soul where he has begun his work he never must nor never will leave it till he has brought it safe home to glory. Bless the lord o my soul dont let me forget thy dear mercies. O let me live thy praise.

March 7-1784. Now the church at wallingsford has dismist mr beech sometime ago on acount of his preaching to them and since he has been very poor and weakly in body to be feared in a consumtive way and o the dreadfuls that have broke out in that church. One man that

[*p. 347*]

use to carry on in our meetings i hear owns he has had twoo children by two young weman. One was his own sisters daughter. He is a married man and i hear he greives and mourns now and pines away. O lord have mercy on him and on us all.

March 29-1784. [crossed out] O it grieves me and makes me pray for him. Last lords day night i awaked and had an uncommon spirit of prayer for him. Methot i see clearly i had acted faith in christ for his soul. I told the lord that i had believed his word and confided in it that he would make him an isaac. My soul cryed to god to fullfill his word on which i had hoped and firmly believed. Lord jesus do come quickly o let my aking heart and weary eyes see thy salvation for my husband and children before i lay my head in the grave. Make no tarrying o my god.

Aprel 4-1784. This lords day i was in hard pain in my back. I believe god dont afflict willingly but if need be. I look upon my sins with horror and loathing myself for them. Lord remove the cause and the affect will cease this day. I have lost the comforter what shall i do lord i have nowhere else to go but unto thee. Thou hast the words of eternal life o how is my soul tormented threw my ears with [crossed out]. One day them words was good to me that joseph spake to his brethren—you ment evil against me but the lord ment it for good.[32] O i do hope in gods mercy that even this shall work for my good according to his promis but o methinks it is a hell above ground to hear such language. O how it weekens my body lord have mercy is anything twoo hard for thee. If thou wilt forgive much will they not love thee much.

May 7-1784. In the evening there was an earthquake o lord make us to turn to thee for jesus sake.

[*p. 348*]

May 9 1784. This day i was in much pain in my back and weak in my joints. I can hardly go alone but the lord is exceeding kind to me. I am for the most part easy anights and can sleep. O i wondred this day that i was out of hell. O father bless the rod.

June 10. I newly have had twoo choice book brought me o how great is the lords kindness to me. I have had great comfort in reading them in my solentary condition scarcely ever any to converce with about the things of god and i hant been to meeting on the sabbath above this six months by reason of illness of body. One book is a discourse upon justifycation[33] the other is the spiritual magazene or the christians grand treasure.[34] O how the latter sets fourth the love of god in electing and everlasting love before the world was made. The book upon justifycation shews how we are justifyed by christ alone all works cut of and flung to the ground. Glory to god for putting such treasure into earthen sinful vessels. Glory to god that has said lo i am with you to the worlds end. Glory to god for being self moved before time began to give his own son for a covenant a representative for the elect. Glory to god for such helps in my journey. O father pray do help me to live to thee.

June 17-1784. Having been sorely tried the week past [crossed out] o how am i punisht threw my ears. God is just in suffering it to be so for i did not take his word for my rule and guide in my marrying. This lords day morning i retired poured out my heart to god for the down powering of his spirit. O that his arm may be reveald and then sinners will bow like willows. O methot i see it so easy with god. O when shall satan be bound o how long shall he have that userped title the god of this world. O i see my own sins o how ignorent i am of god how little do i see of his purity holiness faithfullness and mercy and his boundless lovelyness. O that such a being should ever have a thot of mercy towards loathsome dust and ashes. But o what a sight will it be to be with jesus to behold him where he is not a man of sorrows but in his exalted glory. Behold my soul thy worthless name enrold in lines above. See jesus heart vew there a flame never changing love. Amazed i stand i stand amazed at love so rich and free to one so vile. I well may gaze at love so great to me.

[*p. 349*]

August 14-1784. I awaked with this word in my mind—god is gone up with a shout and the lord with the sound of a trumpet.[35] O the praises the acclamations of angels and archangels they attended the triumphs of our saviours death they sung the trophies of his resurrection. No heart can conceive what joy there was in heaven when the glorious conquerror ascended and i do hope are long to follow jesus where he is gone to prepare a place for me. As one says christ by his resurrection carryed home to his fathers court the princely imperial honours of war and o he ever lives making intercession. Ah here i carry a pained weekly body tryals without and war within but dayly i find the lord is conqueror still threw all the wars that devils wage o he ever lives making intercescion that where he is we may be also with him & then no strokes no frowns no croosses will there be till wee shall rest in blest eternallee but lo the pain and smart will then be gone and nothing but a skeene of love comes on.

Yesterday morning early this word was upon my mind i that speak in righteousness mighty to save o it gave me fresh hopes that the lord would soon pour down his spirit here. O that word mighty to save seems good to me while i rite. O i do hope yet for my poor husband that is now 64 years old since my jesus is mighty to save. Altho he is old in trespasses and sins there is none twoo bad for jesus o he is almighty to save. I found these words in isaiah 63 first all the verce is lovely—who is this that cometh from edom with dyed garments from bozrah. This that is glorious in his apparel traveling in the greatnes of his strength. I that speak in righteousness mighty to save. O did jesus travil for sinners with his garments dyed in blood in the greatness of his strength mighty to save. Above twenty years ago this word use to sound in my mind—he shall see the travil of his soul and be satisfyed.[36] O i thot how great that glory must be in and threw this world that would satisfy the heart of christ jesus. Lord help me to live to thee and if called to it to suffer anything for thee i beg i beg. Suffering grace will brighter grow then gold thats in the furnace tried: only jesus would i know and jesus crusifyed.

The week past i have seen much of the spirit of the devil o lord he is come down in great wrath. O let his time be short. Pled with god today for the downpouring of his spirit and strong desires for my poor husband and calvin. O god do make them know themselves and know thee. I dont feel so distrest for jonathan for i have given him to god thirty years ago and there i am enabled to leave him [crossed out]. O thou that brought israel out of egyept with thy strong arm and bowd manasses stubborn will and converted a blaspheming saul a magdalene possest with devils hast thou not caused these mighty acts of thy mercy and grace to be recorded for me to plead for my husband and children. O father turn and look this way o pyty the souls thou has made.

[*p. 350*]

One day are i was aware i believed and see with comfort that as sure as jesus christ got the victory and arose from the dead so surely shall they that are in him have a glorious resurrection glory to god. O glorious hour o blest abode i shall be near and like my god and flesh and sin no more controwl the sacred pleasures of the soul. My flesh shall slumber in the ground till the last trumpets joyful sound: then burst the chains with sweet surprise and in my saviours image rise.

January 1785. O how loud is god calling to us by a great many sudden deaths among us of late. O lord help me to watch that if thou should come and call me at midnight or at cockcrowing or in the morning i may be in an act of readyness. O let me lay my head in jesus bosom then all is well i can sweetly sleep: my body falling to the dust i leave with thee to keep.

Last week i heard a poor man tell how he got a lawyer at new haven to go into court and plead a case for him and that day the lawyer had a child lay dead in his house but however he went to the court house and spoke but a few words which did no good for the poor man lost his case.

Now some time after the lawyer sews for five pounds hard money and got it only for speaking a few words. Now it grieved me when i heard of it and in the night these words run in my mind james 5-1—go to now ye ritch men weep and howl for the miseries that shall come upon you &. O i see that not only ritch extortioners must perish but all self righteous sinners that think they are ritch and increased in goods and need nothing. O god almighty reveal thy arm lord how does this land mourn under these judgments. O pluck them as fire brands out of the burning for thy name and mercy sake.

I think i have not been to meeting on the sabbath in above a year by reason of pain in my back and weakness. Last saturday just at night i was very sorrowfull and are i was aware my soul went out to god with strong desires that he would give me a heart to serve him altho i am as one alone on earth. Methot i longed to serve god from my very heart root and i see i could not serve god aright except he gives me a heart. Lord help me to be content with such things as i have for thou hast said i will never leave you nor forsake you. O pour down thy spirit lord have mercy on me and on thy world for jesus sake. O when shall thy glory cover all the earth.

Aprel 1785. It is now above eleven years since mr beech moved our meetings to wallingsford on the sabbath and some time ago they dismist him as to his preaching to them and his own mother and brothers and sisters cant tell how to hear him by reason of their thinking that he allows of rong conduct in his family. He has had turns of bleeding and seemd in a consumtive way. It was thot sorrow of mind occasioned it but now he is better and rides out to preach but dont seem to live near to god. O that he had not moved our meetings to wallingsford. Pray fill him again with thy spirit o god.

Last week a prophane wicked neighbour sewed jonathan out of spite and i believe toock from him above thirty shillings rongfully i believe but o how it rallied the power of sin in me. I felt so covetious and wicked i knew not what to do. O what a heart have i lord. O i pray thee do kill the power of sin in me pray make me content with such things as i have wheather mercifull or aflictive. But o how i pytied poor jonathan my child that had met with so much sorrow. O lord bless thy rod. God says of the wicked my rod in their hand is my indignation. O lord let us bear the rod and who has appointed it it is thy hand to punish us for sin. O sanctify thy dispensations to us all for jesus sake.

Aprel 3-1785. Now my soul was fild with horror because of the wicked that foresake thy law. I cant wright what i underwent for several days. O how was i tormented threw my ears. Mr. Flavel says he had rather god would strike him dead while alive then to torment him threw his ears with the dyalects of the damned. O my god deliver my soul from hearing such language as i do. O god have mercy on me.

[*p. 351*]

Now one morning early this word come on my mind it kept runing threw my mind—he will lift up an ensign to the nations.[37] Now wee all know the office of an ensign is to wave his red colours to make a grand shew in an army but i believe our god is acoming to lift up his glorious ensign jesus not only here but among other nations. O is he agoing to spread the gospel among the nations the purchase of his crimson blood glory to god o come quickly get to thyself a great name. If i am laid in dust when it comes i trust there will be joy in heaven if over sinners sure over nations.

Aprel the last day 1785. Now a heavy tryal indeed come upon me i almost sunk under it. My week body and mind was almost overcome for about a fortnit but i kept trying to plead for mercy i see none but god could help. I was now strongly possest with athism and unbelief as if god had no regard to my prayers. It is in vain to pray says unbelief god wont hear you he is angry at your sins. This word one night was some help to me—o thou that dwellest in the clefts of the rocks let me see thy face let me hear thy voice for sweet is thy voice and thy countenance is cumbly.[38] O methot altho i am like one inclosed round with high rocks so that i cant get out yet i must keep praying. Let me see thy face let me hear thy voice. But o this is astonishing for sweet is thy voice and thy countenance is comely. O does god love to have his children pray when they have been sinning against him and his rod upon their backs. Now this was an outward trouble not convenient to be mentioned. However i kept pleading night and day lord have mercy on us tho all means seemd to fail that i used but i was brought at last to pray that if it was not for the glory of god to grant me my petition that it might never be granted. And now what shall i say what shall i wright o glory to god glory to god glory to god he has given me my petition that i asked of him. Here i may set up my ebenezer hitherto hath the lord helped us i can bear witness for he is a god hearing prayr. O christians dont give out pleading altho he may seem to you a hiding god a killing god o cast thy burden upon the lord and he shall sustain thee.

Lord when i count thy mercies ore they strike me with surprise
not all the sand upon the shore in equal numbers rise.

May 11-1785. Met with hard treatment from the wicked took my byble and went into the barn spent part of the afternoon in reading and prayr till my weekly pained body faild so that i went into the house before night. I had desires to god for the conversion of my husband and children and that god would powr down his spirit upon the whole world. And i think it was the same night before morning this word was upon my mind—take unto you the whole armour of god.[39] Now i see that the whole armor of god was all the graces of his spirit and why so. The next words tells that you may stand in the evil day and having done all to stand. O how it run in my mind and made me cry to god that he would work his own graces in my heart for i can do nothing but sin. Now it loockt likely some sore tryal was acoming. I cryed lord fit me lord prepare me for thy whole will dont let me dispise thy chastisements nor faint when rebukt by thee. One day when i was about my work are i was aware there was a great melting come upon me. I cryed o jesus interceed for me. O remember me now thou art in thy kingdom and hast got threw all thy sorrows. Ah lovely saviour o fire my heart to love thee supremely continually o let me honour thee in life and at death. Take the sting away let perfect love cast out fear. Of late i have seen a lovelyness in the death of believers precious in the sight of the lord is the death of his saints christians that have habitual grace as one says. Yet it is like water at the bottom of a well which will not ascend with all our pumping till god pour in his exiting grace and then it comes so our all is from god.

[*p. 352*]

May 1785. I received a letter from a cousen at newark in the jerseys. She wrights that there is a revival of religion there god is powering out his spirit upon the young people. O there is here and there a little cloud like a mans hand. I hope it is a sign of abundance of rain. Lord do come quickly. Is it not time for thee to work for the wicked have made void thy law. On the lords day i felt much pain and weekness but o my soul was boarn down with athism and unbelief. Now i have lost sight of gods everlasting love in christ jesus which was from eternity and will last to eternity. I say now i have no sence of it. An author speaking about this everlasting love says there is three union relations to christ before faith commences in the soul. The first is election union the second is representation the third is regeneration union. Out of all these arises a fourth union viz a vital union which is the souls cleaving to christ by faith. I have loved thee with an everlasting love therefore with loving kindness have i drawn thee love is before drawing and isaiah 5-4-5 thy maker is thy husband the lord of hosts is his name and thy redeemer the holy one of israel.[40] Mark this desolate and barran wife was the gentile church long before the gospel had reacht her or she had a open being in the world. Yet her maker was her husband and christ says as the father has loved me so have i loved you and sure god the father loved christ from eternity. O i do hope the lord will for his faithfullness sake come to my poor soul and strengthen my faith that is now carryed captive by sin and devil. Now i have changed but glory glory be to god that dont change therefore i am not consumed. O lord send the comfortor hant thou promised it. Thou has said i will see you again and your heart shall rejoyce. I hope in thy mercy come lord jesus.

Now it is about a year that my husband and i has lived in a little house built up with the timbers of our old house which they pulled down and wee had a large house raised august 3-1784 in the same spot where the old one stood. Now i shant pretend to tell what i have gone threw with a weeak body and hard pain in my back all day and all the care of so many work folks. About a year and a half i have had all alone the summers back. I have had three and four cows to milk and butter and cheese to make and none to help me. I have often thot of cruel pharaoh that compeld the israelites to make brick without straw.[41] Ah how many times has it sent me to the throne of grace. I daily see them words fulfild that took hold of my mind before i marryed viz thou shall pass threw it hardly &.[42] A good god is just in all this it is i that have sinned but i must acknowledge gods mercy to me. Calvins wife has come a few times and has helpt me several days at a time when i had a great many to do for [crossed out].

[*p. 353*]

June 16-1785. This day calvin and his wife and child moved into our new house. They have lived above twoo years in a hired house about a mile off but o god pray do give me such a heart as thou did give solomon that i may dedicate this house to thee forever. Pray father do make it a house of prayr a house of god and every soul that shall dwell in it a tent for the holy ghost to dwell in forever. O lord grant thy worship to be set up in this house. O pyty my poor husband and children pray convert their souls for jesus sake.

June 26. I being openly reproacht by the wicked the lord helpt me to bear it with patience. I beg father forgive them for they know not what they do. O lord come quickly and bind satan. O how long are it shall be when nothing shall hurt nor destroy in all thy holy mountain. Now my sins and corruptions torment me o my predominant sin i cannot master it i have been tortured with it of late. Ah when i ransack my heart o how poor my store is i have nothing to call mine but sin. O i am almost ready to invi them that have got safe home to glory but o the lord shall redeem israel from all his iniquities. O lord give patience to wait.

Jesus our preist forever lives to plead for us above
jesus our king forever gives the blessings of his love (watts).

This verce has of late been very sweet to my soul but there was an objection arose in my mind viz how can it be said that jesus forever gives the blessings of his love when he often hides his face from his children plunges them into sore sorrows and tryals loos of dear relations sudden deaths reproach disgrace the temtations of satan and o how many crosses and disapointments sometimes left to sin the werst of all so that i could not reconsile that word—forever gives the blessings of his love. But o it soon turnd in my mind and i see with comfort that all is in love to gods chosen ones when he corrects when he chastises when he seems like a killing god o it is all in love and faithfullness. If need be you are in haviness and if ye indure chastisment he dealeth with you as sons. O my soul kiss the rod lord help me to patiently and sweetly indure chastisement. O bless every twig of the rod ah methinks i see some families amongst us that have that wo unto them for they have received their consolation. It seems god wont be at the cost to spend one rod upon them wo to poor christless souls that have it reserved for them to bair in hell. O lord pray pyty poor sinners o pour down thy spirit come lord jesus.

Lord how many are my foes in this weak state of flesh and blood
my thoughts they daily discompose but all my trust and hope is god.

July 31-1785. This lord's day a melting soul i had before the lord in prayr. I was enabled to hold up christs sweet words to mary when she was at the sepulcher that morning jesus arose from the dead. He said to mary woman why weepest thou whoom seekest thou.[43] O methot i sought jesus tho sorroing o for such a comforting word from thine own mouth my father and your father my god and your god. I sung the 61 psalm and 63 first part they seemd good to me a little glimps i had. O lord call me as lazarus out of my grave o may thy spirit guide my feet in ways of righteousnes make every path of duty strait and plain before my face.

[*p. 354*]

August 14-1785. It being the lords day. Jonathans wife miscarried they sent for me and that they was afraid she would not live [crossed out] it being not half a mile [crossed out]. What i unwent for him ah my nine months burden and ten months travel. O my care and prayrs for him before he was born. O the indulgence when he hung on my tender breast. Ah the days and nights i have spent in sorrow to bring him up. O the wearisom steps [crossed out]. I do hope in thy mercy lord for him now. God being gracious i soon had a still mind and felt patient. All my tryals here are sweet cups to what i deserve but o i went down to see the dear child. When i come in there was twoo docters and a number of women round her holding drops at her nose and other means to keep breath in her if possible. The doctors seemd to have but little hopes of her life. I told her to keep begging as long as she had one breath. I asked her if she had a heart to pray. She said she had some and these words kept running in her mind—purge out the old leaven and become a new lump.[44] I begged that she might live if it was gods will o that she may live to god. Yet if not o fit me for it and all nearly concernd.

August 20. My daughter in law is better and glory to god o that mercies may lead us nearer to thee. Her heart seems affected with gods mercy. She told me before she was sick them words ran in her mind—prepare to meet thy god o israel.[45] O lord be gracious to her pray do now make her watch against a light carnal mind that seems to be her predominant sin. O make her live to thee.

September 4-1785. The other night i dreamd that god did not delight in cruelty in his afflicting but if need be. I awakt and it seemd comforting to me next day. Lord thy stroakes are fewer then my crimes and lighter than my guilt. O my sins i am made to posses the sins of my youth. O my sins are more than the hairs of my head they cant be numbred i look on them with horror. I think it was one sabbath night i went with a company of young men and women into a neighbours corn. There we stole watermillions a great number and brought them away to eat and when we had done gloried in it and thot it a cunning trick. O lord i thank thee that i was not flung down into hell in the very act o how much precious time have i spent in pride and vain company o my numberless sins before conversion. And yet a gracious god followd me by turns with convictions and i remember that one sabbath evening i went to a near neighbours where was a young man and a young woman and worried at them to go with me to a house about a mile of where the old man and woman was dead and the young people lived by themselves to have a merry meeting. Now this young man was very loth to go but i got him out at last. O what a merry time wee had and this young man has told me since how consernd he was then about his soul and

[*p. 355*]

when the work of god come among us in the year 1741 he was i believe converted and became an eminant christian and of late years a deacon to mr elisha pain[46] a separate at meacox on long iseland. And o my unsubjectedness to my parents. God says with what measure you meet it shall be measured to you again.[47] Ah just it is that my childrens ways should grieve me and o my sins since conversion they have been so great i cant tell with tongue nor pen. O the wounds christ has had in the house of his friend. O methinks i have many times in the nature of it joynd with judas. Let christ go for the vain trifles of this world. O how many times have i joynd with peter in the nature of it when he denied christ. I ought to spoke for jesus and his cause when i did not and another effect of it is i hant lived to him as i ought. O who can number all the stars or sands upon the shore then may they count my numerous sins my scarlet crimson sins. O jesus wash me in thy blood.

No blood of birds nor blood of beasts
no isop branch nor sprinkling preist
No running brook nor flood nor sea
can wash the dismal stain away
But jesus my god thy blood alone
hath power sufficienct to atone.

And glory to god that in and threw christ he can forgive the greatest provocations. O them words—i am god and will not execute the fireiness of my wrath.[48] Ah his compassions fail not therefore the sons of jacob or christians are not consumed. O how glorious is the riches of free grace in god threw jesus christ before ever the world was made and to be displayed now in time. Glory to god that was self moved before time was to have his delights with the sons of men. Bless the lord o my soul and

Sing of the wonders of his love which gabriel plays on every cord
let all on earth and all in heaven sing hallelujahs to the lord
Twas the same hand that spread the feast that sweetly fourst me in
else i had still refused to taste and perisht in my sin.

September 18-1785. My son calvin had a son born a likely child. Ah god is a god hearing prayr. O that the parents and children may be converted that they may bring up their children in the fear of the lord. O that the worship of god may be set up in their family. A few nights ago i went to bed sad had a distresing pain in my back in the night these words was on my mind—he shall change our vile bodies and they shall be fashioned like unto his glorious body.[49] O how good was them words to me that night and next day and for some time after to think altho i am 64 years old a poor frail pained body almost worn out with pain and sorrow. Ah how vile by reason of sin but yet is there a time acoming that this vile body shall be changed all sin done away and be built up again at the resurrection without sin like christs glorious body. O how kind is the lord to me poor lothsome me. This day i abhor myself and repent as in dust and ashes. While i am wrighting my tears makes me stops a while and mourn. O methinks never any in heaven or hell that has abused mercies abused the ritch grace of god as i have done. O what a mercy is here that he should make his word so precious to me. Glory to god for jesus christ who dyed for sinners such as me. Amazed i stand i stand amazed at love so ritch and free to one so vile i weel may gaze at love so great to me.

[*p. 356*]

I have of late experienced that god is a god hearing prayr o how great is his kindness towards me. I have thot that when i come to heaven i should have cause to sing louder of free grace than manassah saul or mary magdalene. O my aggravated sins against light theirs was ignorance.

A guilty poor and helpless worm on thy kind arms i fall
be thou my strength and righteousness my jesus and my all.

There is twoo things i have often wondred at. One is altho i keep sinning against god yet he dont cease to love me. The other is why he should chuse me and leave thousands leave kings goveners and captains o it is sovereign love.

Wonders of wonders lovely lord to set thy love on me
i know the only reason was thy love alone was free.

On the sabbath i was poorly in body so that i went and lay down on the bed which i dont unless of necessity on the sabbath. Well i fell into a slumber and dreamd that i see a tree that was old and dry fit for nothing but for the fire but out of this old tree there run white lovely gum. I thot it was a pyty it should be lost i would go and try to save it. So i awaked. Now it was opened to me thus. Altho i was like this tree which i see fit for nothing but to be distroyed it was so old and dry yet the free grace of god in his dealings towards my soul was precious and lovely and i must not let them be forgot nor lost and so i got up and went to writing. Of late the 45th psalm has been precious to me. I see glory in it especially the 3-4-5 verces. It made me cry gird thy sword upon thy thigh o most mighty with thy glory and majesty and in thy majesty ride prosperously because of truth meekness and righteousness and thy right hand shall teach thee terrible things. O that thyne arrows of conviction may be sharp in the heart of the kings enemies. O that wee may all fall at thy feet and keep us there. O when shall it once be. Come lord come love come that blest day. My earnest expectation shovel our sins out of the way those hills of separation. Come lord and never from us go. Earth is a tiresom place. How long shall wee thy children mourn the absence of thy face.

November 11-1785. This morning i waked with this verse on my mind. It kept running threw with power. Now i feard god was going to take away some in my family. It made me cry lord fit me for thy holy will. O bring me near to thee my refuge in trouble. The verse was this—

When desolation like a flood ore the proud sinner rools
saints find a refuge in their god for hees redeemd their souls.

O let what will come lord help me to flee into the ark christ and i shall be safe then. I shall love thee love thy works love thy providences and love all thy ways. O what is all earths created injoyments to this. I had my heart refresht many times of late by reading a very old book sent to me by a freind. There was many tytles to it one was properties of forgiveness. He shews the true nature of gospel forgiveness in christ as a free act of gods sovereign will. Bless thee lord o my soul.

[*p. 357*]

December 18-1785. Now a few days after that verce the 11th of november was so powerfull on my mind. My husband told me that there was a hard lump agrowing on his cheek. He went to a docter but means seems to do no good. The lump grows bigger fast it is now as big as a doller and very hard. But o but o how secure and stupid his mind is he goes on in his old ways of sinning not the least reformation. All my labouring with him seems to no purpose. This day i told him i could down on my knees on the floor to him if he would reform his predominant sin. O now i have many arrants to the throne. Lord will it not be more for thine honour to convert him than will redound to thy name if he is lost forever. Lord i believe if thou wilt thou canst make him clean and o that thou would heal his body. If it is for thy honour pray fit me for thy holy will whatever it may be. O fit me for what is acoming. I hope to find a refuge in my god. I hope he has redeemd my soul.

Desember 25-1785. This week past [crossed out] my punishment is just the lord dont wrong me when men of spite against me joyn. They are the sword the hand is thine. O christians take warning by me i have been repenting above forty year ah i did not marry in and for the lord. The night before last i went to bed sad but yesterday morning while i was geting out of my bed this word hung on my mind—o my soul thou has troden down strength. Now i was not sure that it was in the byble. I kneeled down and begged of god to shew me the true meaning of these words and my ignorance made me weep. Then i thot i will go and sarch for the words for they seemd lovely to me but where to look i knew not. I took the byble. It darted into my mind it may be it is in deborahs song and o how it plesed me to find it. Judges 5-21 the whole verse is thus—the river of kishon swept them away that ancient river the river kishon. O my soul thou hast troden down strength. Now as deborah with barak led out the army of israel and got the victory over their enemies she says that ancient river swept them away. O methinks that ancient river holds forth the everlasting love of god the father in and threw christ jesus to his chosen ones that sweeps away or destroys sin and devil our enemies and will finally give us the compleat victory. And this word o my soul thou has trodden down strength this comforted me that i hoped threw christ i had got an intrest in his love and favour. I hope by and by to as it ware to set my feet on the neks of my enimies i mean my sins also. From this word i had a secret hope for my

[*p. 358*]

husbands conversion. O that the strong man armed may be cast out o that he that trod the wine press of his fathers wrath alone would tread down or destroy the enemies of this poor mans soul[50] not for my prayrs above this forty years but for his own name sake but o how precious is gods word when he send it. I told my husband this day that one verce of schripture was worth more than loads of silver well refined and heaps of shining gold.

Jenewary 7-1786. A few mornings ago before i was up this word was in my mind—what do ye more than others.[51] Now it sat me to examine myself and i see as to my external duties i did no more than a hypocrite might do but o it is god and christ that i seek in my duties and when i dont find the lord in them they are all emty things. I see so much sin mixt with my best duties. They look loathsom to me as for instance o how busy has the devil been this winter when i have been praying to throw things into my mind that i never thot of before. O this body of death that gives satan so much advantage. Now i see that a hypocrites duties are all selfish ither to still concience or to pay god something and buy heaven or to escape hell or to be seen of men. O lord jesus pyty them for thy names sake o pyty them that are going down to hell with a lie in their right hands. Self-righteousness is a secret lurking sin. O how many times do i see it atrying to creep into my heart but glory to god that dont suffer it to govern. God be mercifull to me a sinner.

The seeds of sin that bitter root in every heart is found
nor can they bair dyviner fruit till grace infect the ground.

Written out of quarts in cheredion—if thy mother be a widow give her double honour who now acts the part of a double parent. Remember her nine months burthen and her ten months travel. Forget not her indulgence when thou didst hang upon her tender breast. Call to mind her prayrs for thee before thou yet come into the world. Remember her secret groans her affectionate tears her broken slumbers her dayly fears her nightly frights. Relieve her wants cover her imperfections comfort her age and the widows husband will be the orphans father again. Close thine ear against him that would open his mouth secretly against another. If thou receive not his words they fly back and wound the reporter. If thou receive them they flee forward and wound the receiver.

[*p. 359*]

Again a word unspoken is like a sword in thy scabbord thine own. If vented thy sword is in anothers hand. If thou desire to be held truly wise be so wise as to hold thy tongue.

Jenyary 13-1786. There was a croos laid upon me and i sunk under it. O the corruptions of my heart how i felt them o my pride o my wicked thoughts how they kept runing threw my mind. For several days i could hardly think of anything else but of my trouble and i feard this was but the beginning of worse evil acoming. I prayed for help but to no purpose. My sore ran and ceased not but o i was in a hurry to be delivered. Now it is my custom to read every night and morning and one night i was going to bed i took my byble and opened it begged for a blessing cast my eyes on first of chronicles 5 chap 20 verce—and they ware helped against them and the hagarites ware delivered into their hands and all that were with them for they cryed to god in the battle and he was intreated of them because they put their trust in him. But o how good was these words to me. I see that i was in the war in the mistery as rayly as the reubenites the gadites the manasites[52] was in the history and that my help was in god alone. And my trust must be in him alone for victory over his and my enemies. I see that the kingdom of heaven suffers violence and the violent take it by fource. Heaven must be gained as israel took canaan[53] and all by faith in christ jesus who is the captain of our salvation. I remember i once read a story of a general who heard one of his soldiers cry out upon the fresh onset of the enemy. Now wee are ondone now wee are ruined. He called him traytor and told him it was not so whilst he could weild his sword. Jesus christ the believers captain that has redeemed our souls has also undertaken the leading and conduct of them threw all our deficulties. Our duty is to watch fight and contend. His work is to take care of the event. Lord help me to forever commit the keeping of my soul to thee. Glory to god for help and strength from thy word once more. O fit me for what is acoming. Those words have been sweet to me—thy mercyes are new every morning and fresh every moment. Great is thy faithfullness. O what shall i render to the lord for his great kindness to vile me.[54] Glory to god for jesus christ that he dyed to be a covenant for his people for justifyication sanctifycation and their perseverance to glory.

Powers of iniquities may rise and frame pernicious laws
but god my refuge ruiles the skies he will define my cause.

[*p. 360*]

Jenuary 22-1786. The sabbath day o lord indite for me for i cant put my experiences into language i am an object of pyty. Almost all this day i heard such prophane language [crossed out] till in the night anuf to make one tremble. It was a cold stormy day and i was poor in body. I could not flee from it as i use to do but o i was so boarn down with sorrow to see gods holy sabbath broken and prophaned that i was so weak i felt several times as if i should have fainted away. My mind was overcome and drownded. I was all in confusion. My peace was gone and my soul as it were trembled for fear of gods righteous judgments speedyly. I could say with david they laid things to my change that i knew not to the spoyling of my soul. I know the devil was the author of it all but o for want of faith i sunk in deep waters where there was no standing. I did not give up my hope but all lookt dark and dolesom. I went to bed that night but could not sleep in a great while. My soul was so shut up i could scarsely say lord have mercy on me. I lay and greived and mourned at my heart root that god was dishonoured upon his holy day. It seemd to me as if i should never get over it and o my sins that called for such a severe rod grieved me to the heart and so i see i had a hand in it too. The wicked are called gods sword and gods rod but o when no eye pityed nor arm could help my soul thou o my god did think of me in my low estate. The 8th of romans was good to me but especially from the 35th verse to the end. O what shall a poor worm say o how good was gods word to my poor distrest soul. O blessed words for thy sake wee are killed all the day long wee are accounted sheep for the slaughter. Nay in all these things we are more than conquerers through him that loved us. Now i believed that threw christ i was made to get the victory over sin and devil and not only so but sweet comfort flowed into my soul and so i was made more than a conqueror. And now i was enabled to look back to my past experiences. O how many such salvations has god wrought for me when i was in

[*p. 361*]

great straits and now i believed i should win threw all to the last cast threw him that loved us—loved us from eternity and will love to eternity the father and son are both ingaged to love believers as you may see in this 8th of romans—and nothing shall separate us from his love. The angels cant nay the holy angels dont want to. Glory to god for christ jesus. O what shall i render to the lord for all his kindness to me vile me. O sanctyfy my soul and make me live to thee i beg i beg.

Tortured and ract with horrid fears he said peace and be still
the blessed dove come from above with the ollive in his bill.

I was refresht this week by hearing that the work of god was begun at wallingsford and mearidon that twenty was converted they hoped in ten or twelve days after it begun. Now youths and children are flocking together to serve the lord. Glory to god that he hant forsook our world. O let this be a sign of abundance of rain. O when shall thy dominion spread from sea to sea and from the river to the ends of the earth. Gird thy sword upon thy thigh o most mighty with thy glory and thy majesty and in my majesty ride prosperously because of truth meekness and righteousness and thy right hand shall teach thee terrible things. Thine arrows are sharp in the heart of the kings enemies whereby the people fall under thee.[55] Come lord jesus come quickly.

February 15-1786. In the night about bedtime there come in a man and told me that the officer was now agoing to carry jonathan to new haven prison. Now it struck me with great surprise. I cryed out what has he done not knowing but what he had done some aufull crime and so was agoing to prison for it. But this man presently told me it was for his rate and jonathan said he believed he had paid it once and so refused to pay him for he thot there was a mistake. But o how it sent me to god. I cryed to him for mercy and that the rod might be blest and i thot how week natured my child was and it was cold and snow on the ground i did not know but he would perish in the cold prison in the night. Well i prayed till past midnight. My bodyly strength seemd to fail me

[*p. 362*]

i see i must go to bed. This consideration helpt me that my child was in the hands of a good god. The judg of all the earth will do right. Now i felt more calm. I layd me down and slept. The lord made me to dwell in safety. Next day i heard that jonathan payd the man and they setled the affair. Lord i thank thee that thou art better to us than wee could have expected. O give us hearts to live to thee. O keep keep me keep my children from presumtious ways of sining. O bles my seed while sun and moon shall endure. Bless the lord o my soul.

The tender mercies of the lord how precious they be
when i am grieved his bowels move and loudly plead for me.

March 10-1786. There come a man and told me that the separate meeting was almost broken up at the lower end of wallingsford christians divided. He said there was but one of the church that could bair to hear mr beech preach. Ah now i thot of them words in isaiah 5 chap 5 and 6 ver which use to run in my mind after mr beech moved our meetings from north haven to wallingsford which i ever thot was wrong and he has never seemd to prosper since. The words are these—i will tell you what i will do to my vinyard i will take away the hedge thereof and it shall be eaten up and break down the wall thereof and it shall be trodden down and i will lay it waste. It shall not be pruned nor digged but there shall come up briars & thorns. I will also command the clouds that they rain no rain upon it. O methinks this word is now fullfild upon that people. O it makes me pray with this word on my mind. Lift up thy feet unto the perpetual desolations even all that the enemy hath done wickedly in the sanctuary. Come lord jesus come quickly.

Lift up thy feet and march in haste aloud our ruin calls
see what a wide and fairfull waste is made within our waals.

Now one morning early this verce was on my mind and kept running threw my mind o how good it seemed—

Jesus shall see a numerous seed born here to uphold his glorious name
his crown shall flourish on his head while all his foes are cloathed with shame.

O now now i had raised hopes that the lord was agoing to carry on his work here in north haven. Lord thy church waits with longing eyes thus to be ownd and blest. O that i may not set thee a time but give patience to wait thy time. O how many precious souls do i see that seem to not know their right hand from their left in the things of god. Dear youths and children singing and dansing over hell as if nothing aild them. O jehovah reach down thy arm of mercy.

[*p. 363*]

March 26 1786. This day i was distrest with pain in my back. I can wright but little. I have often thot of that proverb namely if you take pleasure in that which is evil the pleasure vanishes and the evil remains. But if you take pains in that which is good the pains vanishes and the good remains. But o my poor body lies like a wait on my soul lord help me.

My restless soul shall nare give ore untill thy bowels move
ile not be driven from thy door till thou shall say i love.

March 29-1786. In the night i dreamd i was from home a great way off among strangers. I thot the day was aspending away the night like to be dark no moon to give light. Some seemd to weep with me. I knew not where i was to lodg that night but i thot i wil press forward on my journey. And so i awaked with these words on my mind—he that ordereth his conversation aright shall see the salvation of god.[56] True it is i am in a strangers land but i hope traveling towards my heavenly home threw many deficulties sins and sorrows traps and snares. But i find my lord is conqueror still threw all the wars that devils wage he gives me a resolution to follow on let what storms there will be in my face. And now methot i see what it was to have my conversation ordered aright. I must love the lord supremely give him my heart and then streams of obedience will naturally flow out of my heart into my life and conversation. O lord help me to honour thee in life and at death (the night of death coming).

Since ime a stranger here below let not my path be hide
but mark the road my feet should go & be my constant guide.

1786. This spring there was a uncommon croos come upon us. We was rewarded evil for good by a wicked man. Now i sunk under the tryal. I had a ancious mind. I laboured and prayed but could not give up the case to god and there satan got the advantage of me. O how was i tortured adays and nights. I could sleep but little in pease. It lasted about a fortnit till i was worn out almost but one morning before i was up this verce was usherd into my mind glory to god—

[p. 364]

Tho in some rugged paths you tread
and numerous foes your steps surround
tread the thorns down
pres threw the host the greatest fight the greatest crown.

But o i was helped in my mind tho no signs of any outward deliverance. Now i see i must press forward follow the lamb whithersoever he goes threw thick or thin smooth ways or rouf. I must tread down the thorns of worldly cares and snares i must press threw the host the world the flesh and the devil and all in the strength of him that is glorious in his apparel who come from edom with dyed garments from bozra traveling in the greatness of his strength and treading the wine press of his fathers wrath alone and has got threw and is gone home to his god and my god to his father and to my father (his vesture was dipt in blood).[57] Now i see altho the enemies of our souls may disturb and vex us jesus has traveled in the greatness of his strength from his cradle to his throne and all to conduct his elect every one of them threw all their defyculties here and to bring them safe and sound home to his and their heavenly kingdom and then and then

No strokes no frowns no croosses will there be
all wee shall rise to blest eternalle.
But lo the pain and smart will then be gone
and nothing but a skene of love comes on.

In May 1786 i was taken very ill with the rumatis in my hip. I had much of the cramp with it and was in great pain. I could help myself but little. They was fourst to turn me in my bed. But now i felt a still mind all the while & could speak well of gods name. One day them words was good to me—i follow after if i may apprehend for which also i am apprehended of christ jesus.[58] O i see that never any soul could apprehend christ savingly till he had apprehended them took them into union with himself and strength to follow after comes wholey from christ and as one says the heart of christ is as much set for serving his people now he is gloryfyed in heaven as he was when dying on the croos. O it was a comfortable consideration to me. O that the lord should ever have

[p. 365]

a thot of mercy towards me as to my soul or body. Now in about a week i grew better. O how sweet is ease after hard pain. But o how precious is gods word to my soul when he sends it and it is all for the sake of him that was at gathsemane[59] when he was sore amazed and very heavy all for the sake of him that was in the lonesome dark cold wilderness forty days and forty nights tempted of the devil. Ah he that made and owned all things must fast so long and be hungry and have his soul exceeding sorrow even unto death and at the mount of olives he prayed till he sweat blood and at calvary he dyed for such as me he says i that speak in righteousness mighty to save. O sinners pray do come to him now he is mighty to save you by and by he will be mighty to punish you forever.

Twas the same hand that spread the feast that sweetly fourst me in
else i had still refusd to taste and perisht in my sin.

Now methinks i must record one thing. On the lords day when i was writing this last on a chest in my chamber there come a filthy snake and crawld along close by my paper. I toock up a bit of wood that lay by me and killd it on the chest. Now i thot the devil is in a rage at me because i record these things. O my god do hasten his binding time. O cast him into hell and set a seal upon him i beg.

1784 [*sic*] june 24. Poor john allin went into the river to swim and several young men with him and he was drowned. Lord sanctify this sudden death to us all but especially to his relations and to our young people. O how often does death come as a thief in the night. O make this world sound in their hearts be you also ready.

A few days ago my husband come home and told me the best cow wee had was dead. She got cast in a ditch and dyed and left a young calf. But them words come to my mind—the cattle upon a thousand hills is mine.[60] But o how still my mind was. I see god had took nothing away but what was his own and i was content. All things here are transitory and fading but o that everlasting love of god in christ cant be lost. Glory to god for giving jesus christ.

[*p. 366*]

July 2-1786. Now every day i am sorely distrest with pain in my back. Many times when i kneel down to pray i cant but a few moments my pain will be so hard that i must go and lye on the bed. I have used much means but to no affect and o how little pyty have i from one that should be my best friend but the lord does right. O wreached one that i am to need so many rods. O it has been a great grief to me that i love my jesus no more but he helps me still to follow after. O lord apprehend me that i may apprehend thee. Make no tarrying o my god lord jesus come quickly.

Ah lord thou seest my frozen heart how little little love
i owe thee all scarce pay thee part drop softness from above.

July 9-1786. Was refresht by reading the book of esther. Now hammon at this day is trying to destroy mordicai but i do hope the time is near when mordicais shall wright as it liketh them. And o that sinners would say and do as esther did fast and pray. She says i will go in unto the king and if i perish i perish. O see what a kind welcom she had. O lord hold out the golden septre jesus make them touch that their souls may live. When i was abed and asleep my soul cryed to god for the downpowering of his spirit i thot i longed to see his power and glory come lord jesus come quickly.

July 16-178[]. I dayly see the vanity and emtyness of all worldly things and yet sometimes mear trifles will discompose my mind. O wreached one that i am because of this body of death. O my unbelieveing hard heart. This word has been pleasant to me in solomons song 1-16—behold thou art fair my beloved yea pleasant also. Our bed is green o that marriage bed of love between christ and the believer when vewed is always flourishing and in its greenness and newnes. O set me as a seal upon thine heart as a seal upon thine arm. I want that love that is strong as death. I want that love that causeth those that are asleep to speak—

Farewell vain world in whom is not my treasure
i have injoyd in thee but little pleasure.

[*p. 367*]

July 30-1786. I see a sinner run upon the bosses of gods buckler and sin as with a cart rope. My heart was payned within me. Now it held till next morning and then i powerd out my soul i trust into the bosom of the lord. I thot i got near to complain and plead and the lord toock off my burden. Glory to god my help springs from the croos and grave of christ jesus for he has received gifts for men even for the rebellious that he may dwell among them ah lovely saviour.

Rage earth and hell come life come death yet my song shall be
god was and is and will be good and mercifull to me
Blest above streams is jordans flood that toucheth canaans shore
ile sing thy praise in jordans streams in cannan evermore.

August 13-1786. The lord sees fit that the biggest part of my time i am in the field of battle. I have admired to see some christians seem as if they had no tryals. I have often thot of them that could not follow david to battle they was so faint and they were made to abide at the brook besor and when david returned who was a type of christ these faint ones went forth to meet him and he salluted them and made it a statute and an ordinance that as his part is that goeth down to the battle so his part shall be that tarryeth by the stuff they shall part alike (first samuel book 30 chap). Christ is every believers portion and them that dont seem to be in the battle yet they abide by the stuff. Jesus is surety not one true believer shall be lost and when christ returns he will salute them with a come you blessed of my father inherit the kingdom prepared for you before the foundation of the world.[61] O that everlasting love and when david returnd from battle he sent presents of the spoils that he had won unto his friends. So our jesus sends to his friends presents of the spoils he won when he got the victory over death and hell.

[*p. 368*]

He the true david israels king blest and beloved of god
to save us rebels dead in sin paid his own dearest blood
Jesus our preist forever lives to plead for us above
jesus our king forever gives the blessings of his love.

August 20-1786. I hant been to meeting upon the sabbath in about twoo years by reason of pain and weakness and but very seldom see a christian to converce with and i have not heard a prayr in about twoo years but i do hope the lord means it all for good to me. He often makes his word precious to my soul and opens it to me. I find great comfort in reading gods word o precious word i commonly read twise a day or more with beging a blessing. O how sweet it is when the seals are loosed and the book opened. Then my meditations are sweet how new it seems & as one says o christians love the word for by it you was converted. I have no cause to complain the lord is exceeding kind to me altho i have no meetings none to take me by the hand no nehemiah to go before me. Yet the lord lives and blessed be my rock he is the god of my salvation.

When nature sinks and spirits droop thy promises of grace
are pillers to support my hope & there i write my praise.

September 3-1786. My predominant sin got hold of my mind tho none knew it but myself. O methot i felt like a devil o i think there never was such a wicked heart as mine. Lord when shall i lay aside every wait and the sin that does so easyly beset me. But o my sins they keep me busy at the throne of grace and now the devil was busy to cast wicked thoughts into my mind. It lasted several days but o when there was no eye to pyty nor arm to help a kind god took off my burden incensibly and i was delivered.

If are i go astray he doth my soul reclaim
and guides me in his own right way for his most holy name.

[*p. 369*]

September 17-1786. This morning it being the lords day i went into the feild to pour out my heart before the lord for myself my husband and children and for the world that the lord would pour down his spirit. Yesterday i begged of the lord that i might have a good day tomorrow and it turnd in my mind that if i was out of hell it would be a good day but o the presence of christ makes my days more than good o then they are extrordinary good days. But it seems the lord is weaning me from frames and feelings and larning me to trust and live upon his naked word which is truth itself firmer than heaven and earth for

This earth that stands so fairm shall from her senter move
sooner then god unchangable shall take away his love
The sun can sooner seace to shine out of the sky
than christ can seace to favour them that do on him rely.

September 24-1786. Yesterday morning i felt something of that well of living water[62] that jesus said should be in believers springing up to eternal life. I now had my hope raised by reading that portion of schripture about joseph and his brethren but they did not know him in their anguish and sorrow. So the dissiples did not know christ when he was coming to them on the water but thot it was a spirit and cryed out. So saints now in their tryals dont know their spiritual joseph is dealing with them and all for their good but i hant time. And o this incouraged my heart that when the famine come on joseph opened all the store houses and sold corn and i am sure of this here is a spiritual famine all around new haven and now wont our spiritual joseph open his stores of grace and mercy and give us bread without money and without price. O i do hope in his word.

Thy glory to thy servants show make thy own work compleat
then shall our souls thy glory know and own thy work is great.

[*p. 370*]

September 31-1786. Tuisday last i went to one of the neighbours to hear mr frothingham of middletown preach. His text was these words—for as he thinketh in his heart so is he.[63] He shewed what thoughts would naturally flow from a carnal heart a worldly heart. He seemd to have some assistance. He said he had got a witnes that this sarmon would not be lost. Now when meeting was done there come a woman i believe a christian to me and took hold of my hand and cryed and said how little do i love christ and cant love him as i want to or to that purpose. I told her how reasonable it was for us to love christ when he had come from heaven down into this wreached world left his glory and dyed for poor sinners. How unreasonable it was not to love him no more and o how lovely he is in himself perfect purity and holiness. Jesus was lovely at this birth that made a multitude of the heavenly host praysing god and saying glory to god in the highest on earth peace on earth peace goodwill towards men. Jesus was lovely in his life perfection in our nature. He was lovely at his death father forgive them for they know not what they do. Jesus was lovely in his ascencion he lift up his hands and blest his disciples and while he was blessing he went home to his glory. And how lovely will he be when he comes to judgment then he will be admired by all them that believe in him. O methinks i have longed for that day. O wreched one that i am that i love god the father son and holy gost no more three persons and one god. I mourn that i can not mourn no more that i love no more. Lord come down touch the mountains and they shall smoak. But o how lovely was jesus at his resurrection. One would have thot he had had anuf with poor sinners. But o he rises with a heart as full of love as ever. He appears first to mary magdalene a town sinner and bids her go to his brethren and tell them that i ascend to

[*p. 371*]

my father and to your father to my god and to your god and he sent word to poor mourning peter this good news that he was risen.[64] And o how glorious and lovely was jesus at his ascension. As one says all the bells in heaven rung when jesus carryed home the trophies of his victory. David says god is gone up with a shout even our god with the sound of a trumpet and glory to his name that now he ever lives to make intercession.[65] But o i have often thot how lovely and tenderly jesus spake to mary that morning he arose. Woman why weepest thou whom seekest thou. Jesus said to her touch me not for i am not yet ascended to my father as if he had said mary you must not think to have heaven on this side heaven here you must betryed. Heaven is only reserved for heaven where i and mine shall soon be in glorious possession and have heaven in perfection. Go mary tell my breethren of it but o how lovely will christ be to all eternity. But o my hard heart how little do i love this lovely jesus when i ow him my all i scarce pay him part. Lord drop softness from above. Sure i am christ is the brightest dyomond in all the ring of glory. And it will be heaven anuf to see christ and be forever with christ. O to see that head that was once crowned with thorns in heaven. Wee shall be spectators and admirors of his glory and have time to thank him to his face for what he is in himself and for what he has done for us rebel worms. I have read of a myrter when he was going to the stake his wife and children was weeping by him. One asked him if he loved them. He said yes if i owned the world and could dispose of it i would give it all to live with them tho in a prison but when compared with jesus christ i love them not.

Rouse up dull heart awake and sing tis day you sleep
how can the suns aproach makes joy to spring tis clear how can you weep
Each pritty bird can pleasant be altho their portions small
o what unthankful wretch am i who droop and yet have all.

[*p. 372*]

October 21. There was a woman that lived near me that got offended at me for a mear trifle. She talked very proving to me. At last i got secretly angry tho i never spoke one misbecoming word to her but i went burdned with it several days. At last i thot i will set her a good example i will acknowledg it to her tho my nature was all up in arms against it. Well an opportunity presented itself and i told her i was sorry that i was vext at her talk i was wrong it was wicked i did not allow myself to have anything against any. Now she seemd very pleased but did not acknowledg anything. But o i found a blessing in it my burden was took off glory to god.

Can i expect to go to heaven on flowry beds of ease
while others fight to win the crown and sail threw bloody seas
If christ make known unto his own what theyll receive at death
theres not a saint but would faint and breath his dying breath.

November 1-1786. This morning before i was up this word was upon my mind—can thine heart endure or can thine hands be strong in the days that i shall deal with thee & (exekiel 22-14). This word was heavier and heavier threw the day in my mind. I thot a great tryal was near but o how did my soul cry to god that day for mercy. Now my child that lives with me went to new haven this same morning and at night he come home poorly and grew worse and worse. Wee got a docter to him he judged that he had the long fever and the bellyous fever and something of the knervis fever but o how my heart trembled at gods word

[*p. 373*]

that followed me several days. Now i know not but the lord ment to take away my child by death and i was sensible that he was in his sins no union to christ. But o now if ever i prayed without seacing on my knees in the barn in the cow house & up in my little chamber it was for his soul that he might be converted unighted to christ. I felt sometimes as if i could deliver up his body but his soul i traveld for. I prayed wherever i went. For some time my apetite to food was almost gone. I fasted and prayed. I dare not pray for his life only so that if it was for the glory of god that he might live that he might take care of his wife and his twoo poor little babes and be a support of his aged parents. But o how glad should i be to have him live if it was gods will. But o my cry was to be resind to gods will who does all things well who has a right to govern his own world as king. Now he grew weak fast his fever increased. On the 15th morning this word come to me before i was up—all thy children shall be taught of god and great shall be the peace of thy children.[66] This gave me a dear hope for his soul. Upon the 17th day the bellyous fever broke and a considerable quantity of frest blood sundry times come from him. The second docter that come to him said he would die he thot so the first of his coming to him. Now his wife told him what the doctor said then he cryed out lord have mercy on me. When he come to face death his

[*p. 374*]

soul was in distres and now he kept on praying for several days and nights sometimes so loud that wee could hear him out in the other room pleading for mercy and when people went in to see him he would hold on pleading and seemd to have no regard to creatures. One morning he told one of the watchers he had been toyling all night but his comforter did not come once or more he said he was afraid hell would be his portion. Once i was speaking to him of the willingness of christ to save great sinners for he had been telling what a great sinner he was. I told him of manasah i intreated him many times to get an intrest in a ritch christ. He said how happy should i be if i had called his father and i to his bed and asked us to forgive him. I told him i did freely forgive him but said what profit will that be to you if god dont forgive you. Now he grew so weak i dare not say but a few words to him at a time and as his weakness come on his sences failed tho they never wholly failed him only by turns. One day i heard him say meaning his sickness this is for sin sin has done all this. Mr beech come to pray with him he asked him if he had anything in perticuler he wanted him to pray for. He answered that i may have an intrest in christ. Now he was so ingaged for mercy that when one spoke to him he said you hender me.

[*p. 375*]

O how loth now was he now to be hindred from improving every moment of time but now the long fever and the nervis fever seemd to increase. His body weakned fast but this word was upon my mind i have heard thy prayrs and have seen thy tears and i will heal him i will heal him kept running threw my mind that day but wheather it was his soul or body or both i knew not that was to be heald but it gave me a supporting hope. O this word upon the 19th day of his sickness was good to me tho as to appearance he seemd to grow worse. But o how my heart went to god for him. But upon the 22 day of his sickness in the morning i think early this word siezed my mind and lay so heavy that day that i was knoct down by it viz concerning my sons and concerning the work of my hands command ye me—isaiah 45-11. Now i could hardly pray at all for i was afraid i had been trying to command god because i had been so earnest with him for mercy. But o how it grieved me to the heart that my joseph spake roughly to me and took me for a spy as if i had been trying to get mercy by my industry. Now i told a woman of this schripture that greived me so she said she believed it was from the devil but be it so or not i began to sarch the schriptures there. Jesus i see taught his disciples to pray and wee are to pray without ceaseing and wee read of great and wonderful things in the old testament and new god has done for his children in the way of prayr

[*p. 376*]

And now i was helpt up again and begun to plead as before but it was many times darted into my mind that if my child did dye i should sink under it but these verses helpt me to hope in my god. It was the 25th day of his sicknes.

How can i sink with such a prop as the eternal god
who beairs the earths huge pillers up and spreads the seas abroad
When we have been there ten thousand years ashining like the sun
wee have no less days to sing and praise then when wee first begun.

And the consideration of a promice that i believe god gave me above forty years ago when i was in travil with my first child helped me viz i will be with you in six troubles also in seven. Well thot i then god wont leave me if my calvin dyes. Now wee had sent away all his grave cloaths to the neighbours to be done up expecting soon to want them to lay him out and now the nervis and long fever was hard upon him and he grew weeker and weeker but o the anguish of my soul for his soul for he now seemd more and more stupyfied. But o how large did the purchase of christs blood look to me. I often told the lord that if he would but only touch the bier as he did the dead young mans the only son of his mother that he would be heald immediately. O how easy i believed it was with god to heal him soul and body he could do it.

[*p. 377*]

He toucht the hand of peters wives mother[67] [page folded] or a word but i begged hard for submission let the case be how it would. I thot god alone has a right to govern his own world. I poor dust dont know what is good for myself but upon the 27th day of his sickness this word was very precious to me—surely he hath boarn our sicknesses and carryed our sorrows.[68] These words kept running threw my mind with much sweetness. Now i thot christ has took the case of my child into his own hands has now undertook for him. O how i hoped for my childs soul and body. It was a lovely support to me but upon the 29th day of his sickness this word was powerfull upon my mind isaiah 59-16—and he saw that there was no man and wondred that there was no intercessor therefore his own arm brought salvation unto him and his righteousness it sustained him. O how comfortable now my mind was. I went to bed that night and slept comfortably and in the morning had a still mind altho my child seemd to decay fast. I see that if the lord spared him it would be free grace. His righteousness it sustained him o precious words to me. Now i hoped for his soul and body both that the lord would convert his soul and raise him again to health when docters nor none could help. The lords arm would bring salvation o free grace. But o i feared i should never render to the lord acording to the benifit received when he was pleasd to heal him. I thot of hezekiah.

[p. 378]

But now upon the 30th and the 31 day of his sickness it was thot the fever was at the height. The docter said he could be no sicker his case will now soon be decided and he ordered and four fowls was split open alive and one after another laid on his feet and he kept beging to go home and was very deef and breathed hard.[69] O to see him look up at me and say what shall i do. I told him he must come to christ he is ritch in mercy but he was much shattered in his sences. The docter would say when he went away i will come again tomorrow if i dont hear nothing from him. It seemd he expected to hear he was dead before tomorrow. O now i had no room to hope but in gods word of mercy that was a choice support to my soul. And o how kind was a good god to me all the time of his sickness for i for the most part slept comfortably. Also this word was on my mind very comforting to me—fear not daniel for from the first day that thou didst set thine heart to understand and to chasten thyself before thy god thy words were heard and i am come for thy words.[70] But o when everybody thot he would dye and his fever raged so and his weaknes seemd to gain upon him for many days sometimes when i looked upon him and begun to consult with flesh o what a trembling hart i had for a few moments. But then again the word of god was my stay and staf to support me and so i lived along till i think about the 50th day of his

[p. 379]

sickness he seemd to grow a little better tho slowly and seemd pleasd when he found out that he was at home and said why did not mother tell me of it before. And now in a few days he begun to sit up and seemd to be gaining but he was soon taken with terrible fits of the ague and a hard fever followed for near a fortnit more he now seemd weeker than ever but in the use of means his fits was broke and he grew better again. Now upon the 66th day of his sickness these words kept running in my mind—for wee are alike in christ jesus. I think these words come to me before i was up in the morning and o how they followed me that day. Now the night before i was consernd for and begged for my childs soul. Now i did not remember any such words in the byble and the next night as i was by the fire pondring and thinking of them this word flowed into my mind—for wee are his workmanship created in christ jesus.[71] Paul says wee he means the ephesians were in christ jesus as himself was. O this gave me a raisd hope for my poor childs soul. Glory to god o how kind is the lord to me. One word from a god of truth is anuf to make me believe forever. But o how many blessed words has he spoke to my sinfull sorrowfull heart about my childs salvation both soul and body altho i know not that he is yet adopted into christs family. O that everlasting love before the world was made his delights was with the sons of men. O what am i or what was my fathers house all sinners that god should be self moved to shew such undeserved mercy. Glory to god on high.

[*p. 380*]

And when my son begun to walk about he seemd affected with tears at gods goodness to him. I asked him what he thot god spared him for he said i dont know but it is to fill up my cup. And o how many mercies wee had in his sickness. Wee was all spared when some of our neighbours was sick with the same fever and some dyed but wee are still alive. My jonathan had it lightly but he is got well again and o how kind the lord made the neighbours to us. I think my calvin had about 70 watchers. O lord reward them with an intrest in thy love and favour. And he had i think 27 blisters drawd on his body in his sicknes glory to god that has spared him a little longer. O that wee may go and sin no more as we have done lest a worse thing come upon us. Lord sanctify it to us all. Thine own arm has brought salvation glory to him that has boarn our griefs & carryed our sorrows. He carryed them from his cradle to his croos and dyed for us he rose again ascended and ever lives to make intercession for poor sinners. Now these words ran for some time in my mind. I thot i could joyn in some measure with the heavenly host and say glory to god in the highest and on earth peace good will towards men.[72] O me thot jesus christ had good will towards my child and did not cut him off in his sins. Also this was sweet to me—god so loved the world that he gave his only begotten son that whosoever believes on him shall have eternal life.[73] Ah so loved the world so loved sinners as to give his life a ransom for them. O lord pray do prepare our hearts i beg i beg to live thy praise while life lasts.

[*p. 381*]

When thou with rebukes doth correct man for iniquity thou makest his beauty to fade like a moth. Now when calvin come to grow better he scarcely lookt like the same person he was before his sickness. It seemd awfull to look upon him his eyes sunk his flesh gone he seemd scarcely nothing but skin & bone. He seemd almost like lazarus raisd from the grave. Lord let it be for thy honour and glory thou didst give me to beg for it when i thot he was upon the verge of eternity. The lord killeth and maketh alive he bringeth down to the grave and bringeth up—samuel 1 book 2 chap 6 ver. Oh what shall I say what shall i write i am a poor guilty sinner. O lord do make us render our hearts lips and lives to thee. O make us live in some good measure answerable to the benifits received. O make us to come forth out of the furnis like gold puryfyed seven times more fit for thy use. O my god i thank thee for raising both my children from a sick bed to health again and above all for a comforting belief that is founded upon thy word for the salvation of both their souls glory to god in the highest and on earth peace good will towards men. O the love of god in christ jesus towards poor sinners. Ah sweet words—peace and good will towards men. Now i have great cause to bless the lord for this affliction and as to my hope for my jonathans soul i have writ it on page 27th and in calvins sickness it was confirmd to me again.

He frees the souls condemd to death and when his saints complain
it shant be said that praying breath was ever spent in vain. (Bless the lord o my soul).

[*p. 382*]

Calvin was in his 33th year of his age when he had this fit of sickness. I can say as david it is good for me that i have been afflicted.[74] O how many nights in the time of his sickness did i lye down in my bed like one beat out with the exercise of my mind all day and drop to sleep like a stone till morning—out of the eater come fourth meat & out of the strong came forth sweetness.[75]

Why should the wonders god has wrought
be lost in silence and forgot
No treasure so inrich the mind nor shall thy word be sold
for loads of silver well refind nor heaps of shining gold
Jesus our priest forever lives to plead for us above
jesus our king forever gives the blessings of his love.

Geneway 24-1787. This word was upon my mind with glory and sweetnes—when you shall see this your heart shall rejoyce and your bones shall flowrish like an hearb.[76] O now i was full of hopes that god was agoing to pour down his spirit that i should see and hear saints rejoysing in the lord and sinners crying out what shall i do to be saved. O come lord jesus come quickly.

March 12-1787. In the morning before i was out of bed this word took my mind—sharon shall be a fold of flocks.[77] Now i had raisd hope that this sharon was north haven. Lord pray gather thy flock here like the sheaves in to the floor come lord jesus.

March 14. Now wee met with many loses crosses and disapointments as to this world but this word run in my mind—seek first the kingdom of god and the righteousnes thereof and all these things shall be added unto you. Ah it is better to have

[*p. 383*]

gods promis than to have a house full a barn full or the largest estate in the cyty of new haven that may soon all be gone but gods truth and faithfullness cant never fail. Glory to his holy name.

Lord when i count thy mercies ore they strike me with surprise
not all the sands upon the shore in equal numbers rise.

March 16-1787. I was poor in body a pain in my back i went and lay down. After awhile i dropt to sleep and dreamd about the sufferings of them worthies in the 11th of the hebrews and wakt with it on my mind. But alas how unlike to them me thot i was for a tryal immediately come up on me from the wicked but how poorly did i bare it. O for that faith these heroes had. Lord learn me to suffer if calld to it. God be mercyfull to me a sinner thy stroakes are fewer than my crimes and lighter than my gilt. I beg for more patience.

Aprel 8-1787. I was in much pain in my back but had a heavy tryal on my mind and all by reason of my not being subjected to god. I thot i was like that ship that paul was in[78] where twoo seas met but yet i hopt i should not be run aground as that was. We read that the four part stuck fast and remained unmovable so my hope in god remaines but the hinder part of the ship was broken with the violence of the waves. O how i dishonoured god in my heart tho none knew it but god and myself. But i kept beging for help lord have mercy on me o for patience to wait thy time. Now i was worried day and night with the devil and sin—deep calleth unto deep at the noise of thy water spouts all thy waves & thy billows are gone over me. O my god my soul is cast down within me therefore will i remember thee from the land of jordan &.[79] O how many times has the lord helpt me when in depths and i do hope in his mercy.

[*p. 384*]

One night this word was some help to me—thou art more glorious and excellent then mountains of pray.[80] (Lord i have read thy blood was spilt to wash away the sinners gilt). Now when i had waded many nights and days in sorrow the lord helpt me and took off the sharpness of the tryal tho i was not quite delivered but it seems dying away glory to god.
Aprel 22-1787. O the sin i see and heard yesterday sunk me down o how little i felt of the spirit of christ [crossed out]. But the lord is kind to me i slept comfortably last night. This morning this word was early on my mind—kings of armies did flee apace and she that tarryed at home divided the spoil.[81] Ah now i had raisd hopes again that the lord was going to carry on his work and i do hope to divide the spoil. And i talked to the poor creture that sind so i asked him if he was not sorry. I told him i was sorry that there was so much sin but he did not own that he was sorry but this is all the fruit of my ill doings. O christians god tells you and i tell you by woofull experience marry in and for the lord. O god be mercyfull to me a sinner but o them words—kings of armies did flee apace. Now i do hope the lord is about to sanctify my sinfull soul. I hope the devils is agoing to be bound tho as kings of armies of sins lusts and corruptions. I hope babylons ruin is nigh. O jesus rain king here in thy own world in my heart and all hearts. But i am in such pain in my back i must stop.

The world and sin will ever vex will trouble & molest
but i will trust my soul with christ to bring to heavens rest.
And then i can sit down and rest me and dry me. Hannah Heaton

[*p. 385*]

May 28 1778 [*sic*]. This morning i had a sort of a fit. I was taken at once with an extreem pain in my bowels and sick at my stomok. I was outdoors and i lay down on the ground and sweat with the anguish. I found i grew weak fast i knew not but i was adying. At last i thot i will try to get into the house and with deficulty i did and lay down on the bed and my husband gave mee some camphire and i soon grew better. Now i had nothing special upon my mind i was overcome with extremity of pain. O sinners repent now in health. O what an unfit time it will be for repentance when the agonies of death are upon you.

July 2-1787. This day i was made to possess the sins of my youth. O i look on my sins with horror down to my cradle. O my sins they made me to weep before the lord on my knees. O my god be mercyfull to me a sinner. I can say as bunyan did o my sins my sins they kill me with care they daunt me with dread they consume me with sorrow they pine me with pain they eat me with grief they overcome me with bigness they press me with weight they overlay me with load too many in multitude of too deep a dye too wilfully consaited too often repeated twoo much followed and practised. I have not been anough sorrowfull for them heaping dayly sin upon sin. I have loved sin twoo much and thy ways too little. O my sins they press me down as a heavy burden. Lord heal my soul and hide my sins in the wounds of thy perced side. I do believe god many times sorely afflicts his children for past sins that

[*p. 386*]

he has forgiven and the lord does right. It is for their good to kill pride to keep them humble to make them live upon him their only help & refuge. Nathan told david the lord had put away his sin in the matter of uriah[82] but yet the child must die and the sword was never to depart from his house. God forgave their iniquities but he said i will punish them for their inventions. I lookt out at the door & see a company of chickens scratching and picking in the mud. How they pratled it took my mind to see how content they was with what god had provided for them. I see that all gods creatues answer the end of their creation better than poor sinfull man unthankfull for the mercies wee have and yet discontent because we hant greater mercies. This word come to my mind—what is man that thou art mindfull of him or the sons of men that thou visitest them.[83] O free grace in christ jesus. In the night my sins greatly troubled me especially my marrying contrary to gods word against the warnings of a faithfull minister of christ a few days after i first knew the lord. O it will be just if the sword never departs from my house. O heavenly father let moab be my wash pot[84] a means to purge our sin.

Each pritty bird can pleasant be altho their portions small
o what unthankfull wretch am i that droop and yet have all.

But o my wicked heart i am full of wounds bruses and putrifying sores and my wounds stink and are corrupt. O what can be more loathsome than a putryfyed corrupt

[*p. 387*]

stinking wound but sin is alas still [page folded]. Sin strikes at lovely puryty itself. Oh i do hope and believe it is out of a gracious respect and love to god that i so abhor myself for sin. I do believe god for jesus sake has forgiven me but methinks i cant never forgive myself—who can number all the stars or sands upon the shore—my sins my sins are multitude my soul thy sins are more. I dont remember that ever god dealt with josephs breethren[85] for their selling of him till they went down to egypt to buy corn which was above twenty years after they sold him and altho joseph forgave them and told them not to be greived nor angry with themselves for it was gods hand. Now it seems they could not forgive themselves for after jacob their father was dead they went and fell down before joseph and begged forgiveness again. O lord ever give me such a sweet christlike spirit as dwelt in joseph. He says to them fear not i will nourish you and your little ones and he comforted them and spake kindly to them. But o at this day it seems our spiritual joseph is speaking roughly to his brethren but i do hope the day is near when they shall be humbled anuf that our glorious joseph will reveal himself to them and open all his store houses as joseph in the famine. O lord bowe thy heavens and come down.

Come lord and never from us go earth is a tiresom place
how long shall wee thy children mourn the absence of thy face.
O come quickly i beg i beg.

[*p. 388*]

I am still sorely afflicted with a distressing pain in my back. The lord dont rong me but he helps me yet to think and speak well of his name altho he is shaking me over the grave. But o what a wicked worldly heart have i. God is dayly shewing me the vanity and emtyness of all things here below and yet meer trifles will discompose my mind. I think it is thirty years or more that wee have kept bees and this spring we lost them all. But o how ancious i felt tho i beged hard for submission and after a while i got still about it. And i think in june my husband took a swarm upon shares and upon the 5th of august my little grandchild of almost twoo years old went out to the bees and a great many stung him. The child cryed as hard as he could about an hour and a half with the pain and then went into convoltion fits. His father went near four miles for a docter and wee feard every moment would be its last but the docter come and the lord blest the means so that the child grew better and the next day run about the house. This was a undeserved mercy but o how did i beg for the childs life if it was gods will for it was i that had sind. But these sheep what have they done. O how plain did i see the cause. Now i thot that god answered me out of the whirlwind. O father i thank thee for the childs life. Pray make it an heir of thy heavenly kingdom. Lord bless the rod to me o make it like aarons to bud and blossom and bring forth the peacefull fruits of righteousness.

O when shall i from sinning cease & wait on pisgas hill
untill i see him face to face then shall my soul be still.

[*p. 389*]

September 14-1787. This day poor in body low in mind my faith seemd almost gone i went this night to bed sad distrest with pain so that i could not sleep. It seemd as if the lord took no notis of me and now that come into my mind concerning nathaniels being under the fig tree.[86] It is thot he was there praying and jesus lookt upon him and dear nathaniel soon after had a vew of his lovely face and the lord witnest to his sincirity so that he cryed out thou art the son of god thou art the king of israel. Now this incouraged me to pray and to press forward toward the mark of the prise christ jesus. O lord help me to live upon thy word when frames are denyed i beg i beg.

I live upon thy heavenly truth nor shall thy word be sold.
for loads of silver well refind nor heaps of shining gold.

September 21-1787. I hant been to meeting upon the sabbath in above 3 years. I am gods prisoner. I think often of joseph that was in prison or a servant thirteen years whose feet they hurt with fetters. He was laid in irons untill commandment come from the king then he was losed and free and as i have red made lord over egypt forty years. But believers shall be made eternally free. They shall be made kings and priests unto god when the commandment comes from the king of heaven to break this mas of corruption and sin and bring our souls out of prison and leave the iron fetters behind. Our king jesus is not only

[*p. 390*]

lord of egypt and goshan but of heaven and earth and hell. O will it not make up all when wee shall get with our jesus. Joseph bid his brethren tell his father of all the glory that they had seen him to have in egypt but o the glory that god has prepared in heaven for his chosen ones eye hath not seen nor ear heard neither hath it entred into the heart of man to conceive the things that god has laid up for them that love him.[87]

When i at length get grace and strength to strike them heavenly notes
ile praise him twoo as angels do with their sweet warbling throats.
Lord when shall wee like angels be to travel threw this air
and all thy host travil this coast and meet together there.

October 7-1787. This sabbath morning i prayed and sung a himn had some refreshment of soul then i talked to my children charged them to get an intrest in christ. I asked calvin what he thot god spared him for when panting for breath on a sick bed. I told him it was to try him a little longer. Said i god is coming again. I told him of the barran fig tree it may be when god comes to deal with you again he will say cut it down why cumbreth it the ground. I said to them all that hant an intrest in christ must go down to hell never to come out again. Now he and his wife seemd to have a hearing ear o heavenly father have mercy on them. I thank thee for a special hope for my twoo children that are my own.

[*p. 391*]

I also told calvin how i begged for his life when he lay sick if it was for the glory of god. And now i feard he grew worse and worse more and more careless and how fraid i was when he was sick that it would be so. Now he said nothing but seemd willing to hear. Lord i leave him with thee speak the word and the work shall be done. I have and do believe that thou art able to when shall it be according to my faith. Matthew 9-29 lord have mercy on him.

October 11-1787. This day my daughter in law rhoda brought home my great byble which i bought at new haven. My heart went out to god that i might improve it to the honour and glory of god. My husband asked me where will you keep it. I told him in my bosom for wee still live in our little house. But alas what shall i say this part of our zion lies i think with their souls cleaving to the dust. Saints are called the temples of god but o i think the money changers have got into these temples and what shall wee look for but a scourge of cords sore judgments if the lord dont power down his spirit. Lord when shall it once be. Dont let the blood of souls be found in the skirts of zion pray father of mercies come quickly.

Come lord and never from me go this worlds a wicked place
how long shall wee thy children mourn the absence of thy face.

O when shall our deborahs awake awake and utter a song to the everlasting i am.

[*p. 392*]

October 31-1787. I was weak and poor this morning but i wrapt up went out in to the field to be alone to pray. I had desires for myself and others. O that i may see the salvation of god with mine eyes before i lay my head in the grave. The week past my heart has been almost ready to break to see and hear so much sin committed. Gray hairs are here and there and they seem to know it not. I am dayly tempted to dispair of gods mercy to my soul and body but especially the support of our bodies that we are coming to want. But this blessed word has been upon my mind—seek first the kingdom of god and the righteousness thereof and all these things shall be added unto you. And i do believe that if god is kept uppermost in my heart a faithfull god will take care of me both soul and body. A kind god has been my supporter 67 years and shall i dispair now no no no with the lords help. And this word of late is good to me—trust in the lord at all times ye people pour out your hearts before him.[88] O here is incouragement to pray and to trust in the lord let what will come. Altho i by turns a long while have been tempted to fear temporal death the pain of dying but it is all for want of faith and trust in god that can make a dying bed feel soft as downy pillers are while i lean on his breast. O let me praise thee while i live & praise thee when i dye and praise when i rise again and to eternity.

[*p. 393*]

February 1788. O the enmity in one against me for reading and praying. I have had my spectacles took away and hid from me and o the cruel persecuting spirit in poor sinners. I wright these things to warn others to marry in and for the lord. Where there is no agreement in the things of god how can they walk together. These sweet words have been on my mind by turns for several days—unto you that fear my name shall the sun of righteousness arise with healing in his wings.[89] O tis supporting and strengthning to me and whither it is my soul or pained back that is to be heald or both i know not i must wait to see. I have a raisd hope in gods mercy. I dare not pray for ease only so that if it is for the honour and glory of god he would have mercy on me and heal me for wee dont know what is good for ourselves. The lord knows what is best for us ease or pain. Sometimes i think it is hardly worth while for god to heal my body that is so near the grave. I am now 67 years old. O father help me to live more in obedience to thee. O let my loins be girt and my lamp burning when calld from this earth.

I could renounce my all below if my creator bid
and run if i was cald to go and dye as moses did.

[*p. 394*]

February 26-1788. Confined and shut up with the wicked under much pain and weakness i seem as much alone as if there was not another creature on earth. As to the things of god i am as a specled bird among the birds i am an eye sore to them all. I cant go to no meetings i have no nehemiah to go before me none to take me by the hand a stranger in a strange land. Sometimes these things grieve me but glory to god that i find grace in his eyes that thou should take knowledge of me seeing i am a stranger ah on a double account by nature and by providence. A few nights ago i dreamd that i was talking to a person i told them this schripture belonged to me—return to thy rest o my soul for the lord hath dealt bountifully with thee.[90] O lovely words. Here i see that god threw christ was my souls resting place. I awaked and pondred on these words. Me thot the lord invited me into this rest. Here i find no rest as noahs dove for the sole of her foot.[91] Father put fourth thy hand pull me in to the ark jesus as noah did the dove. O draw me and i shall run after thee for thou hast i believe dealt bountifully with my soul. I sung the 15th himn in the book called the penitential cries called a song of praise for the gospel. It was sweet to my soul glory to god on high hosanna.

March 11-1788. Calvin sind and was disobedient to his parents but o how it grieved me to think what great things god had done for him in raising him from the

[*p. 395*]

brink of the grave and did not cut him off in his sins but restord him to health again. O to think how ill he requites the lord o lord touch his soul as thou did the lepers body and say i will be thou clean—mark 1-41. Lord if thou wilt thou can make him clean. O lord draw near to my distrest soul. Thou has i believe deliverd me out of egypttion bondage and carryed me threw the red sea when pharaoh and his host was at my heels (the devil and sin) and didst drownd my sins in the red sea of thy blood and has helpt me in many wars changes and distresses. And now i have just got threw the wilderness help me o lord god threw jordan. O let me go threw dry shod into the promist land where there is no sorrow nor crying.

O land me on the eternal shores my jesus to behold
those crowned heads that dazel bright more than the shining gold
Where cheribims & ceryphims streatch fourth their charming wings.
o there they talk of nothing elce but lovely glorious things
The wicked there from troubling seace the weary are at rest
there i shall see my jesus dear and lean upon his brest.

May 29-1788. This lords day morning i was scarcely awake this word was on my mind—sun stand thou still.[92] Now joshua was a type of christ and in his gods name he bid the sun stand still while he destroyed the lords and his enemies. Now these words seemd to be opened to

[p. 396]

me thus that the lord was carrying on the work of sanctifycation in my soul adestroying his and my enemies in my heart (i mean my sins and corruptions) altho the sun of righteousnes as it ware stands still. The most of my time now i hant them sweet frames and feelings that i had when i was young. Now the lord seems to be larning me to live upon his word of truth and faithfullness. Frames are sweet but fadeing and failing but the word of the lord endureth forever. Stoddard says saints have anuf to live upon when frames are gone. Lord let me find thy word & eat it.

Thy word like silver seven times tryed threw ages shall endure
and they that in thy truth confide shall find the promis sure.

Oh what a wonder it is that the lord can carry on his work and uphold the being of grace in a soul where there is so much sin and opposition as i find in mine.

Turn aside a sight to admire i the living wonder am
see a bush that burns with fire unconsumd amidst the flame
See a stone that hangs in air see a spark in oceans dwel
kept alive with death so near i am i am out of hell

I am full of guilt and shame my heart as black as hell i see
i the cheif of sinners am but jesus dyed for me
Meanest follower of the lamb his steps i at a distance see
i the chief of sinners am but jesus dyed for me.

[*p. 397*]

November 26-1787 [*sic*]. Glory to god that has not yet forsaken our world for wee hear that in one place down the country there has been lately four hundred hopefully converted and a few days ago i hapned to see a letter a woman sent here to her mother from the nine pardners. She writes how wonderfull the work of god is there souls flocking to christ. O lord when shall that day come that all nations shall bow to the mighty god of jacob. O lord put zion into travil for the birth of souls for thou has told us in thy word that as soon as zion traveled she brought fourth her children. O lord make bare thy holy arm in the eyes of all the nations and make all the ends of the earth see the salvation of our god—isaiah 42-10. O how i want to have the whole world bow at the feet of a lovely jesus that was dead and is alive and lives forevermore. And i am glad that he has the keys of hell and death glory to god.

December 16-1787. I received a great temporal mercy in answer to prayr. O how kind is god to me oh help me lord to improve this and all thy mercies to thy honour and glory. I awaked in the morning with this word on my mind—proverbs 31-9—many daughters have done vertuously but thou excellest them all.[93] Here i see altho i am a poor sinner and carry a wicked heart dayly which costs me a sea of sorrow and repentance but yet the grace of god in my heart excelleth all moral vertues. Christ in you the hope of glory excelleth all all all.

Twas the same love that spread the feast that sweetly fourst me in
else i had still refusd to taste and perisht in my sin.

[*p. 398*]

Jenuary 20-1788. We read that when israel took his journey to go to joseph when he come to beersheba[94] he offered sacrifices to god for it seems that joseph was alive. And it is now above a year since the lord had mercy upon my child and raised him from a sick bed and restored him to health again and did not cut him off in his sins. But i do believe and hope for the conversion of both my children. O make me like israel to offer the sacrifice of prayr and praise to thee o let me see thy salvation in their being turned to god. Lord sanctify me o fit me for thy holy will i beg. These words of late have run in my mind—thou are glorious in holiness fairfull in praises doing wonders.[95] Lord let me see thy salvation. I am still poor in body a distressing pain in my back. I decay fast i shall soon go the way of all the earth but o i fear the passage. But i shall not if perfect love casts out this natural fear and the sting of death is taken away. My god be mercyfull to me a sinner.

Claspt in my heavenly fathers arms i shall resine my breath
and loose my life among the charms of so divine a death
Thy loves a sea without a shore spreads life and joy abroad
o tis a heaven worth dying for to see a smiling god
Shew me thy face and ile away from all inferior things
speak lord and here ile quit my clay and streach my airry wings.

[*p. 399*]

Now i think i must tho in much pain write a few hints of calvins behaviour towards his aged father.

In 1788 in aprel my husband sold a yoke of his own oxen and it was noised about that he was agoing to sell a peice of wood land but it was false for he never once attemted to sell it. But calvin and his wife and some of their friends consulted privately and went and postted my husband. We read that jezebell suborned false witnesses against naboth to stone him and all to get his vinyard.[96] And now the next day after it was done we heard of it but o how it struck our minds. My husband was very angry and o i felt all nature rallyed for several days and nights. Our appetite to food was almost gone wee could sleep but little. I myself was brought very weak and low. O to see children rise up against parents and put them to shame and sorrow. But ah the sin against god was above all. O how did i bare him with sorrow and brought him up tenderly took much pains in his education o my prayrs and tears for his soul and body and now wee are rewarded evil for good. But o it drove me to god. I prayed day and night for a still resind will and that we might all be turnd to god. I begged of god not to give us up in to the hands of men that lookt dreadfull. I thot of david i wanted to fall in to the hands of my god for very great are his mercies but alas god now shewed me what a heart i had. I see if the lord should let my corruptions loose and i be holden with the cords of my sins—proverbs 5-22—

[*p. 400*]

i should have a real hell in my soul. O i felt a heart like the devil tho none knew it but god and myself but o i kept crying to the lord day and night tho sometimes the mouth of my soul was stopt up. At last this word seemd to run in my mind—he calleth his own sheep by name and leadeth them out.[97] Now this word was a support to me. I hoped the lord would bring us out of this snare of the wicked. Ah had they done it for any breach of law i had held my peace. And now the 146 psalm was sweet to me. Had i time and strength i would write how every verse was opened to me. O it was a word in season to my soul. The lord preserveth the strangers that was i but the way of the wicked he turneth upside down. The lord shall reign forever even thy god o zion unto all generations praise ye the lord glory to his name forever. Now i see this word fulfilled—gad a troop shall overcome him but he shall overcome at the last.[98] Glory to god on high that hath regarded the low estate of his handmaid. When the waves arise thou stillest them. This is the lords doings and it is marvellous in my eyes. Praise the lord everything that hath breath praise him heaven and earth praise him saints and angels. O that i was with you in heaven falling down before the throne and the lamb. O jesus the most sparkling dyomond in all the ring of glory. Bless the lord o my soul.

[*p. 401*]

And now i think it was about a week after my husband was advertized one morning there come into our house three men of calvins freinds and prest my husband to give calvin part of his land and they said if he would they would go to the selectmen and take off the advertisement and set up a paper of recommendation (and get fifty siners) and that there was mistakes reported. Now my husband was so full of grief that he yielded to apoint a day the next week and thot he would do it and then they went away but before the appointed day come we see the snare (gods hand was in it) and word was sent to the men not to come for he would not do it. He said must he give away his land that he workt so hard to buy to pay his enemies for abusing him. O the mercy of god that did not suffer them to have their wills upon us. For several days when wee first heard the news my husband would groan and cry like a child. He slept but little his appetite to food almost gone he would sweat with anguish his body was much weekned worldly sorrow works death. Lord make us truly to mourn for sin the cause of all our sorrow. I believe it is but seldom for tryal. O give us both that spiritual sorrow that worketh for life and peace for jesus sake. O lord bless the rod or it will do no good.

[p. 402]

But o how kind the lord has been to me. He gives me a quiet mind his word is precious to me. This verse flows and runs in my mind—happy the man whose hopes rely on israels god—psalms 146.

He made the sky and earth and seas with all their train
and none shall find his promis vain.

And now when these bonds had lain upon us almost three months one of the squires of the town went to the towns men and a meting was warned. And now when the day come my husband seemd sunk down not knowing what the enemies might bring in against him. But i felt a still mind yet and told my husband not to be discouraged for i did believe god would deliver us. Now he went to the meeting.

June 27-1788. There was a considerable number of the princypel men of the town. The twoo squires was by and not one enemy dare shew his head. Now they examined the matter as much as they pleased and had proof that it was all deceit and lies and so skyre peirpoint[99] writ a clearing. I was poor in body went to bed before my husband come home that night. I did not want to ask him about it. I was easy slept comfortably. Next morning he told me

[p. 403]

they had cleard him. Then i sung the 145 psalm intytuled mercy to sufferers or god hearing prayr. Glory to god he has turned the way of the wicked upside down. God has not suffered this child of belial to be hid that contrived so privately but by an unexpected providence it is spread abroad. Wee know who they be five men and one woman. Lord have mercy on them. O make them know themselves and know thee. And o that my poor husband now he is at liberty from these temporal bonds lord pray do deliver him from the bonds of sin and satan. O make him to know christ jesus i beg i beg. Surely god is a god hearing prayer. The lord liveth & blessed be my rock and let the god of my salvation be exalted. And o my dear calvin what shall i say to thee. Had not jezebel with her friends consulted to stone naboth in order to get his vinyard i believe thou would never have done it but i hope yet in the mercy of god that is boundless. I do hope for thy precious soul tho thou hast done as bad things as thou couldst. Yet there is none twoo bad for christ. Magdalene full of devils nor manasseh a wiszerd was not twoo bad for christ. Your tender mothers prayrs are going up to god for you day and night. O methinks while i wright i long you should come to christ then you would know that he is a great forgiver. O my dear child is this the fruit of your sickness winter before last. Ah i want without fear that you would be worse then

[*p. 404*]

ever if the lord heald you but god can i know bring great good out of this seeming evil. He is a god glorious in holiness fairfull in praises doing wonders. I do hope yet in thy mercy for poor calvin. Lord grant that out of the eater there may come fourth meat and out of the strong sweetness. He that is our god is the god of salvation and unto god the lord belong the issues from death and thou alone dost wondrous things—psalms 68-20. Holy and reverend is thy name. O lord bow thy heavens and come down touch the mountains and they shall smoak. Lord grant that i may see thy salvation. O let my heart rejoice in thee and made my bones to flowrish like an herb—isaiah 66-14. And now when my husband and i had lived in our little hut about four years but a few rods from the new one it was very uncomfortable partly under ground and very damp and there was twoo rooms done in the great new house just for calvin and his family to live in and altho a joyner we had payd yet no more might be done to the great house on purpose we suppose to keep us from living in it. And now our son jonathan and his wife pytying us in our suffering condition often envited us to come and live with them. Now my husband was amind to go but i declind it. I told him i wanted to know the mind & will of god before i went.

[*p. 405*]

My daughter rhoda told me that she see it clear from these words—2 of kings 7-4—if wee sit still here wee die also. Now therefore come let us fall unto the host of the syrians. If they save us alive we shall live. If they kill us we shall but die. Now this seemd something to me but i wanted to see for myself. And now for some time i kept praying to god to know his will. I told my husband that i had done so many rong things i dare not go till i see it to be gods will unless he forst me to go. And now after a while as i was up in my little chamber as i kneeled on the floor to pray these words are i was aware dropt into my mind with a small still voice—if they persecute you in this cyty flee ye into another.[100] And now i soon told my husband that i was willing to go. Also these words was running threw my mind—genesis 46-1—and israel took his journey with all that he had and came to bersheba and offered sacrifices unto the god of his father isaac. Now i hoped the lord would go with me that i may offer the sacrifice of praise to god. Now i prayed lord if thy presence go not with me carry me not hence. And now upon the 5th of july 1788 we moved to our son jonathans a large comfortable house not half a mile from where wee had lived 39 years. Now when calvin see the team was come to carry away our goods i believe his con-cience smote him for he asked his father not to go. But it was now twoo late. Ah poor calvin. When he was about

[*p. 406*]

twoo or three and twenty years old his father bought and gave him forty acres of good land at waterbury but against our minds he soon sold it and spent it by trading. A little while before calvin spent his estate i dreamd that a fine dove come into our house and flew round. Its wings seemd flowered with yallow gold. I cryed out so loud in my sleep for calvin to cetch it that he heard me up in the chamber in his bed and cald to me. I dreamd he hastned to cetch the dove but it flew threw the roof of the house and was gone so i awakt. Now i soon see this dream fulfilled. The money he sold his farm for was this dove but it soon flew away and was gone. But ah he is my dear child still. O father pyty his precious soul. O let me and mine be thine tho we have had but little comfort together here. O let us live together in heaven with thee at thy right hand never to part. And now glory to god for his mercy to us wee live comfortably. Jonathan and his wife is exceeding kind to us. Lord reward them o bring them into nearness to thyself. They have no child lord reward them as thou did the shunamite woman in elijahs day.[101] But o my poor calvin how can i leave thee altho thy delilah[102] with her friends has inticed thee

[*p. 407*]

and led thee astray to thy great damage to both soul and body when thy father had layd out sixty pounds wee believe or more to build us a new house on his own land. It must not be finisht for fear that i believe wee should live in it with you. And besides this how have you robbed us of our money and the produce of the earth only just what wee eat and drinkt and wore. But for all this o my dear child i hope yet in the lord for you if you will come to christ. Gods mercy is such that he can overlook the greatest provocations and take delight to make you happy. But if you will not repent and come to christ your portion will be

Eternal plagues and heavy chains tormenting racks and fiery coals
and darts to inflict immortal pains dyed in the blood of damned souls
There guilty gosts of adams race shriek out and howl beneath his rod
once they could scorn a saviours grace but theave insenst a dreadfull god.

O lord pyty them that dont pyty their own soules. O thou that wept over jerusalem and at gethsemane thy soul was exceeding sorrowfull even unto death. O thou that sweat as if ware great drops of blood and dyed for poor sinners pray have mercy mercy mercy i believe thou art able.

[*p. 408*]

August 28-1788. Mr wildman[103] of danbury come and preacht in this house. His text was isaiah 45-17—but israel shall be saved in the lord with an everlasting salvation. He shewed in many perticulers the sartainty of believers eternal salvation in the lord jesus christ. I gave him a little money lord bless him o let this young man grow in grace and in the knowledg of jesus christ our lord. O make him wise to win souls to thee lord pyty thy poor world. Now one day as i was fearing trouble that was likely might come upon me as i was going to pray these words dropt into my mind—mercy shall compass him about.[104] O it was supporting and comforting to me. I hoped in gods mercy and was comforted. This morning my heart went out to god i had freedom to tell him my wants. O sanctify me purge out sin pyty my poor husband and children. O have mercy upon this frozen world. Pray bind satan destroy antecrist by the spirit of thy mouth and the brightnes of thy coming. O take to thyself thy great power and rein. Make thy dominion spread from sea to sea and from the river to the ends of the earth. Lord jesus come quickly i beg. Now there was a report that my calvin was going to trouble us again but i followed god by prayr for mercy and up on the 30th of october he come with

[*p. 409*]

a officer and tacht his fathers estate or his body for a thousand pound called a book det for work that he had done after he was of age. Now my jonathans wife lookt every hour when she should send for the weemen but o this aufull sight for a child to tach the land and body of his aged father. Sat she and i atrembling her pain come on she sent out the four part of the night and was extream bad supposed accasioned by the fright. Her first docter thot her child was dead and was ready to think she must dye twoo but they sent for another docter and they together delivered her the next day about twelve oclock with a living child like to do well. But now they had sent for a third docter supposising he must take away the child by peaces but just before he come to the dooor she was delivered. Glory to god he that is our god is the god of salvation and unto god the lord belong the issues from death—psalms 68-20. But to return. Now my husband never said a word to calvin but just before the court set went to coronal hall of wallingsford for advice and got him to come down here to my son jonathans and he advised what to do and just at night he sat away to go home but he hapned to

[*p.* 410]

see calvin and come back and brought calvin to his father and laboured much with him to acknowledg his sin. He also pleaded with his father to receive him again if he acknowledged. But o how was my heart melted. I went to calvin and took hold of his hand and cryed over him. I said to him you are my child still o that god would give you repentance but poor child he was not free to acknowledg that night and so he went away wee suppose to consult with his advisers. For the coronal had told him that he would sertainly be beat in the law and that the whole court would be against him and he would be put into jayl. And now it being late in the evening the colonal concluded to stay all night but in the morning calvin come again and acknowledged to father and mother as much as his father desired and begged our forgiveness. And now his fathers spirit fell he said he could not help but forgive him. And for time to come he promised obedience to his father and he received him again as a child and calvin soon come again and brought the original writ and gave it to his father. Lord thou dost work wondrously. O lord god of abraham of isaac and of israel our fathers keep this forever in the imagination of the thoughts of the heart of thy people and prepare

[*p.* 411]

their heart unto thee—1 chronicles 29-18. And now this morning the dear coronal sat away to go home but o how pleasd did he seem to be that he was an instrument to make peace between father and son. But o how unexpected this was for the coronal had drawn up a large acount for my husband against calvin to be carryed into court and he himself was going to undertake the affair. But gods mercy was spread over us and compast us about glory to his name whose hand was in it all. O lord bless coronal hall with the choisest of heavens blessings o make him shine in heaven like the stars forever & ever. O methinks i can bair witnes for god that he is a prayer hearing god for many times did i go to god and begged that he would come and turn the heart of a father to the child and the heart of the child to his father and the lord has done it in some good measure. O my god turn our hearts to thyself truly for jesus sake. And now about a fortnit before this tryal come one day as i stood by my bed this word come to me with power—to us a son is born unto us a child is given and the government shall be upon his shoulder—isaiah 9-6. O methot was this child jesus born for me did the father give his son for me and the government of all things is on his shoulder. It was so when he dyed on the croos for he raisd himself from the dead and the government of all things is on his shoulder now

[*p. 412*]

and will be untill the day of judgment and then the kingdom will be delivered up to the father and god will be all and in all glory to god for this sweet scripture it was supporting to my soul. Now i hoped i should win threw all since the government was on the shoulder of my jesus and also a few days before the forenamed trouble come upon us these words in the first book of samuel 14 chap and 15 ver—and there was trembling in the host—i say these words run in my mind. Now again i had a comforting hope from this consideration that as the lord saved israel from the hand of the philistines by jonathan and his armour barer[105] when they made the host to tremble so i did hope the lord would save us from our philistine enemies tho i knew not by what means. But if it had not been the lord who was on our side when men rose up against us then they had swallowed us up quick when their wrath was kindled against us. Then the waters had overwhelmed us the stream had gone over our soul. Blessed be the lord who hath not given us as a pray to their teeth. Our soul is escaped as a bird out of the snare of the fowlers the snare is broken and wee are escaped. Our help is in the name of the lord who made heaven and earth—psalms 124. O now methot i see gods hand in every step that was taken for our deliverance out of this trouble.

[*p. 413*]

Lord when i count thy mercies ore they strike me with surprise
not all the sand upon the shore: in equal numbers rise

O let me by thy croos abide thee only thee resolved to know
the lamb for sinners crucified a world to save from endless woe
Who jesus sufferings share: my fellow prisoners now
ye soon the crown shall wair on your triumphant brow
Rejoyce in hope rejoyce with me
we shall from all our sins be free
Bound down with twice ten thousand ties
yet let me hear thy call my soul in confidence shall rise
Shall rise and brake threw all
What did thy only son endure before i drew my breath
what pain what labour to secure my soul from endless death
Beneath my load he faints and dies
i fild his soul with pangs unknown
I causd those mortal groans and cries
i killd the fathers only son
O that i might as a little child may follow thee and never rest
till sweetly thou has pourd thy mild and lovely mind into my brest
Here ille set my ebenezer hither by thy grace ime come
and i trust by thy good fafour shortly to arrive at home
Him eye to eye wee there shall see our face like his shall shine
o what a glorious company when saints and angels joyn

[*p. 414*]

O what a joyfull meting there in robes of white arrayd
palms in our hands we all shall bair and crowns upon our heads
(glory to god on high)
Ye hearts of stone come melt to see
this he endureth for you and me
Christ sufferd all our sins forgiven
and on his blood we swim to heaven.

December 26-1788. The only child of jeames tood was scalded and dyed in a few hours. Lord sanctify it to the poor mourning parents. This child was about three years old. Lord make this town to hear thy voice in thy providences o pour down thy spirit. A few days ago i see a man do a wicked deed he being under gods rod. O how it grieved me. These words come into my mind—why should you be stricken any more ye will revolt more and more.[106] Now i spoke of it and said is this a time to sin. My heart aked to see him sin against god. I went to the barn to pray and while i was going threw the cow yard something are i was aware come behind me and kickt me down on my face. In a moment i got up as soon as i could and lookt behind me and i see it was the old mare that kickt me. It was a wonder that i was not killed in the spot but the lord preserved me. I was hurt but a little blessed be god for his mercy and kindness to me the chief of sinners. I do believe the devil was inraged at me for going there to pray. Lord help me now to live to thee in my last moments. We have heard of three in about the cumpass of a year that have murdered themselves one at wallingsford and twoo at new haven. Lord have mercy on thy world oh that thou wouldst the heavens rent in majesty. Come down stretch out thine arm omnipotent and seize us for thine own.

[*p. 415*]

One or twoo days much disturbed in my mind because of croos providences. O what a body of indwelling sin i felt o wretched one that i am i kept pleading with god for a resind will. I felt a proud heart right in opposition to god but these words was powerfull on my mind—altho he slay me yet will i trust in him.[107] O lord give strength to my resolutions. I am helpt in some measure but not delivered o my god my god dont forsake me. I hope in thy mercy o jesus draw near to my soul i beg i beg.

Conquerror of hell and earth and sin
still with thy rebel strive
Enter my soul and work within
and kill and make alive
More of thy life and more i have
as the old adam dies
Bury me saviour in thy grave
that i may with thee arise.

Oh wicked heart will thou never be still till i get to heaven that in deuteronomy 33-26 to the end—there is none like unto the god of jeshrun who rideth upon the heavens in thy help and in his excellency on the sky. The eternal god is thy refuge and underneath are the everlasting arms and he shall thrust out the enemy from before thee and shall say destroy them &. I say these four verses was sweet to me i see glory in them they kept running threw my mind several days. O there is none like to my god o does he ride on the heavens to my help when i need it and in his excellency on the sky o lovely puryty. I do hope this eternal god is my refuge and that i find and feel the everlasting arms underneath to support and maintain the seed of grace

[*p. 416*]

which his own right hand has planted in my heart in spite of all oppositions from world flesh and devil. But o he has promised to thrust out the enemy and say destroy them and methinks the 28 verce points to the glorious day israel shall dwell in safety &. Happy art thou o israel who is like unto thee a people saved by the lord but i hant time. O methink these verces are all full of glory & sweetnes glory to god on high bless god o my soul. Oh here i have a weak poor pained body a sinful backsliding heart to carry strong oppositions from without and within. I have been tryed tryed much of late about going over jourdan death. But i do find christ is conquerer still threw all the wars that devils wage. O my god take the sting of death away which is sin.

O let perfect love cast out fear and then shall i from all below
insensibly remove my soul its change shall scarcely know
Made perfect—first in love—and then with ease
my soul threw death shall glide
into its paradice and thence on wings of angels
ride triumphant threw the skies.

May 24-1789. Now i had not been to meeting upon the sabbath in three year. Now the breethren have sat up a saparate meeting about twoo miles from us for the church at wallingsford is broken up. Now this day i sat away to go to meeting. The beast i rode on was very apt to stumble. I was afraid but i had not got but a little way these words come to me—he that keepeth israel doth neither slumber nor sleep.[108]

[*p. 417*]

Now i said to myself i shall be safe and i was for the beast went and come very well. The meeting seemd pleasant i do hope something of gods spirit was moving. I had a pain in my back all day and after i go home i was taken vomiting was poor all night and next day. Lord fit me for death for it is near. O make me live actually ready with my loins girt and lamp burning. Why should i fear deaths grim looks since christ for me did die.

June 7-1789. There was one that threatned me positively that he would take away something of my worldly intrest and i thot he did but in a few days one come and told me that it was not so. These words come to my mind—he suffered no man to do them rong.[109] Then i thanked the lord for his care and mercy. But soon after another heavy tryal come on. But it drove me to god i went on pleading with god for submission let it turn how it would and in twoo or three days the lord shewed mercy. It seems to be the lords will that i must get all by prayr. Glory to his name he is a prayr hearing god.

July 5-1789. This day wee have lived with our son jonathan a year in the best room in his house. They are very kind to us blessed be god for his mercy to vile sinners. The lord has lent them a lovely babe. It is now 8 month old o that like hannah wee might give it to the lord and now [crossed out] ever since [crossed out] obedience to his father. Bless the lord o my soul.

[*p. 418*]

July 25-1789. This day john ward was drownded in the river a young man about 16 years old. My son jonathan had hired him to live with him this summer. He went to new haven with the team and coming back fell off of the bridge into the river by what means wee know not and was drownded. The next morning which was the sabbath a number of men went and found him and they brought his corps to our door in a horse cart an awfull sight. They carryed him home to wallingsford to be buried. He had lived with my son about six weeks. He was very prophane and wicked it seems the lord cut him off in the midst of his sins. Now there was another young man that lived here this sudden death seemd to take great hold of his mind. Now he seemd willing i should talk to him about his soul but before he could not bare it. O lord grant he may never find rest till he finds it in christ jesus. Pray bless it to all his near relatives and to us. O make our dear youths take warning by this aufull sudden death. And now my son jonathan was left alone as to his farming work and where to get a boy he knew not but in about eight days after this lad was drowned there come a boy to us that ran away from new york. He says he is twelve years old. His father and mother is dead and he had none to

[*p. 419*]

take any care of him but a father in law a very drinking man. He use to beat him when in his cups and he suffered for food. He seems to be a boy of truth a pritty lively child very obedient and pleasant wee all love him.

His wicked father in law never sent him to school altho schools anuf in york. When he come here he did not know all his letters but now he larns and loves his book. I have many times admired the providence and mercy of the lord in this thing. Lord bless the child o make him an heir of thy heavenly kingdom pray make us do by him as we can answer it before thee with comfort i beg.

August 22-1789 a few mornings ago i awakt was poor. I thot what if i should be past labour and could not help myself what would become of me. But it turnd in my mind that god would still be the same to take care of me. I reacht my hand and tooke my byble that lay near my bed. The first words i cast my eyes upon was in isaiah 46-4—even to your old age i am he and even to hoar hairs will i carry you. I have made and i will bair even will carry and i will deliver you. O how sutable and good were these words i dont remember i ever saw or heard them before. O will the lord god of heaven and earth the sea and dry land and all that is there in who is purity itself bair and carry me a poor sinner and deliver me glory to his holy name.

1790–1793

[*p. 420*]

A few nights ago when i was in bed i was mourning that i loved christ no more. These words come to my mind—the lamb is worthy that was slain to receive power and riches and wisdom and strength and honour and glory and blessing.[1] O methot he was worthy indeed. O i do hope there is a time acoming that i shall love my jesus and be swallowed up in him to eternity. Ah lovely jesus o give me patience subject me to thyself for thy honour sake. Now some time after i felt my sweet peace withdrawn them words run in my mind—you have slain the prince of peace. Now my heart was grievd to think that my sins had slain my jesus not only on the croos but now by my sins and my unbelief. But these words come to my mind—i am not sent but to the lost sheep of the house of israel.[2] O me thot jesus come to save just such as i am this was supporting. Now i dreamd i was traveling on the road with a great multitude and i heard a voice say to me pray pray pray. I lookt up to see who or what it was that spoke to me and i thot i see a little creature all in white and it flew upward out of sight and

[*p. 421*]

was gone but i still heard the voice pray pray pray. Now this was opened to me thus let the world do as they will do thou hold on praying. That come to my mind—let these go their way follow thou me.[3] Now this quickned me to prayr for some time. O i do wonder and admire sometimes that ever the lord should have a thot of mercy to such a creature as i am so tainted with sin and leave others. O he loves because he will and if he will love a creature who can hinder him. Glory to god on high peace on earth and good will towards men.

Aprel 8-1790. This winter past i have been poor and low a distressing pain in my back and now the pains of death go hold on me with all its terrors by turns for several months. I found i could not put my trust in god about my dying. At last this word come sweetly—if i go and prepare a place for you i will come again and receive you to myself that where i am there ye may be also.[4] O this was supporting i thot never man spoke like this man. What king or prince on earth did ever tell any of his poor

[*p. 422*]

subjects that he would prepare a place for them and come again and receive them to himself that where he was there they should be also. But o sweet saviour never was words like thine. I am comforted glory to god on high peace on earth and good will towards men bless the lord o my soul.

1790 aprel 31. This day my husband was taken sick with the plurasy and he dyed on the fifth day after he was taken which was may 4. He seemd loth to talk about his soul. Once i asked him if he was afraid to die he said yes. I asked him if he had a heart to beg for mercy and i thot he said not much. He told jonathan that there was but twoo of them to settle things and not to differ about this poor good for nothing world. Mr beech come and prayed with him in his sickness. I many times told him the necessity of having a christ or there can be no salvation but he is gone i must leave him the judge of all the earth will do right. O lord be the widows husband and the orphans father. His age was 70 years we lived together 47 years and a half.

[*p. 423*]

These words has been with comfort in my mind—continue in well doing seek for glory immortality and eternal life.[5] Bless the lord o my soul for quickning grace. The lord is kind to me since the death of my husband he gives supporting mercy. And my children i live with are exceeding kind to me and it is all from the lord glory to his name. Pray father reward them all with an intrest in thy love. A few mornings ago as soon as i awakt these words come to me with a small still voice—the labour of the olive shall fail.[6] Now these words when i began to consider of them i was fild with concern. I feard i was going to be past all labour and should be a burden to my children and i find that i decay fast. O lord grant that the inner man may be renewed day by day for jesus sake. Dont forsake me now i am old and gray headed and my strength faileth. Be thou the strength of my heart and my portion forever. Blessed be god that has said i will never leave you nor forsake you i have loved you with an everlasting love. Bless the lord o my soul help me to praise thee.

[*p. 424*]

June 1-1790. I was refresht by hearing that the work of god was begun at westfeild some hopefully converted and many crying out—what shall i do to be saved. I cryed lord make thy dominion spread from sea to sea and from the river to the ends of the earth come lord jesus come quickly.

1790. I have lately heard from long iseland that the work of god is begun in many places on the east end. O lord grant that the showr may reach poor north haven that seems to sleep in security. Here the righteous are sad the wicked mad while thou withholdst thy spirit. O god spread out thy hands of mercy here like a man that spreads out his hands to swim to pull in souls to thyself i beg. Now i was poor and weak distrest with pain in my back these words come and run in my mind—take my yoke upon you learn of me. O methot i felt more willing to take up my croos. Lord help me to live and walk in thy spirit that i may not fulfill the lusts of the flesh i beg i beg.

1790. Now above a year i have had sore eyes and in aprel there was in the night a thunder storm there was one very hard

[*p. 425*]

clap. I thot the lightning seemd to hurt my left eye and i soon found it was blind. I was much distrest in my mind with fear that i could read no more. I got up went to the fire lit a candle and toock a book. I found i could see to read with my right eye. Blessed be the lord for his mercy who might justly have toock away the sight of both my eyes because i have improved them no more to his honour and glory. Lord bless every twig of thy rod.

October 1790. My son jonathan had another daughter born. The midwife was with her not much more than an hour before it was born a perfect and living child. O the mercy of god when wee was fearing she would be as bad as she used to be. Surely the lord is a god hearing prayr glory to god on high peace on earth and good will towards men. O that we might live thy praise. Before my husband dyed for half a year or more i heard in my mind the sound of death acoming it seemd very near. I often told my husband of it. I also told my daughter of it that i live with. I thot because i was so unwell it looked likely i must dye first but after he dyed my distress was took off my mind.

[*p. 426*]

The lord is mercyfull and kind to me he gives me supporting mercy but only by turns. These thoughts come into my mind that i have prayed for my husband above 47 years i believe not faild one day and sometimes many times in a day and yet got no evidence of his being united to christ. But this consideration helpt me moses prayed that he might go into canaan but god would not let him go in. Elijah prayed that he might dye when jezebel was after him but god would not let him die.[7] So i thot if god hant heard my prayrs he hant ronged me. The judg of all the earth will do right. The lord is still gracious to me since my husbands death. I am well provided for. My children are very kind to me. Their kindness to me sometimes melts me into tears to think that the lord should be so kind to me and move them to

[*p. 427*]

love and pyty me so vile a sinner. Lord i thank thee pray father reward them with an intrest in thy love i beg i beg.

1790. About the middle of december i was taken with an inflamation as the docter cald it in my poor blind eye. I was in distressing pain and sick with the extremity i had in the ball of my eye. It seemd sometimes to me as if it would split or start out of my head. I was very weak and low also my old pain in my back raged harder than it use to do. One spell i lived almost without sleep. It is now above four months since i was taken. My eye is better but not well but my back remains painfull. The lord has given me all along some measure of patience. I have been enabled to justify the lord and condemn myself. O it made me pray and search myself. I often thot shall i receive good at the hands of the lord and not evil i that have been so vile. These words one day dropt into my mind and kept running with joy—shall you draw water out of the wels of salvation.[8] O it was supporting to me

[*p. 428*]

and all along now and then sweet schriptures would come to me to support me. O the kindnes of the lord. These words was good to me—even unto hoar hairs will i carry you. O methot does the lord bear me up with his sweet promises and carry me along. Glory to his name. Also that in deuteronomy 33 the four last verses they abode upon my mind for some time. O they was very precious to me. I se everything in them four verses that my soul wanted. Lord never was words like thine none so pure heavenly and divine. O lovely god the father god the son god the holy ghost i want to love thee with all my heart soul streng and mind. O i do hope the time is coming that i shall swim in the sea of endless love and glory. But i now can wright but very little because of my eye.

1791 may 16. About half after ten at night there was a terrible earth quake it was on the sea we hear as well as on the land. Some heard a loud noise like thunder. O lord do i beg cause this to wake up this poor drowsy world pray turn us to thyself for jesus sake.

[*p. 429*]

One night in bed it was cast into my mind thus dont christians sometimes sin themselves out of the promises as israel did. God promised them canan a type of heaven but by their sin they lost it all but only caleb and joshua the rest dyed in the wilderness. A little while i was sorely distrest for i felt myself a great sinner but by and by i see that canaan was only the promis of a temporal blessing that all spiritual blessings are secured in christ jesus. This come to my mind—i will visit your iniquities with stripes but my loving kindness will i not take away nor cause my faithfullness to fail[9] i have loved you with an everlasting love &. And jesus says—i am the good shepherd and know my sheep and am known of mine and they shall never perish neither can any man pluck them out of my hands my father gave them to me. Neither is any able to pluck them out of my fathers hand and as the father hath loved me so have i loved you.[10] Glory to god for everlasting love bless the lord o my soul.

[*p. 430*]

July 12-1791. This morning i think before i was awake these words was whispered into my mind—i will save you with an eternal salvation and none can pluck you out of my hands. O what a wonder it is that ever the lord should concern himself to speak a comforting word to my poor sinfull soul. O help me to live to thee the few moments i have to be here (but o i am in so much pain i must stop).

August 16-1791. I met with a sudden tryal but o it almost overcome my weak body. I cryed to god for mercy and o how gracious was he to me. I thot of gods promises and i soon had my mind stild. That night i went to bed slept comfortably—bless the lord o my soul and all that is within me bless his holy name.[11] O sanctify every twig of thy rod to me help me to trust in thee at all times in life and at death for jesus sake.

September 13-1791. Some of my relations came over from long iseland to see me and brought the news of my aged mothers death. She lived in wonderfull nearness to

[*p. 431*]

god i believe was a witness for him and at her last appeared not to have the least fear of death but wanted to have it hasten. Her last words was she lifted up her hands and said come lord jesus come quickly and fell asleep as quiet as a lamb. Now when i heard of it my heart cryed glory to god for his mercy and kindness to my dear mother. Lord i thank thee for godly parents. O sanctify me & make me to be in an act of readiness always that i may depart in peace and joyn with them and all the redeemed giving glory to god the father son and holy ghost to all eternity. My mother cook dyed in june 1790. She had lived a widow 36 years and dyed in the ninetieth year of her age. My cousen told me that there was a minister apreaching at my brothers where my mother lived and he spoke these words some are galled [illeg.] like they care for none of these things. This word struck her mind so powerfull she fainted away.

[*p. 432*]

It seems that she was overcome with grief because sinners set so light by jesus christ. Her soul seemd to be swallowed up in him. These words run in my mind—be you followers of them that threw faith and patience inherit the promises.[12]

September 1792. My twoo sons went over to long iseland upon worldly business. Now when they had been gone about a week wee heard that they was seen on the water coming home loaded in the hole and upon deck and we could hear no more of them. For several days people thot or feard they was lost. This struck my mind them words that rebecka said to jacob run in my mind—shall i be bereaved of you both in one day.[13] O it drove me to god. I thot how job lost all his children and yet could bless the lord. I was sore afraid that if they was dead i should sink under it and dishonour god. Now this was supporting i hopt i had a portion in god that cant be lost. Now i was also afraid if that the lord brought them home alive i should not live answerable to the benifits

[*p. 433*]

received i have so often misimproved my mercies. Now i thot how aaron held his peace when his twoo sons was slain. I begd for submission but o to our surprise that day fortnit they went away they come home. Glory to god lord i thank thee pray make them to know jesus christ and him crucifyed. O make us live in some good measure answerable to the benefits received.

fabruary 23. The wife of capt samuel munson of this city hanged herself. O it drove me to god for myself and others. Lord sanctify it to us all especially to the husband and children. Pray make it a means of their being turnd to god the lords voice cryeth to the city &.

Aprel 15-1793. Thes words was sweet to me—never man spake like this man.[14] O sweet jesus no words like thine none so pure sweet and divine who bought salvation for the poor and bore the sinners shame. O purity itself in whose mouth was found no guile. Also this word was good to my soul in the morning before i was up—unto you that fear my name shall the sun of righteousness arise with healing in his wings. It made me hope in his mercy that he was coming to heal the wounds that sin has made in my soul. Now before these places of schripture come i got so discouraged. I tried but could not pray my strength was

[*p. 434*]

gone glory to god that has strengthned me. Then this word come—here we have no continuing city but look for one to come even a heavenly whose maker and builder is god. Bless the lord o my soul. Now i cant write no more methinks (i am in such pain).

July the last day 1793. Mr marshal of canaan come to visit me twice and prayed. Each time my heart went out to god with him. I believe the lord sent him here to tow me along. Afterwards i felt comforted and strengthned. Gods word was sweet to me he gave us good counsil. Lord set it home. O lord i thank thee that thou should take so much notice of me and remember me in my low estate. O make me to live a life of faith upon the son of god and dye in faith. O let me expire in thy bosom in thy petarnal arms for jesus sake.

1793 in august. Lydya beecher was found drownded in the river and a child was murdered at hartford. The man is in jayl that did it in a raging fit of anger with a hammer and another child is murdered six years old by a man at cheshire in a fit of anger killd it with a ho. He is now in new haven jayl. Lord have mercy on us bow thy heavens and come down. O hasten the day when violence shall be no more heard in the land nor destruction within our borders. O make our walls salvation and our gates praise. O come lord or wee are ondone. Oh keep me keep me gracious lord and never let me go.

[*p. 435*]

Never let me go till i up boarn on wings of love reach the mentions of the sky and take my seat above. Sometime after i was tried with a body of sin and death i felt for several days bound down under it but i kept beging of god to help me and one night these words come to me—the weapons of our warfare are not carnal but mighty to the pulling down of strongholds and the evil imaginations of the heart.[15] Then i see what them weapons was. Read ephesians 6 from the 13th to the 18 ver—wee must be like good soldiers who have their instruments of war when they are going against the enemy to pull down their strongholds and fortifycations. Glory to god that all our spiritual armour is in the hands of the captain of our salvation jesus christ and above all taking the shield of faith wherewith ye shall be able threw god to quench all the fiery darts of the wicked. Soon after this dropt into my mind—for we know that if this earthly house of this tabernacle were dissolved we have a building of god an house not made with hands eternal in the heavens.[16] Now my sore burthen was toock off. The lord helped me against his and my enemies lord i thank thee for sweet peace in my soul. O sweet saviour pray dont let me be holden with the cords of my sins i beg i beg unless it be for thy honour and glory and my good. O how love i thy law it is my meditation day and night. And as to street of cheeshire i mentioned before that killd a child with a ho now after he had lain in jayl for some time he hanged himself in prison and is gone to his place. Glory to god that has kept me or any from such presumtious ways of sinning lord have mercy on us.

[*p. 436*]

I dreamd these words and sung them in my sleep—he shall come down he shall come down. The whole verse is thus—he shall come down as rain upon the mown grass as showers that water the earth.[17] They were good to me. It apeard to me that christians now was like mown grass but o how a rain of righteousnes will make them flourish again. I cryed and do cry come lord jesus come quickly but alas there was soon a croos laid upon me and instead of taking of it up i sunk under it. I was holden with the cords of my sins i was like jonah in the whales belly i went down to the bottom of the mountains i felt my will up i was unsubjected i thot i was like demos and judas. I cannot tell the distress of my mind. For about 5 or 6 weeks it would be upon my mind night and day and when i prayed tho none knew it but god and myself i believe the devil had a great hand in it to distress me. Now i kept beging for mercy and this consideration helped me that the lord never had left me in any tryal but always had helpt me out again. And these words come to my mind—surely he shall deliver thee from the snare of the fowler and from the noisom pestilence 41 psalm 3.[18] A sweet support this was to me that god in his own time would deliver me from the snare of the devil and from the pestilence of sin. And now i do hope the snare is broken and i am escaped. Lord i thank thee for thy tender mercy and kindness to me. O how was i forst to fight all day upon the sabbaths while this temptation lasted to keep out sinful distresing thots. I abhor and loath myself. Lord have mercy on me pray keep me from sin or i am ondone

Here the law of sin and grace will jar
both dwelling in one one room
The saints expect perpetual war
till they are sent for home.

[*two pages blank and one missing*]

APPENDIX

LETTERS TO AND FROM FAMILY AND FRIENDS

1750–1792

[*p. 440*]

Now i am sorry that i have not written coppies of the letters that i have sent to my freinds for above 30 years but i could not spare time i being alone as to any womankind in my family but myself. But o lord thou can bless them where they be altho written by an unworthy sinful worm. Thy truths are truth still glory to god. Now i think to write over some of the letters my father sent me before he dyed. This was writ in the year 1750 march 4.

Southampton long iseland

Dear children our love to you and all that love the lord jesus in sincirity grace be multiplied. I received yours with the joy of my heart. Dear child stand fast in the liberty wherein christ hath made you free. Run away christians & refined hipocrites they both will speak to you fair things of mens enventions but believe them not. They say lo here is christ he is in the secret chamber but remember christ tells you his coming is like lightning which cant be hid. But now the religion of this day is for a hidden christ. Therefore jew & gentile have agreed to keep him in the cepulchre and if any are obbliged to praise him they must as it were go under ground where they maynt be heard lest it make disturbance. But glory to god i see a glorious day acoming and then the rocky hearts of saint & sinners shall rend asunder.

[*p. 441*]

Then jew and gentile shall long for and embrace a rison jesus. O lord hasten on the latter days glory. I bless my god that gives me to see the waters of the sanctuary is rising & that the day is near that no cananite can dweel in that glorious house the lord is about to build. For the love of god will be so great that they cant dweel in it it will so burn them. Dear lambs of north haven seek ye a rison jesus. O lord onvail thy glory to thy lambs they will know thee again they will remember the calvery groans and garden tears & bloody sweat. O send the holy ghost that shall bring all things to their remembrance whatsoever thou hast spoken to them. O sanctify them by thy truth &. Fear not scoffing ishmaels. The lord will make you a feast at your weaning from the world & self. Abraham made a feast when isaac was weand. Dear christians in north haven i think i have some feeling of your dificulties but all is nothing when king jesus doth appear to the soul for perfect love casts out fear. Dearly beloved walk closely with god and nothing can hurt you hannah. I am glad the lord is pleasd to let you see something of the latter days glory. Be not high minded but fear. The christians at brighapton have a saparate meeting and have much of gods presence. O the goodnes of god to his poor things. There is a large saparate meeting at southhold:ocaboag. Mr richard howel[1] is leader. We have sweet meetings sometimes with the dear indiens. The lord is going to do great things. Fear not little flock i long to see you all. Farewell in the lord. *Jonathan Cook.*

[*p. 442*]

March 24-1749. Dear children our love to you hannah. We have heard that you are like to go home. At first my proud hart could not bare it but my god made me willing & now through grace i can bless my god & your god. Let him take you away when it pleaseth him. O sweet jesus that hath redeemd our souls ah dear child i feel a longing to go along with you and it wont be long before wee shall be with him whoom our souls love. O lovely god father son & holy ghost jesus hath taken away the sting. Jesus hath prayd that our faith fail not. I pray god to be with you and uphold you by his free spirit be faithful to souls. The lord is at hand jehoshaphat is feasting with ahab and the false prophets while micaiah is in prison.[2] Last sabbath the lord was with us give my love to gods children especially to them that come out into the open feild of battle. Stand for your lord fight under the banner of jesus but with a meek & humble spirit and o my son heaton fly fly into the arms of a compasionate saviour. O my sun my sun is the great and dreadful god your enemy or is he reconciled to you through christ. If not wo wo be to you have you not grieved away the spirit and brought a more dreadful curse upon yourself. I long you should praise the lord. Hannah if he will not do you praise the lord. Wee are loth to praise him till we behold his beauty and hear him as it were groaning out his soul with bitter agonies

[*p. 443*]

upon the croos and by faith behold the bloody clodders fall and all for me. Dont you think it will be glorious by & by when you and i and your mother and the rest meet in heaven. O how sweet how sweet it will be there wee shant have no sin to pester us but shall look on jesus with open face. What matters it then what we meet with here. *Jonathan Cook*

October 5-1749. Dear children our love to you. We are well. As common the children of god here are now most of them in bondage. Their hearts condemn them & they have no confidence towards god. Poor jehoshaphats have made a leauge with ahab and are feasting with the fals prophets whilst micaiah is hated for the truth. O that the set time to favour zion might come. Dear child sit you up the watch tower and hear what the lord will say to you. For if you are christs you are led by his spirit and that leads into all truth and the truth makes free from original bondage but if you have not the spirit of christ you are none of his. Now the spirit produceth in them that have it meeknes patience self denial &. Not rendering evil for evil but blessing for cursing. If you would live thus you would find a spirit of prayr. Your hart would not condemn you before god. Hannah remember always that the woman was the first

[*p. 444*]

in the transgression & not the man and o my son my son heaton what shall i say to you. I am afraid you will perish forever. O fly fly for your life to the city of refuge make haste death hell eternity is at your heels. Dear child i love you or i would not be in such ernest for your best good. Farewell in the lord. *Jonthan Cook*

March 19-1748. Dear children our love to you. We are in health blessed be god & o that these may find you so. I received yours and am glad to hear that you have any feeling of your nothingness but dear child you say that christians are underlings here in north haven but doth god make them so. No no. God makes them kings & priests. It is they make themselves so by their unbelief. Was the three children in daniel underlings because wicked opposers flung them into the fire. No they ware through christ more then conquerors. So shall you be if you live upon god. Dont tell me you cant tell me you wont and then you will speak true. Dont look to none but to god to help you. Most are looking to ministers but they have played the harlot. Read hosea 2 in the 4 verce—God saith he will not have mercy on her children not those children zion begot when she walkt with god but since she hath joind with the world for they are the children of whoredom. But i believe god is agoing to do great things for zion. Read the 50eth

[*p. 445*]

of jeremiah and the 47 of ezekiel. The waters of the sanctuary are rising & will rise to the neck and then christians will swim. *Jonathan Cook.*

1747 june 4. Dear children our love to you. It is a sickly time here. John topping benjamin howel silas sayre is dead & many are sick with the plurise. I dont know but god is about to redeem zion with judgments and no matter which way if god be glorifyed. Now prepare to meet thy god o israel. O christles souls what shifts will you make before the burning throne of an angry god when you shall know what a dear saviour you have refused. But comfort ye comfort ye my people saith your god. Pure religion runs low and i think it began to go down at the sanctuary. O christian ministers what are you adoing. I believe god has many things against you and i am sure of one because you suffer that woman jezebel which calleth herself a prophetis to live and to teach & seduce the people and cause them to commit fornication and to eat things offered to idols. Dear children get into the arms of the dear jesus and then let earth and hell do their worst you will be safe. Dear christians in north haven my heart is with you. O you little remnant contend not for forms but for

[*p. 446*]

faith. Pray without seacing and in everything give thanks. Beware of the leaven of the farices which is hipocricy. I see a glorious day acoming but there is a dark cloud between. The devil is come down not in great wrath but in great peace. Beware beware of him try them that say they are jews with the truth and you shall find them out. O be faithful to your dear lord for he that confesseth him before men him will he confes before his father. There is a great fight acoming the dragon & his angels are labouring to destroy the saints but michal and his angels[3] will appear in the best time and defeat the helish host. Fear not little flock. O that the son of righteousnes would arise with healing in his wings. Lord jesus come.

Jonathan Cook

Aprel 16 1745. Dear children our love to you. Hoping you are in health as we are praised be god for it. Tell sister beech to hold out a little longer the fight will be over. I long to know how it is with you and hannah. Do you love jesus. O grace abounding to the chief of sinners. Give my love to the christians at north haven. My heart has been of late with you i feel your tryals. You are as sheep having no shepherd. Yet fear not little flock its your fathers pleasure to give you the kingdom.[4] Keep you own language. Be of one heart & one mind. Meet often together and

[*p. 447*]

sing and pray if there is no brother to lead. Dear sisters do you meet and pray together. I dout not you will find your harts warmed with the love of god. Fear not reproach. This is my mind about unconverted ministers that they are the devils ministers and not christs. My sheep says christ know not their voice that is not to approve of it. O if you have a hart pray for them. Wee have had much of the presence of god in our meetings blessed be god for it. O little bethlehem poor in walls rich in firneture pray for me. And o my son heaton are you converted do you say no then god says his wrath abideth on you. O my son what do you wait for fly to jesus or you will quickly be bound down with the black chains of eternal dispair. O that i could write to you both as children of god. Hannah keep yourself from idols throw yourself into the arms of a compasionate jesus. None goeth a warfare for christ upon his own charges. Read the second book of chronicles the twentieth chap beg god to give you the spiritual sence thereof. The lord has done great things for me whereof i am glad. Fareweel in the lord jesus glory glory. *Jonathan Cook southhampton long iseland*

The august following after this letter was written my honoured father went over to lime and was with mr whitefield and he rit to me thus blessed be god for it.

[*p. 448*]

Southhampton geneway 26-1758. Dear child my love to you and yours hoping these will find you in health as they leave us. We are never to forget any outward mercy. But our dear christ is the mercy of mercies the life of all life the perl of great price the treasure hid in the gospel field.

And this is all that a child of god can raily call his own for all our outward injoyments are but lent blessings. Our bodies and souls our dear lord hath purchased to himself with the price of blood. Wee have given all to him therefore we are no more our own so then we are to glorify god in our bodies and spirits that are his. But our beloved is ours and we are his and this treasure tho hid from the world is the believers riches and as touching the life of my own soul. Tho i live in widowhood and so in that state am very lonesome yet i am not alone for through divine grace i do injoy my spiritual husband which is better infinitely to me than any earthly friend. Ah the love of christ passeth all knowledg o let our prayrs meet often at the throne of divine grace tho absent in body a few moments more and we shall have done with the world & shall land on the eternal shores i trust of compleat hapynes. And now i must tell you there has been a glorious work of god at mecox of late. The children do cry hosanna to the son of david. Dear child dont you sometimes loose your evidences for heaven. Well then you look back to see whether your experiences ware right and what holines there is

[*p. 449*]

in you and finding none you are cast down and then you go to work and want to live a holy life. My dear child if you would live a holy life live dayly upon christ. Paul tells you to walk in the spirit & you shall not fulfill the lusts of the flesh.[5] How can you live upon the food you eat last week. I remain your loving mother till death. *Temperance Cook.*

Southhampton on long iseland 1755 august 31. Loving cousen we are all weel excepting myself. I know not but my work is almost done and the time nigh when i must pas over jordon. O that i may find my feet stand on them stones on which the priests feet stood that i may go over dry shod. Dear cousen i daily find i stand in need of sensible need of help from god through many tryals here in this thorny maze. Christ has told me true that without me you can do nothing. The lord is yet very gracious to mecox for he yet holds up the standerd of truth tho some are offended at it but o let us not yield ourselves to obey the world in pleasing their forms & customs while they are set to destroy pure religion out of the world. Dear cousen hannah be not afraid to shew yourself a bold soldier for the cause of christ for the which so few has courag to speak in this day. For i must say i never spoke one word for christ in all my life and had not a blessing in it. O let us watch and

[*p*. 450]

pray for we are in an enemies land. My wife gives her love to you wishing you strength in the lord. Tell your husband from me i wish he knew the reality and sertainty there is in religion and the worth of his soul before the harvest is past & summer ended. Cease not dear cousen to warn sinners of death and hell. Let them think what they will of you for the truth wont die here but go to the judgment seat of christ with us. Dear cousen wee have a remembrance of you as of one we love in the lord with the dear flock of jesus. In your place mercy and peace be multiplyed to the end. Great is the mistery of eniquity that is working at this day by which means many of the dear saints are entangled for satan through the deceitfullnes of good words pretends great good to the church of christ while he would advance men to take such a charge and care of the church as to give her laws and prescribe rules for her to walk by while christ is secretly pusht off the throne and his laws laid aside as being not sufficient for the churches protection and instruction. And while union & love is pleaded for it is on such conditions as shall sertainly divide had that people believed the promis of god they would never have builded a babel & if wee do but believe the promis of jesus that he will send down the holy ghost the comforter that shall lead in to all truth & bring to remembrance all whatsoever he hath said unto us wee need not to build with nor upon pope nor counsels blackguards nor associations creators that god hath not made. Dear sister i believe the word & spirit of god is suffishent to inform us both in faith and practice and i am sure they that are thus led know something about that peace which christ gives not as the world giveth that will support in life & at death. Dear sister you remember how our redeemer took the field here in

[*p*. 451]

Dear brother deavenports day in 1741 when both armies contended openly under their own kings coulours and how satan was obbliged to give way and the testimony of the witnesses prevailed. But now satan comes by flattery and some of christs dear lambs are bewilderd and have lost their way and begin to think it is because they are no christians and while they are hereing the law which commands perfection they think tis presumtion to think of claiming a priveledg in the gospel or covenant of grace. But thou o lord can make us consider hear & live. Farewell in the lord jesus. *John Cook*

Southhampton october 1-1758. Loving and dr sister. I had oppertunity to see a letter you sent to your mother which refreshed my spirit and others that heard it. You desired i would send you a letter which i am glad to do and think i can write to you as to one i hope knoweth something what that legacy is that jesus christ left for his saints that in the world they should have trouble but in him they may have peace. Dr sister do wee not pray to god to wean us from the world to purge away sin to refine our hearts for himself and yet when god is in wisdom taking his own way to do the work how are we startled and often ready to think all these things are against me. Sometimes i find it is good for me to look back and call to mind

[*p. 452*]

the covenant i first made with god. Dr sister was there any reserves made by us when wee closed with christ. Did we reserve husband wife children or any injoyment for us to dispose of as we would or did not we give up even our lives to be unfainedly at gods disposel. Weel if so why should wee be discouraged at all the triumphs of the wicked since god has given us such a blessed promis to rest on that all things shall work for good to them that love god. Their triumping is but short for consider how many we have known in our day that took a great deal of pains to worry christs lambs that are sommoned forth before that god that has said vengance is mine & i will repay it.[6] Pyty pyty them my cousen and remember such were we and should have been now had god let us alone in the way we chose. O let us stand astonisht and bless god who has chosen us to be of that small number that hath entred in at the streight gate when so many millions do decend in the broad road down to everlasting burnings. The christians last sabbath had a blessed meeting at meacox and ant cook your mother was here and had a refreshing time. She bid me tell you in my letter that she was weel in soul and body. Do write to us as often as you can. Last fall there was a blessed shower here and at shennecock many was

[*p. 453*]

converted. I think i may say i am sure of it of both english negroes and indiens chiefly young people and some do yet retain a relish of divine things. I must say tho unworthy i never had such a sealing time in all my life before. I find it is one thing to walk in fellow with god and his saints in judgment and another thing raily to walk in the spirit being a living witnes for god in the truth. And here is where a great many christians seem to stay and where i stayd a great while but wondering much in my mind about many things not seeing my duty and indeed afraid to see the path god had markt out for my feet trying to get such a religion or to walk so that the world would not fight with me but with shame do i confes how many times i denyed the cause of christ. Dr sister i can from my own experience in these things both bless god for the strength he gives you and would incourage your hert to yet yield to the teaching of god in his word by his spirit both to live and speak his truths. Let the world rage and blind guides preach never so much against standing in the testimony of god. They are as easily known as a tree is known by the fruit altho they think they are hid. O lord how much doth this land mourn under this heavi curse and what is still worse the people do chuse

[*p*. 454]

them that will lead them to hell and many of thy dear saints have drunk of their pisonus doctrins but they shall vomit them up again. Hasten the time o lord when thy saints shall learn not to trust in a friend nor put confidence in a guide. Dr sister glory to god tis anough to fear and love god and wee have none other to fear and he gives his peace that the world knows nothing of. Let us be willing to stand all alone from creatures for god will not forsak us if we will follow him. Run my dear sister run fight with meeknes and courage and i will run with you. Our work is almost done. God is hastning the end. See him in this war rooling the wicked by thousands into eternity. The fig tree blossoms the summer is near rejoice ye saints & be glad be not dismayed. Read with comfort revelations 19—god is making war with the beast and he shall be taken. O sinners there may you read what is to become of your captain and of you o cousen theophilus heaton.

Come in come in while you may list under christ while there is room. I want to write more but the bearer hereof is moveing. We all give our love to you all farewell in the lord. *John Cook.*

Southhampton long iseleland july 24-1761. Loving cousen my respects to you & yours. As to religion tis a low time in general but o my cousen i can say great is the

[*p*. 455]

goodnes of god to poor dispised mecox. He condecendes to pour out his spirit on his servants and on his handmaids and they prophesy or speak forth the truth recorded in gods word. I long to write perticulerly to you about many things but hant time. Moloncoly it is to think of the imaginary

worship that is now raised in poor north america a religion envented by fallen mortals and set up by the athority of man and is called the gospel and the religion of god while the spirit of the gospel of jesus is driven out of doors because there is a power in it that pulls down the alters of baal and cuts down the groves and spoils the glory of that kingdom that is sat up in opposition to the redeemers. But take courage all ye saved ones to stand fast in the faith once delivered to the saints and fight against babylon but with the weapons that king jesus hath harnesed us with which is the truth with a meek & quiet spirit denying ourselves and all ungodlines remembring vengance is gods and he will repay it. Be strong my dr sister and suffering brethren in the lord for altho the bush is kept on a light flame it shall not be consumed because of the good will of him that dwells in it. God has promised and will not go back that the beast and his followers shall be taken & cast alive into the lake that burns with fire and brimstone. O that sinners would hasten out of his kingdom and get into christs. My wife and her mother sister pain our pasters wife and the rest of the sisters give their love to you in the lord. Do come and see us if not do write to us as often as you can. From your loving cousen. *John. Cook*
O my god keep me at they feet. Amen.

[*p.* 456]

June 28-1773. Dear cousen it glads my hart that my deer redeemer shews his power in vindicating his truth and supporting his feeble lambs to contend for the liberty of the gospel and tho the world says we have lost the battle in argument while overpowered by the wisdom and athority of this world yet glory to god wee gain the battle in the name of our redeemer by paciently suffering in a just cause and by gods grace we may brook threw the powers of darknes and put on them crowns triumphant crowns of glory which our dear redeemer has purchased for us if wee faint not. My brethren in north haven be not offended at the croos. The servant is not greater than lord. Love the truth and those that stand in it earthly powers must fall christs church must live they shall posses the good land the grace of our lord be with you all amen. *John Cook*

September 24-1774. Loving cousen it is a very low time here in religion i mean the power of that religion that is of christ that weans from sin and makes the heart chearful in god. But tis a florishing time with antechrist for his priests are adorned with the very extent of the fashion & the people count it a very great honour to their gospel to see them so living glut-tonus lives for then there is a good example for nature to follow. Their religion is calkilated exactly to suit nature for tis wholly independent and unsertain. Indeed to speak the whole in short tis inconsistant nonconnect and

[*p. 457*]

unsertain these three tryanguler pillers greatly bare up their building. Tis inconsistant because they profess to know god but in works they deny him. Tis nonconnected because there is many heads and every one cries lo here and lo there. Tis unsertain because their faith is a general faith or perswation without the seal of the holy ghost. O my soul come not thou into their secret unto their assembly mine honour be not thou united. But now my sister their perplexity is come. God has sent preachers to declare his glory that they cant forbid nor imprison. The earth says i am burdned with mans sins and i testify for god that i am greatly forbidden to yield my strength for man. The clouds says i am forbid to give my watry blessings (how is the earth dryed) for man hath kicked against heaven and waxed fat the wide and watery ocian. Methinks says i hear the blasphemies of those that trace the watry paths and there denying the holines and purity of my maker and tis god that gives man dominion over the fish of the sea therefore now at the command of god i withhold my fish from man. Our happy constitution that hath long under god yielded peace to our nation says since man has forsaken god and has taken the reigns out of my makers hands i testify for god that the troubles man shall feel shall make man know that without god he can do nothing. Yet dear sister be not dismayed at these signs for our god knoweth who are his and will love them to the end. Dear cousen it glads my heart

[*p. 458*]

that my redeemer shews his power in vindicating his truth and supporting his feeble lambs to contend for the liberty of the gospel. And tho the world says we have lost the battle in argument while overpowered by the wisdom and athority of this world yet glory to god we gain the battle in the name of our redeemer by paciently suffering in a just cause and by gods grace wee may brook threw the powers of darkness and put on them triumphant crowns of glory which our dear redeemer has purchased for us if we faint not. My dear brethren at north haven be not offended at the croos the servant is not greater than his lord. Love the truth and those that stand in it. Earthly powers must fall gods church must live they shall posses the goodly land. The grace of our lord jesus be with you all. Amen. *John Cook.*

This fall there come twoo young women over from long iseland to see me related to me they come from the place where i lived and brought me the foregoing letter. I believe they was both christles. I felt a pyty to their souls. I warned them to flee from the wrath to come. I told them how little they thot what a day was acoming. It lookt likely we should see each other no more till we met in eternity. I reminded them of the warnings that they had had. Ah they have seen god working there. O god do convert their souls. Now i sent a preasent to my honoured mother and the following letter to cousen *john cook.*

[*p.* 459]

October 30 1774. Dear brother i can say with you it is a florishing time with antechrist. The jehoshaphats are feasting with the ahabs while the dear micaiahs are hated for the truth while they bare witnes for the bleeding cause of christ. And o how many of the children of god are deceived by our standing ministers while they have larnt to preach the truth and yet live in the fashion of the world. They have no croos to take up and not only to take up but carry daily. O sweet croos jesus carryed it. I see their religion exactly to suit nature. Ah dear brother let us not halve christ but be willing to take croos and all. Afflictions are in the covenant yea many tribulations if wee will live godly in christ jesus and walk in his rules abyding in the vine. Here the righteous are sad the wicked mad while god withholds his spirit. God brought many sore judgments upon the jews for their sins but they would not be reformed. The last was their quarrelling and fighting amongst themselves and that proved their final overthrow. And o how just it is now for god to destroy our nation and say lo i turn to the savage heathen but i hope yet in gods mercy. Of late i have had my soul rest here that god lives reigns and governs the judge of all the earth will do right. Dear brother earth death nor hell cant take away the portion of gods children. Here i carry a weekly body which is often a clog to my soul. Here i see sinners going down the road to hell laughing and dancing as if nothing aild them as if there was no god no heaven

[*p.* 460]

nor no hell nor no day of reckoning acoming and o the body of sin and death i sometimes feel makes me groan for deliverance. But i believe sometimes that i in all these things am and shall be more than a conqueror through christ jesus that dyed for me. Here i want meetings but i hope to meet you and all christs lambs in that general assembly and church of the first born and see jesus face to face and be like him. Here my meetings have been broken up twice but there will be a meeting that will last to eternity glory to god. Give my dutiful love to my dear honoured mother. Tell her to keep her loins girt and her lamp burning actually ready to go into the mariage supper of the lamb whenever her lord shall call. But o what shall i say to my dear brothers. Tell them to haste flee for their lives stay not for the the avenger of blood is on his way. Dear cousen i want you and our brother peter to come here if the lord will. There is some here that i believe would be glad to see you. Write to me as often as you can you know not what good it may do. By and by wee can write no more. We have loud preaching at this day. Prepare to meet thy god. O north america—gad a troop shall overcome him but he shall overcome at the last—read leviticus 26. I feel my heart inlarged. I want to write more but hant time. Now give my love to all. Farewell in the lord.

Hannah Heaton

A few words was written to me in a letter which my mother sent me. They was writ by a man my eyes never see viz the saparate mininster at meacox on long island.

[*p. 461*]

*M*y dear unknown and yet well known sister according to the common faith and fellowship of the spirit of our dear beloved lord jesus christ grace mercy and truth be multiplied in all hearts and churches. Tell my young fellow sufferrers not to be ashamed to go with out the camp unto jesus bearing his reproach. I know by experience that there is great sweetnes in being imprisoned for the gospel sake. He that hath pauls gospel in his heart and makes it manifest in his life and conversation must expect pauls reward here and hereafter in kine if not in degree. So farewell in our dear beloved lord and saviour jesus christ. So be it written at quoage on my journey westward in the 80eth year of my natural age and the 50eth year of my spiritual age. *Elisha Pain*

A coppy of four short letters which i writ to my brothers and sisters at newark in the jerseys.

*D*ear brethren and sisters grace mercy and peace be multiplied in each of your harts. Wee are as weel as common the lord be praised. Dear brother baldwin you writ to me in your last letter as if you was in darknes as to the state of your own soul. Tho i know not how it is with you now yet pray dont rest till you have made your calling and election sure. If you have believed dont rest till you are sealed to the day of redemption. O think of the foolish virgins that had lamps but no oil and except wee have the spirit of christ

[*p. 462*]

wee are none of his. Now the spirit of christ produceth in them that have it supreme love to god. For what he is in himself fervant love to the brethren meeknes patience humility not rendering evil for evil but blessing for cursing a rejoycing in god in christ jesus an increase of grace and perceverience therein to the end. I mean those that have the sanctifying influences of gods spirit. Dear brother dont be discouraged in pleading hard with god. The great god is threatning our nation and land with destruction but i will turn and overturn untill he whose right it is to reign shall come. O let all that has an intrest at the throne cry day and night. Tho times look dark ile tydings spread that fills the peoples harts with dread. The lord is on his way to bring salvation to zion.

My dear brother jonathan cook has the lord been making you to feel the rod. Pray take care dont dispise gods chastisement nor faint when rebuked by him. I pray gods rod may be blest to you. O that it may become to you like aarons to bud & blossom and bring forth the peacible fruits of righteousness. O brother death is coming to meet you and i ah how can you come up to gods bar without a christ. O think of eternity what horrible company must they have that die christles. Make haste dear brother flee for your life. The

[*p. 463*]

avenger of blood is on his way. O that you may be like jonathan of old to taste of the honey that your eyes may be inlightned.

My dear sister elisabeth how is it with you. Have you forgot the day that you could say my soul doth magnify the lord my spirit rejoyseth in god my saviour. Ah i know not how it is with you now. Do you say you are with the prodigal in a far country. God complains of israel that they forgot him days without number that is they did not live to him but went after other lovers. But behold he says again yet will i not forget you are graven on the palms of my hands &.[7] Dear sister dont rest till you get christ in your heart. Dear sis i hope and fear for you. O that you may be like elisabeth of old to have your soul filled with the holy ghost.

Dear sister phebe how is it with your soul. Are you still with hager wandering in the wilderness and cant find jesus. Behold he says whosoever will let him come & take of the water of life freely again he says you will not come to me that you may have life. Dear sister i myself stood off from christ a great while when i was afraid of perishing forever not knowing what the matter was but as soon as i was willing to except

[*p. 464*]

of christ upon his own terms the match was made for a whole eternity. I trust glory to god o sister that you like blind bertimeus the begger may hear jesus calling you and like him cast away the garment of your own righteousness and come to christ that you may receive your spiritual sight. Ah you will have as kind a welcome as he had. O think of the prodigal when he returned home to his fathers house. O that you may be like phebe of old a servant of the church of christ.[8] O that god would remember the cries of our godly father for his seed tho being dead yet speaketh. O sister think of his warnings. How can any of his children meet him at the great day without a christ. O lord grant that wee all may meet together in heaven never to part. *Hannah Heaton.*

A letter to my mother at quoage on long iseland.

Dear honoured mother. My love and duty once more to you. We are as well as common. God is merciful to me altho i am weakly yet i keep about for the most part. Dear mother it is through many tribulations we are to enter into the kingdom. Jesus has gone before us and paved the way. Here wee are upon the waves but when wee get ashore christ and heaven will make amends for all. Dear mother live near to god

[*p. 465*]

in subjectednes to him and nothing can hurt you. Yet a little while and i hope to meet you where all tears shall be wiped away where there is no sorrow no sin nor no crying. The lord grant that like abraham wee may give up our all to god and all will be well with us and that which is better god will be glorifyed. It is a low time here upon the account of religion but ah the promises of god are sure promises. He has said he will send showers of blessings and altho amarica is now under gods severe rod for sin and how deep wee must drink of the cup of gods displeasure i know not but this i know the lord knoweth them that are his and tho if forsaken of all things below the sky jesus will never forsake the purchase of his own blood. I give my love to my dear brothers if any are yet alive. O wake up wake up hear gods voices in his providences. You have loud calls to come to christ that your souls may live. O precious time by and by it will all be gone. God will call you no more. O come in while you may it looks likely this is the last time that ever i shall invite you to come to christ. O why will you die. Pray mother do forgive

[*p. 466*]

me all that ever you see amis in me which was much much. Give my love to all my friends at meacox. Tell them to keep [illeg.] up watch and pray. O pray that god would make good that word in malachi 4 chap 5 and 6 ver—god lives and blessed be my rock.[9] *Hannah Heaton.*

A letter to a freind at nine pardners.

*M*ay 12 1783. Dear sister let us remember that afflictions are in the covenant for he whom our dear lord loves. He chastens & and he has told us to be content with such things as wee have for i will never leave you nor forsake you tho sometimes for our sins he may seem to hide. Yet he wont never forsake the purchase of his own blood. And now if the lord is pleasd to call you away by your preasent illness are you not willing to go and see that head that was once crowned with thorns. Musculus says that in heaven there are angels and archangels but they do not make heaven. Christ is the most sparkling dyomond in all the ring of glory. Dear sis in heaven wee shall be spectators and admirors of his glory and have time to thank him to his face for what he is in himself and for what he has done in redeeming our ondone souls from the lowest hell. Glory to god sin cant get into heaven. There the wicked cease from troubling there weary travellers are at rest and as one say the believers winding sheet will wipe away all their tears. I have cause to thank the lord that you think of my poor husband and me in writing a few lines to us that you may carry us to throne in faith that i may be sanctifyed sin purged out and live in obedience to god and that my husband may be converted. I think he seems to be more secure of late then ever altho wee have been severely corrected by gods rod for sin. O pray for my poor children that jesus may be their eternal

[*p.* 467]

portion. I have lately heard from long iseland they have a young man a minister that come from ingland. The people flock to hear him and there is a great awakning. O pray pray for the downpowering of gods spirit. O lift up thy feet and march in haste aloud our ruin calls. *Hannah Heaton.*

The coppy of a letter writ to my aged mother at long iseland. August 1783.

Dear mother my duty to you i am poor and low in body by reason of pain and weekness. God is often shaking me over the grave and i want to go home glory to god. I feel my anchor fast. I hope by and by to be above pain. I want to be in the full injoyment of god where sin cant never come all tears forever wiped away. O glorious hour o blest abode i shall be near and like my god and flesh and sin no more controul the sacred pleasures of the soul. Dear mother you must expect soon to pass over jordan o that you may go over dry shod o that like simeon you may get jesus in the arms of faith that you may depart in peace. O live in obedience to god now in your last moments. Do pray hard for the downpowering of gods spirit that he would come and thaw this frozen world. And o that you may give up all your children to god in faith. O that wee all may meet at gods right hand never to part. Pray mother forgive me all you ever see amis in me. Ah how many worisome steps you have taken for me i never can reward you. I pray that god may with his preasence here and in heaven eternally. Give my love to brother and sister. Pray dont rest till you know that christ jesus is yours. O how fast does time run away. Death is coming and it may be sudden as we have seen and the great day of gods wrath is coming. And who shall be able to stand why not one that is out of christ shall stand with him in that day but must

[*p.* 468]

have that hart rending sentence depart from me you cursed. Pray come in while you may list under jesus while there is room. Then christ and heaven and all is yours. Pray dont let us forget fathers warnings and prayrs for his children. O how dreadfull will it be to sink under them eternally if you die out of christ. I believe he will stand with jesus in the judgment day and joyn in sentence with him against even his own children if they die christless. O think much of it now before it is twoo late and the door shut o do come to christ then you will be happy in life happy at death and eternally happy. *Hannah Heaton.*

August 24-1783. Coppy of a short letter i sent to a young man at long island a preacher that come from old ingland as my mother informd me in a letter that people was greatly awakened by his preaching.

Dear sir. I have heard that god is with you in your preaching. Do ask the lord to let you come over to this macedonia and help us here. Christians are asleep o that the lord would make you a hunter to hunt them out of the holes of the rocks where they are hid. Here sinners are assleep in their sins here is multitudes that dont know their right hand from their left in the the things of god. O when shall the young solomons ride fourth upon king davids mule[10] threw our world. O when will the lord spread fourth his hands like a man that spreadeth fourth his hands to swim to pull in souls to himself. O when will the lord gather his flock like the sheaves into the floor. O when shall our jesus ride fourth upon the white horse of the gospel with his bow and crown conquering and to conquer. O that the watchmen and all zion may never hold their peace till

[*p. 469*]

jerusalem is made a praise in the whole earth. Come lord jesus come quickly. Breethren pray for us pray for me. North haven. *Hannah Heaton*

The coppy of a letter sent to me by a woman that was once my neighbour but now lives at the nine pardners.

September 1783. Dear mrs. heaton. I thank you that you have not left off your kindness to me but have taken the trouble to send me a few lines. Now you cant speak to me as formerly you use to do. I have thot sometimes i would not try to wright any more for i could neither write playn or propper but these words would be in my mind—in the morning sow thy seed and in the evening withhold not thy hand for thou knowest not which will prosper this or that.[11] In the morning of my life i use to speak to my friends tho in a broken manner and now in the evening tho in another form tho broken still the lord grant it may prosper. We are to take heed how we sow for whatsoever a man soweth that shall he reap. If we sow to the flesh wee shall of the flesh reap corruption but if we sow to the spirit we are to reap life everlasting.[12] We are every moment of our life from our cradle to our grave asowing. O i pyty them that are spending a long life sowing in such a manner that ere long will have to reap an harvest of eternal wo. Bound to the bottom of the burning main knawing their chains theyl wish for death in vain. If any shall hear these lines that are concious to themselves that they are christless o let them think of the dreadfull sound forever. Jesus that loving lamb that will soon appear like a lion to poor sinners that now refuse his offers. He that hath his eyes like a flame of fire will tread them in his anger and trample them in his fury. He will trample them down where the worm dyeth not and the fire is not quenched.[13] O dreadful words where the worm dyeth not o to be forever dying and never die. What heart can endure the thot. Wee commonly tremble at thots of the death of the body but o what a death is here.

[*p. 470*]

What to be banisht from my life and then forbid to die to linger in eternal pain yet death forever fly. O that he that hath in his mouth a sharp sword would ride fourth conquering and to conquer but if the sinner will not regard that sword of the spirit which is the word of god it seems as if there is no hope for truth itself has said it. If they hear not moses and the prophets neither will they be perswaded tho one rose from the dead[14] (now i do suppose that this true tho a terrible message was writ in perticuler to my husband for when this woman lived by us she use to be distrest for him). But again o why was it that any of us was made to yield to this almighty prince. Was it not because he had his name written king of kings and lords of lords. Glory to his infynite wisdom that he has found such a way to make the sinner submit. Blessed thrice blessed be my god blessed forevermore (twas the same love that spread the feast that sweetly fourst me in: else i had still refusd to taste and perisht in my sin).
Patience Ford pray for me. Lord bless what i write or all will be in vain. O help lord.

The copy of a letter i writ to mrs patience ford at the nine pardners.

Dear sister grace mercy and peace be multiplied to you threw our dear lord jesus christ. O methinks god is calling us at this day to live upon him in a special manner to live upon him and not on christians gifts and graces. O what contentions and divisions there is in zion some that wee thot was eminant falling but o gods electing love from eternity that will last to eternity and all in christ jesus. Do read romans 8—the truth and faithfulnes of a god in christ has been a mighty stay to my soul an anchor sure and stedfast in storms when waves and billows passed over my soul. Ah neither sin nor sorrow shall things present nor things to come shall ever be able to separate us that have believed jesus from the love of god in christ.[15] And i know a vew of these things sets a keen edge against sin. We are here in a thorny maze and changing world christians changing our frames changing but glory to god he is a god that changeth not therefore the sons of jacob are not consumed. O that everlasting love of god the father

[*p. 471*]

before the world was made and all in christ jesus. Dear child clasp the promises sweep the blessings of the everlasting covenant into your own bosom a divine treasure for your immortal mind to feed upon. O what a sight will it be to be with jesus to behold him where he is not as a man of sorrows but in his exalted glory. O lord pour down thy spirit upon us i beg i beg

Behold my soul thy worthless name: enrold in lines above
see jesus heart vew there a flame: of never changing love
No frowns no strokes no crosses shall there be
till we shall rise to blest eternalle
But o the pain and smart will then be gone
and nothing but a skene of love comes on
Amazed i stand i stand amazed at love so great and free
to one so vile i well may gaze at love so great to me.

Now methinks i must tell you i heard your brother benjamin preach a funeral sermon a few days ago at sergant heatons they having heard that their daughter mary is dead she dyed at heartlands. She was taken in travel and dyed in about four hours. She was thot to be a dear christian. With her dying breath gave her husband and children to god. Death was no terror to her. His text was in psalms 49-14. He preacht exceeding well and seemd to have a sence of the upright having the dominion over them in the resurrection morning. Do pray for him and all christs ministers that god would make them hunters to hunt zion out of the holes of the rocks where they are hid. O when shall our glorious solomon ride fourth upon king davids mule. O when shall christs dominion spread from sea to sea and from the river to the ends of the earth. O come lord jesus come quickly. Do pray for me and mine.

[*p. 472*]

Dont dear child let unbelieveing fear press you down. Christ says as the lily among thorns so is my love among the daughters.[16] The thorns of corruption temtation and worldly snares and cares we are almost ready sometimes to think we shall be torn to death but this is for want of faith. Scratch and grieve us they may but thats the worst they can do to us we know says paul that all things work for good to them that love god who are the called according to his purpose.[17] As saith one the heart of christ is as much set for serving his people now he is glorifyed in heaven as it was when he was dying on the croos. O think much of it. He ever lives to make intercession. Weel then if he ever lives to interceed he must plead to eternity for his bride his spouse and he said when here on earth that he knew his father heard him always. The lord lives and blessed be my rock. He follows his chosen now in this wilderness and causes the water to gush out of the rock the chearing and sanctifying influences of his spirit.

Whatever plea my jesus makes: the voice is precious blood
by life by death he undertakes: to bring my soul to god.

I have wanted of late to have the lord pour down his spirit. O when shall we hear the sound of his goings as on the tops of the mulbury trees. O when shall we hear the voice of the angel among the myrtle trees. Come lord make no tarrying.

He left his shining throne above: his bowel yearnd his heart was love
he could not bair to loose his bride: to save my soul the bridegroom dyed. *Hannah Heaton*

The coppy of a letter i writ to my two sisters at newark in the jerseys in 1783.

*D*ear sisters. My love to you. I am weekly and poor in body but hope by and by to be above all payn where sin cant never come. O glorious hour o blest abode i shall be near and like my god and flesh and sin no more controul the sacred pleasures of the soul. O how is it with your souls. Do you love jesus and love holiness because it is gods likenes. Does your duties give you peace or are they all emty things when god is not injoyed in them. And when you get near to god can you say whoom have i in heaven but thee and there is none on earth that i love in comparison of thee. Is sin the greatest burden you have in this world. Do you daily mourn for sin as it is against god or do you mourn for it because it leads to misiry. We are told in gods word to examine ourselves and except christ dweel in us we are reprobates. O my dear sisters i am consernd for you wee have almost run our race cold death is very near to us precious time is to us almost gone. How does death look are you willing to go over jordan to go in to possess the promised land. O blessed land where there is not one sin nor one pain but an otian of joy and glory holiness and love. Musculus says that in heaven there are angels and archangels but that does not make heaven. Christ is the most sparkling dyomond in all the ring of glory and it is heaven anuf to see christ and be forever with christ. O to see

[*p. 473*]

that head that was once crownd with thorns. There saints will have time to thank jesus to his face not onely for what he is in himself but for what he has done in redeeming their ondone souls from the lowest hell. But o my dear sisters if you die christless then farewell god christ heaven forevermore. Farewell saints and angels alas to be forever banisht. How great his vengance no tongue can tell. Pray wake up before the door is shut. I have lately heard from long iseland there is a wonderfull work of god at southhold and at quoag carryed on by a young man a preacher. Dear sisters i take it hard that you dont write to me twas the same love that spread the feast that sweetly fourst me in else i had still refusd to taste and perisht in my sin. North haven. *Hannah Heaton*

The coppy of a letter writ to cousen john cook at meacox on long iseland.

Dear brother. Our dear lord jesus christ has markt out our path threw this world to everlasting glory and it is threw many tribulations that wee are to enter into the kingdom. Our dear lord who is gone to the father went in this very path and has paved the way for us. Dear brother i have heard of some of your tryals. Dont look back but press forward toward the mark of the prize. Jesus is the captain of our salvation and we are bid to endure hardness as good soldiars and a good soldier will follow a good captain when he calls and whereever he goes and wont turn back for a few bullets. Lord give us the spirit in our warfare to fight with meekness and humility. Again christ says as the lily among thorns so is my love among the daughters. The thorns of corruption temtation worldly cares and snares but fear not you shall be torn to death. Scrach and grieve you these thorns may but thats the worst they can do to you. All shall work to our advantage for our lovely captain has said all things shall work together for good to them that love god that are called according to his purpose. Ah we shall soon get home where there is

No strokes no frowns no crosses shall there be till wee
shall rise to blest eternalle
But lo the pain and smart will then begone
and nothing but a skene of love comes on.

O how dear are christs lillies to him he calls them his love he says as the father hath loved me so have i loved you. O the love of god to christ from eternity who was one with him and o the everlasting love of god to the elect before the world was made. I have loved you says he with an everlasting love therefore with loving kindness have i drawn thee. Gods love is before conversion to believe gods love to christ thats easy but says jesus as the father hath loved me so have i loved you. O this everlasting love has been a mighty stay to my soul in storms when waves and billows passed over me this has been an anchor sure and stedfast. The lord lives and blessed be my rock. Ah will it not be lovely by and by to go and see that head that was once crowned with thorns to sit down with abraham isaac and jacob in our fathers kingdom. Renownd musculus that was fourst to dig in a town ditch for his maintanance says that in heaven there is angels and archangels but that does not make heaven. Christ is the most sparkling dyomond in all the ring of glory and it is heaven anuf to see christ and be forever with christ and glory to god sin cant get into heaven. Dear brother now is the time for us to clasp the promises o sweep the blessings of the everlasting covenant into your bosom a divine treasure for your immortal mind to feed upon. Oh what a sight will it be to be with jesus where he is to behold him not as a man of sorrows but in

[*p. 474*]

his exalted glory and wee shall be like him and see him as he is. Oh i have often thot that when i come to heaven i could tell john that when he was upon earth he saw a great wonder in heaven but behold john here is another great wonder the greatest sinner that ever was is come to heaven. Ah the believers winding sheet will wipe away all their tears.

Twas the same love that spread the feast that sweetly fourst me in
elce i had still refused to taste and perisht in my sin.

And methinks i must tell you that the lord is carrying on a glorious work in many of our upper towns. Saints rejoysing in the lord and sinners crying out what shall wee do to be saved. But it is a low time here breethren falling out by the way and some that we thot was eminant falling into horrible sins. O lord have mercy on us o power down thy spirit and thaw this frozen world. O when shall wee hear the sound of thy goings as on the tops of the mulbury trees. O when shall wee hear the angel among the myrtle trees. O when shall our glorious solomon ride fourth upon king davids mule threw our world. O i do believe the lord gave me to see that he was acoming thirty years ago and i was glad in some measure like abraham when he see christs day. O how rejoysed i was to see that the lord was agoing to get to himself a great name and now i do hope the lord is refyning and purging his children now in the furnice that they may bring fourth more fruit. But glory to god tho wee change yet he is the same yesterday today and forever. His compasions fail not therefore the sons of jacob are not consumed. O them soft sweet words our dear jesus spoke to mary that morning he arose from the dead. Go says he to my breethren and say unto them i ascend unto my father and to your father to my god and to your god. O how he condecends to be one with believers my father and your father my god and your god. O what a heart full of love he arose from the dead with. Ah lovely jesus when shall i give thee the glory and praise that is due thy holy name.

He left his shining throne above
his bowels yearnd his heart was love
How loth he was to loose his bride
to save my soul the bridegroom dyed.

And now he ever lives to make intercession. Well if he ever lives to make intercession he will plead to eternity but he will save that soul that puts their trust in him.

Whatever plea my jesus makes the voice is precious blood
by life by death he undertakes to bring my soul to god.
Jesus our priest forever lives to plead for us above
jesus our king forever gives the blessings of our love.

I give my love to your wife and to all my friends. Tell them to make sure of the love of god in christ. The besom is coming that will sweep down the spiders webb. I am poor and week in body a constant pain in my back all day but easy anights supposed be the rumatis. But o this body of sin and death and the sins of others is worse to me than all. Do pray hard for me that my rod may be like aarons to bud and blossom and bring fourth the peacyble fruit of righteousness. Do beg for my husband and children that christ may be formd in them. Pray do send me one letter more.
North haven 1785. *Hannah Heaton*

[*p.* 475]

North haven. June 1 1786. Dear cousen. Grace mercy and peace be multiplied to you threw our lord jesus christ. I have heard that the work of god is decaying at symsbury o cry after a departing god. We read that when jesus made as tho he would have gone further they constrayned him and he went in and tarryed with them. O plead hard for the downpowring of gods spirit. Ah what shall we do if god takes away his spirit. Wee are an ondone world. O let us cry with david lord take not thy holy spirit from me. Let us search out our sins that has caused the lord to withdraw and mourn before him. Plead for true repentance accompanied with real reformation. Watch and pray it signifies but little to pray if we dont watch. They are put together but watching is the first. O flee all youthfull vanities keep yourself out of the way of all temptations live near to god and that will arm you against sin. Walk in the spirit and you shall not fulfill the lusts of the flesh. I have often thot of late what a reasonable thing it is to love christ above everything who has dyed to redeem our ondone souls. But o how precious is he in himself perfect purity and holiness. But o what a sight it will be to behold him where he is to be with him and see his glory not as a man of sorrows but in his exalted glory. O to see that head that was wonce crownd with thorns and as one says in heaven there are angels and archangels but that dont make heaven. Christ is the most sparkling dyomond in all the ring of glory. Ah there wont be one wicked heart not one aluring sin no tempting devil. O well may angels desire to look into the mystery of redeeming love and grace in god the father first who said before the world was made that his delights was

[*p.* 476]

with the sons of men. Do read the 8th of proverbs from the 22 verse to the 32. Glory to god the father for giving his only son to die for us poor sinners. Glory to jesus who said lo i come to do thy will. Saints nor angels cant dive into this mystery god manifest in the flesh. None but infinite wisdom can sound the bottom of this infinite ocean. Glory to god the father son and holy ghost.

Amazed i stand i stand amazed at love so rich and free
to one so vile i well may gaze at love so great to me
Twas the same hand that spread the feast that sweetly fourst me in
else i had still refusd to taste and perisht in my sin.

And as to the people here there seems to be death all round us the righteous sad the wicked mad while god withholds his spirit. O when will the lord come and thaw this frozen world. Dear cousen it is threw many tribulations wee are to enter into the kingdom. Wee that have a jesus must have croos and all wee must go to heaven. As israel took caanan with wars and fightings we are to indure hardness as good soldiers. I read once of a general that upon a fresh onset of the enemy he heard one of his soldiers cry out wee are ruined wee are ruined wee are ondone we are ondone. He calld him traytor and told him it was not so whilst he could hold his sword. Jesus is the captain of our salvation. He has undertaken the leading and conduct of his people threw all their defyculties. He says because i live you shall live also. Ah precious words indeed.

Tho in some rugged paths we tread
and numerous foes our steps surround
Tread the thorns down press threw the host
the greatest fight the greatest crown.

And says our dear jesus if any will be my disciples let dayly take up his croos and follow me in my tryals. I have many times seen this word fulfild. Gad a troop shall overcome him but he shall overcome at the last. I have seen that as jesus got the victory over

[*p. 477*]

sin death and hell for believers and is set down with his father so surely shall believers threw him get the victory. They shall be carryed threw all tryals and oppositions here. Forty five years god has been my salvation in tryals and temtations and carryed me threw with victory and at death a compleat victory for all believers. They shall be made like jesus christ and sit down with him and with abraham isaac and jacob and all the prophets saints and myrters in our fathers kingdom. O everlasting love i hear there is a work of god at the upper end of wallingsford. The children do cry hosanna to the son of david. They are flocking together to pray and praise and to talk of gods wonderfull works. This verse has run in my mind of late. I do hope the lord is on his way coming to power down his spirit (namely)

Jesus shall see a numerous seed: born here to uphold his glorious name
his crown shall flowrish on his head while all his foes are cloathed with shame
Ah the believers winding sheet will wipe away all their tears their crowned heads they pleasently will sing
sweet hallalujah to their lord and king
But lo the pain and smart will then be gone
and nothing but a skene of love comes on
No stroakes no frowns no crosses shall there be
till wee shall rise to vast eternallee.

May 31 1787. The copy of a letter sent to me from my brothers son. Easthampton.

My dear ant. I have the hapyness of writing to you. I am at town following lerning in the academy. I shall endeavour to get what i can and let god do with me

[*p. 478*]

as he pleases. To proclaim the gospel to a lost world looks very aufull pleasing and glorious. I am not determined what branch of business to follow. I want to give myself blindfolded into the armes of my master jesus. I have threw infinite mercy a sweet relish and foretastes of the joys arising from faith in jesus christ in believing my sins are pardoned and that i shall be fild with the fruition of gods glory in the coming world. Last night i heard by a letter that my fathers family was well. My dear aged grandmother whom i left four weeks ago was rejoysing in her saviour and said that death appeared nearer and nearer and she wanted sometimes to have it over and be with her god where she should offend him no more and have power to love him with all her soul. She is often praising with uplifted hands. May her lord carry her at death to heaven in a chariot of love. My sister Esther is i hope a sincere christian. I hope my dear ant injoys a heaven on earth has much of the love of god. I should be glad to see you here but know not as i ever shall but i hope to around the throne of the lamb glory to the sacred three for love that is so infynite as is manifested in the plan of salvation. The salvation of my country both from tyranny and the slavery of satan lies upon my mind. Aufull chastisements have exercised our countrymen but wee must look for more if people remain so impenitent. O lord our god bring on a universal reformation. Some among us seem to be peeping into heaven some i hope just stepping upon the ladder tho not but few.

[*p. 479*]

Mr. buel blows the trumpet powerfully. The lord throws down the walls of jericho. I am at his house. He gives me my board for some weeks. The lord reward my benefactors. Your afectionate nephew in the bowels of the dear emanuel. *Nehemiah baldwin Cook.*

September 7 1787. I writ to him this letter.

*M*y dear cousen. Grace mercy and peace be multiplied to you threw our dear lord jesus christ. I received yours which was refreshing to me and i hope for good to those that heard it. Dear child stand fast in the liberty when christ has made you free. Watch against all sin when thou hast thy best suit on. Remember who made it who paid for it o the love of god in christ jesus before the world was made. His delights was with the sons of men glory to god for everlasting love. I am poor in body under much pain and weakness. I have cause to think i am not far from that place where the wicked cease from troubling and the weary are at rest. O blessed rest in the bosom of jesus where no sin can ever enter. In this town the righteous are sad the wicked mad while god withholds his spirit. O plead hard at the throne for a rain of righteousness that the devil may be bound antechrist may be destroyed by the spirit of crists mouth and the brightness of his coming. At wallingsford there has been a work of god of late. A number has been hopefully converted chiefly young people. O when shall the lords dominion spread far and wide from sea to sea and from river to the ends of the earth. And my dear cousen dont let selfish motives influence you to take upon you the work of the ministry. God says how can they preach except

[*p. 480*]

they are sent and he that enters not into sheepfold by the door but climes up some other way the same is a thief and a robber.[18] I pray you consider it weel. Dont let the blood of souls be found in your skirts soul blood is dreadfull to answer for. Plead with god to know what your gift is and improve it faithfully and if the lord should call you into his work ask him to let you come over to this macedonia and help us. Ah how many poor souls is there that dont know their right hand from their left in the things of god. Give my love to my honoured mother if yet alive. Tell her to keep her loins girt and her lamp burning to lean on jacobs stall jesus christ and then she wont be afraid to go over jordan but will go over dry shod. And tell cousen Esther not to go back as orpah did but to press forward towards the mark of the prise christ jesus. O i do hope yet that amarica will be emanuels land. O lord bow the heavens and come down that the mountains may flow down at thy presence mountains of sin and unbelief. Ah when shall we hear a noise among the dry bones. When shall jesus take the possession of his inheritance amongst the heathen. When shall jews and gentiles long for and imbrace a risen jesus. Come lord jesus come quickly. My love to all relations. Tell them to haste away to jesus before it is twoo late. *Hannah Heaton.*

A letter i sent to a friend.

1789 december 10. Dear sister. Our lord has told us that threw much tribulation wee shall enter into the kingdom. Our jesus has markt out the

[*p. 481*]

path for us and gone threw it himself to glory and says it is anuf that the disciple be as his master and the servant as his lord. O how unreasonable it is to want to get out of this path that jesus has markt out for us and trod himself. Why should i then fear deaths grim looks since christ for me did die the sting of death stung our jesus to death for us. Ah lovely saviour precious in his life precious in his death precious in his resurrection precious in his ascention and will be precious to eternity.

O may we die in faith and then shall we from all below insensibly remove
our souls their change shall scarsely know made perfect first in love
With ease our souls threw death shall glide
into their paradice

And thence on wings of angels ride
triumphant threw the skies
Our souls are in his mighty hand
and he will keep them still
And you and i shall surely stand
with him on zion hill
Him eye to eye we there shall see
our face like his shall shine
O what a blessed company
where saints and angels join
But what a glorious meeting there
in robes of white arayd
Palms in our hand we then shall bair
And crowns upon our heads.

In patience posses ye your souls.[19] Of late these words have for several days run with sweetnes in my mind. Give ear o shepherd

[*p. 482*]

of israel thou that leadest joseph like a flock thou that dwellest between the cherubims shine forth before ephraim and benjamin and manaseh. The church of christ stir up thy strength and come and save us. O methinks my soul cryed to god with these words. Pray dont be discouraged our shepherd wont never leave us but will carry us threw all threw darkness as well as threw light for he never lost one prayr. He says father i will that those thou has given me be with me where i am that they may behold my glory. The mirits of his blood pleads and he ever lives to make intercession and says because i live you shall live also. Bless the lord o my soul tho our shepherd dwells between the cherubims surrounded with glorious angels yet he is pleading for poor sinners and shall not wee plead. Ah look on poor zion let jerusalem come into your mind. O by and by what will they do that are now sipping at babylons golden cup.[20] Why god tells us that they are to pertake of babylons plauges. But now the world is pleasd and says so would we have it. O when shall our glorious solomon ride fourth upon king davids mule. O when shall wee hear the sound of goings as on the tops of the mulbury trees

[*p. 483*]

o when shall wee hear the angel amongst the myrtle trees o when shall zion arise and shine. Do pray hard for the fall of babylon for satans binding and for christs reigning. O lord pyty poor sinners that are running upon the bosses of thy buckler and sin as with a cart rope. O how will they with dives lift up their flaming eyes in hell if they die in their sins. Lord pyty them that dont pyty themselves. O thou that wept over jerusalem and when shall redeemed ones live to him that has loved loved them and dyed to save them. O how long shall jesus be wounded in the house of his friends. Lord bring thy captives out of babylon that they may rebuild jerusalem and say the lord liveth that brought us up out of the north country and give honour and glory to his name. Farewell in the lord our time is short i trust wee shall soon meet where all tears shall be wipt away no sorrow nor crying. *Hannah Heaton.*

August 17 1790. The coppy of a letter to a woman at wallingsford who has lately buryed her husband and is left with four small children. I have had some acquaintance with her.

[*p.* 484]

Dear sister. Grace mercy and peace be multiplied to you threw our dear lord jesus who dyed that wee might live. The lord has been calling you and me in a perticuler manner to live to him. O let us pray hard that our rod may be like arons to bud and blossom and bring fourth the peasible fruits of righteousness. Ah let us truly say as david did i was dumb and opened not my mouth because thou didst it. The lord has a right to govern in his own world and he does all things well and it is much for his honour and glory when we subject to his kingly athority and government. O let us be still and know that he is god. He has bid his children to trust in him at all times and tho he slay me yet will i trust in him. Altho he slay me yet will i trust in him says that holy man. Dear woman wee must learn to trust in a hiding god a killing god if cald to it. I believe god christ and heaven will make amends for all our travel here in time. Bless the lord o my soul. Praise him.

[*p.* 485]

Tho in some rugged paths wee tread
and numerous foes our steps surround
Tread the thorns down press threw the host
the hardest fight the greatest crown.

Dear sister by faith get hold of the promises of god and keep hold. He says i have loved you with an everlasting love therefore with loving kindness have i drawn thee. Here you see gods love to the elect was before drawing or before conversion. It was election union from eternity and will last to eternity. To view this o how sweet and transporting it is to the soul. Yet a little while and the lord will gather his lambs into his bosom. There will be no sorrow nor crying.

No strokes no frowns no crosses shall there be
till we shall rise to blest eternalle
But lo the pain and smart will then be gone
and nothing but a skene of love comes on
When weve been there ten thousand year
ashining like the sun
Weve no less days to sing and praise
than when wee first begun.

Do plead with god that he would pour down his spirit. O when shall zion arise and shine be fair as the moon clear as the sun and terrible as an army with banners. O when shall our glorious solomon ride

[*p. 486*]

fourth upon king davids mule o when shall wee hear the sound of the lords goings as on the tops of the mulbury trees o when shall wee see and hear the angel among the myrtle trees that was answered with good words and comfortable words. O let us cry come lord jesus come quickly and bind satan and destroy antechrist by the spirit of thy mouth and the brightness of thy coming. Do plead hard for poor sinners. How many do i see dayly laughing singing jesting as if nothing aild them. They go on as if they thot there was no god no heaven nor no hell. O when shall souls flock to christ as doves to their windows. O when shall our dear lord be king of nations as well as king of saints. Lord remember thy covenant for the dark places of the earth that are full of the habitations of cruelty. O that he that wept over jerusalem would pyty the souls of our friends and children and bring them all into the number of his sanctifyed ones. And as to this world the lord has told us to be carefull for nothing but in everything with prayer supplication and thanksgiving to make our requests known to god.[21] Ah if wee live to god he has promised if wee seek first the kingdom of god and the righteousness thereof all these things shall be added unto you. God feeds the fowls of the air they sow not neither do they reap nor gather into barns yet your heavenly father feedeth them. Are not ye

[*p. 487*]

much better then they. Read matthew 6 from the 25th to the end. Farewell in the lord.

Twas the same hand that spread the feast that sweetly fourst me in:
else i had still refusd to taste and perisht in my sin. *Hannah Heaton*

May 1 1792. A letter to a aged widow under confinement one of my intimate acquaintances.

Dear sister. We are the lords prisoners but glory to god wee are prisoners of hope. Altho wee cannot serve god in a publick way by active obedience now the lord is calling for passive obedience. These words of late have run in my mind with much sweetness and comfort—to them who by patient continuence in well doing seek for glory and honour and immortality eternal life. O to love jesus is eternal life begun here and this word has helpt my sorrowfull soul—be patient breethren unto the coming of the lord. Behold the husbandman waiteth for the precious fruits of the earth and hath long patience for it untill he receive the early and later rain. Be ye also patient stablish your heart for the coming of the lord draws nigh.[22] Dear woman dont faint nor be discoraged. I do hope you and i shall yet hear and see the goings of our god before our heads are in the grave. A few nights ago i was going to bed this word come to me—occupy till i come.[23] O let us up

[*p. 488*]

and be doing for it is god that says what time you shall reap if you faint not. This winter past when i was poor in body and mind sin trying to get victory these words come with help and strength—with joy shall you draw water out of the wells of salvation. O how sweetly they run in my mind. For some time now i see every time we get a vew of christ by faith and get strength against sin it may be called a well of salvation. Jesus said it shall be in you a well of water springing up to eternal life. Lord what is sinful man or any of his race that god should make it his consern to visit him with grace. The lord is good and kind to me he gives me sometimes sweet cordial meditations on his word. This day i was thinking which famous ministers here was once in scotland and ingland. O how sweet to hear such preaching but it quick turnd in my mind jesus is a priest forever he is made so by oath the lord sware and will not repent. Thou art a preist forever mortals must die but jesus our high preist ever lives to make intercession. O firm comforting words. Jesus our preist forever lives to plead for us above. Jesus our king forever gives the blessings of his love. Pray read for your comfort the 33 of deuteronomy the four last verses.

[*p. 489*]

But o when i look upon christians and poor sinners they at this day seem to prefer this world above the blessed jesus. They seem to join with one voice great is diana of the ephesians.[24] Lord is it not time for thee to work for they have made voide thy law. And let us pray hard that the lord would remember the covenant for the dark places of the earth that are full of the habitations of cruelty that the lord would come and take the possession of his inheritance among which jews and gentiles all nations shall call him blessed. Come lord jesus come quickly. Of late these words have been comforting to me—if i go and prepare a place for you i will come again and receive you to myself. Dear sister dont be afraid of the pasage for jesus says he will come at death and receive his to himself. Our time is short. O let us redeem time pray without ceaseing. Glory to god in the highest peace on earth good will towards men.

O heaven our happy home which is our country dear
what cause have we to long for thee and beg with many a tear.

Hannah Heaton.[25]

Notes

Introduction

1. Studies based on eighteenth-century diaries and letters include Joanna Bowen Gillespie, *The Life and Times of Martha Lauren Ramsay, 1759–1811* (Columbia: University of South Carolina Press, 2001); Laurel Thatcher Ulrich, *A Midwife's Tale: The Life of Martha Ballard, Based on Her Diary, 1785–1812* (New York: Vintage Books, 1991); Joy Day Buel and Richard Buel, Jr., *The Way of Duty: A Woman and Her Family in Revolutionary America* (New York: W. W. Norton and Co., 1984); Carol F. Karlsen and Laurie Crumpacker, eds., *The Journal of Esther Edwards Burr, 1754–1757* (New Haven: Yale University Press, 1984); Michael J. Crawford, "The Spiritual Travels of Nathan Cole," *William and Mary Quarterly,* 3d ser., 33 (1976): 89–126; Charles E. Hambrick-Stowe, "The Spiritual Pilgrimage of Sarah Osborn (1714–1796)," *Church History* 61 (December 1992): 408–21; Elaine Forman Crane, ed., *The Diary of Elizabeth Drinker,* 3 vols. (Boston: Northeastern University Press, 1991). Helpful analyses of autobiography as a genre include Paul Delany, *British Autobiography in the Seventeenth Century* (London: Routledge and Kegan Paul, 1969); Daniel B. Shea, Jr., *Spiritual Autobiography in Early America* (Princeton: Princeton University Press, 1968); Diane Bjorklund, *Interpreting the Self: Two Hundred Years of American Autobiography* (Chicago: The University of Chicago Press, 1998); Margo Culley, ed., *American Women's Autobiography: Fea(s)ts of Memory* (Madison: University of Wisconsin Press, 1992); Suzanne L. Bunkers and Cynthia A. Huff, eds., *Inscribing the Daily: Critical Essays on Women's Diaries* (Amherst: University of Massachusetts Press, 1996); Judy Nolte Lensink, "Expanding the Boundaries of Criticism: The Diary as Female Autobiography," *Women's Studies* 14 (1987): 39–53; Ronald Hoffman, Mechal Sobel, and Fredrika J. Teute, eds., *Through a Glass Darkly: Reflections on Personal Identity in Early America* (Chapel Hill: University of North Carolina Press, 1997).

2. Genealogical information for the Heaton family is found in Donald Lines Jacobus, comp. *Families of Ancient New Haven* (New Haven, 1931), III, 726–30.

3. Benjamin Trumbull, *A Century Sermon; or, Sketches of the History of the Eighteenth Century* (New Haven, 1901). Other studies of North Haven include Sheldon B. Thorpe, *North Haven Annals: A History of the Town from Its Settlement, 1680, to Its First Centennial, 1886* (New Haven, 1892), and Lucy McTeer Brusic, *Amidst Cultivated and Pleasant Fields: A Bicentennial History of North Haven, Connecticut* (Canaan, N.H.: North Haven Historical Society, 1986). On the Connecticut town, see Bruce C. Daniels, *The Connecticut Town: Growth and Development, 1635–1790* (Middletown, Conn.: Wesleyan University Press, 1979); according to Daniels, the land of North Haven (incorporated in 1786) measured twenty-one

square miles and had soil of average productivity. Articles about eighteenth-century Connecticut by Richard Buel, Jr., Richard Bushman, and Bruce Daniels may be found in *Connecticut History* 36 (fall 1995).

4. Marian Yeates, "Beyond 'Triffles,'" *Journal of Women's History* 6 (summer 1994): 150–60.

5. Mechal Sobel, "The Revolution in Selves: Black and White Inner Aliens," in *Through a Glass Darkly,* ed. Hoffman, Sobel, and Teute, 163–85; Charles Taylor, *Sources of the Self: The Making of the Modern Identity* (Cambridge: Harvard University Press, 1989).

6. A family tree, prepared by A. M. Cook in 1915, is available in the Long Island Collection, East Hampton Free Library, East Hampton, N.Y. Limited genealogical information can be found in George Rogers Howell, *The Early History of Southampton, L.I., New York, with Genealogies* (Albany, N.Y.: Weed, Parsons, 1887).

7. Richard L. Bushman, ed., *The Great Awakening: Documents on the Revival of Religion, 1740–1745* (New York: Atheneum, 1969); Richard L. Bushman, *From Puritan to Yankee: Character and the Social Order in Connecticut, 1690–1765* (Cambridge: Harvard University Press, 1967); Jon Butler, *Awash in a Sea of Faith: Christianizing the American People* (Cambridge: Harvard University Press, 1990); Patricia U. Bonomi, *Under the Cope of Heaven: Religion, Society, and Politics in Colonial America* (New York: Oxford University Press, 1986); Harry S. Stout, *The New England Soul: Preaching and Religious Culture in Colonial New England* (New York: Oxford University Press, 1986); William G. McLoughlin, *New England Dissent, 1630–1833,* 2 vols. (Cambridge: Harvard University Press, 1971); C. C. Goen, *Revivalism and Separatism in New England, 1740–1800: Strict Congregationalists and Separate Baptists in the Great Awakening* (New Haven: Yale University Press, 1962); Stephen A. Marini, *Radical Sects of Revolutionary New England* (Cambridge: Harvard University Press, 1982); Peter S. Onuf, "New Lights in New London: A Group Portrait of the Separatists," *William and Mary Quarterly,* 3d ser., 37 (1980): 627–43.

8. Rodger M. Payne, *The Self and the Sacred: Conversion and Autobiography in Early American Protestantism* (Knoxville: University of Tennessee Press, 1998); Roger Sharrock, "Spiritual Autobiography: Bunyan's Grace Abounding," in *John Bunyan and his England, 1628–88,* ed. Anne Laurence, W. R. Owens, and Stuart Sim (London: Hambledon Press, 1990), 97–104; Charles E. Hambrick-Stowe, *The Practice of Piety: Puritan Devotional Disciplines in Seventeenth-Century New England* (Chapel Hill: University of North Carolina Press, 1982). On evangelical psychology, see Philip Greven, *The Protestant Temperament: Patterns of Child Rearing, Religious Experience, and the Self in Early America* (New York: Alfred A. Knopf, 1977).

9. Alan Heimert, *Religion and the American Mind, from the Great Awakening to the Revolution* (Cambridge: Harvard University Press, 1966), 145; Susan Juster, *Disorderly Women: Sexual Politics and Evangelism in Revolutionary New England* (Ithaca, N.Y.: Cornell University Press, 1994); Susan Juster, "'In a Different Voice': Male and Female Narratives of Religious Conversion," *American Quarterly* 41

(March 1989): 34–62; Patricia Caldwell, *The Puritan Conversion Narrative: The Beginnings of American Expression* (New York: Cambridge University Press, 1983); Hambrick-Stowe, *The Practice of Piety;* Stephen R. Grossbart, "Seeking the Divine Favor: Conversion and Church Admission in Eastern Connecticut, 1711–1832," *William and Mary Quarterly,* 3d ser., 46 (1989): 696–740; Lewis R. Rambo, *Understanding Religious Conversion* (New Haven: Yale University Press, 1993); Murray G. Murphey, "The Psychodynamics of Puritan Conversion," *American Quarterly* 31 (1979): 135–47; Richard Godbeer, "'Love Raptures': Marital, Romantic, and Erotic Images of Jesus Christ in Puritan New England, 1670–1730," *New England Quarterly* 68 (September 1995): 355–84.

10. Elizabeth Reis, "The Devil, the Body, and the Feminine Soul in Puritan New England," *The Journal of American History* 82 (June 1995): 15–36; Richard Godbeer, *The Devil's Dominion: Magic and Religion in Early New England* (New York: Cambridge University Press, 1992).

11. Fasting and suffering were experienced by medieval female mystics, such as Catherine of Siena, in their emulation of the humanity of Christ. Rejection of ordinary food expressed the idea that all forms of physical desire were sinful and should be transcended. In contrast, among Puritan writers female sanctity was associated with women's devotion to marriage and motherhood. As a result of the demarginalization of female sexuality, Puritan women enjoyed some freedom from the hostility to female flesh that led medieval saints to self-starvation (Amanda Porterfield, *Female Piety in Puritan New England: The Emergence of Religious Humanism* [New York: Oxford University Press, 1992], 124–27). In New England, both men and women were expected to observe fast days and attend fast sermons, which pointed to the way in which the current generation conformed to their inherited mission and to the ways in which they had fallen short, saved only by providential mercy (Stout, *The New England Soul,* 75).

12. Joel Perlmann and Dennis Shirley, "When Did New England Women Acquire Literacy?" *William and Mary Quarterly,* 3d ser., 48 (1991): 50–67; F. W. Grubb, "Growth of Literacy in Colonial America: Longitudinal Patterns, Economic Models, and the Direction of Future Research," *Social Science History* 14 (winter 1990): 451–82; Cathy N. Davidson, *Revolution and the Word: The Rise of the Novel in America* (New York: Oxford University Press, 1986); David D. Hall, *Worlds of Wonder, Days of Judgment: Popular Religious Belief in Early New England* (New York: Knopf, 1989).

13. Margaret Spufford, *Small Books and Pleasant Histories: Popular Fiction and Its Readership in Seventeenth-Century England* (London: Methuen, 1981); Erik R. Seeman, *Pious Persuasions: Laity and Clergy in Eighteenth-Century New England* (Baltimore: Johns Hopkins University Press, 1999); Susan O'Brien, "Eighteenth-Century Publishing Networks in the First Years of Transatlantic Evangelism," in *Evangelicalism: Comparative Studies of Popular Protestantism in North America, the British Isles, and Beyond, 1700–1990,* ed. Mark A. Noll, David W. Bebbington, and

George A. Rawlyk (New York: Oxford University Press, 1994); Nathan O. Hatch and Mark A. Noll, introduction to *The Bible in America: Essays in Cultural History,* ed. Nathan O. Hatch and Mark A. Noll (New York: Oxford University Press, 1982); Richard D. Brown, *Knowledge Is Power: The Diffusion of Information in Early America, 1700–1865* (New York: Oxford University Press, 1989).

14. Endnotes document biblical quotations identified by the editor. The index points to quotations identified by Heaton in her text. As indicated in the endnotes, Heaton referred to the OT about 200 times and the NT more than 100 times. She refers to nearly all of the OT books and a majority of the NT. Approximately 250 additional biblical references can be found in the index. In contrast, John Barnard, an eighteenth-century Boston housewright, referred almost exclusively to the Psalter in his diary (Seeman, *Pious Persuasions,* 18–43).

15. Bibliographic information about her reading may be found in the endnotes, which cite the earliest dated editions, to show that her reading consisted of venerable texts; she herself states that the Hildersam volume was from lectures presented more than a hundred years ago, and another work was so old that the date was gone [pp. 308, 340]. Charles E. Hambrick-Stowe, "The Spirit of the Old Writers: Print Media, the Great Awakening, and Continuity in New England," in *Communication and Change in American Religious History,* ed. Leonard I. Sweet (Grand Rapids, Mich.: William B. Eerdmans, 1993): 126–40.

16. Cecile M. Jagodzinski, *Privacy and Print: Reading and Writing in Seventeenth-Century England* (Charlottesville: University Press of Virginia, 1999).

17. Isaac Stiles's Arminian tendencies are discussed in Edmund S. Morgan, *The Gentle Puritan: A Life of Ezra Stiles, 1727–1795* (New Haven: Yale University Press, 1962), 61–64. Goen, *Revivalism and Separatism in New England.* The growth of dissent in the wake of a religious revival in northern New England at the time of the Revolution is described in Marini, *Radical Sects of Revolutionary New England;* Brusic, *Amidst Cultivated and Pleasant Fields,* 46.

18. Goen, *Revivalism and Separatism in New England,* 307; Thorpe, *North Haven Annals,* 325; Brusic, *Amidst Cultivated and Pleasant Fields,* 46; Phyllis Flood, *History of the Baptist Church in Montowese, Compiled for the North Haven Bicentennial Celebration, 1975–1976* (n.p., n.d.).

19. A notable account of a Connecticut woman, Elizabeth Backus, who was imprisoned for not paying rates is given by her son, Isaac Backus, in *A History of New England, with Particular Reference to the Denomination of Christians Called Baptists* (Newton, Mass., 1871), II, 98.

20. Benjamin Trumbull, *A Discourse, Delivered at the Anniversary Meeting of the Freemen of the Town of New Haven* (New Haven: Thomas and Samuel Green, 1773), 28.

21. Juster, *Disorderly Women.*

22. Donald A. Stauffer, *The Art of Biography in Eighteenth Century England* (Princeton, N.J.: Princeton University Press, 1941).

23. Mary Beth Norton, *Founding Mothers and Fathers: Gendered Power and the Forming of American Society* (New York: A. A. Knopf, 1996); Taylor, *Sources of the Self.*

24. One such purchase of land is recorded in New Haven Colony Records, Town Records, XXI, 263, MS, New Haven City Hall, New Haven, Conn. For the will, see Theophilus Heaton [Sr.], 1760, Conn. Probate Records, New Haven District, Town of New Haven, MS, Conn. State Library.

25. North Haven, Conn., Tax List of the 1754 Levy, Mar. 1, 1755, MS, Conn. State Library. For the records of Heaton's husband, see Theophilus Heaton [Jr.], 1791, Conn. Probate Records, New Haven District, Town of North Haven. Spouses never became owners of real property, although widows were entitled to life-estates in one third of the land their husbands owned at death and a room in the house. Widows, along with blood relations, did inherit ownership of personal property. They were entitled to one third of all personal property remaining after debts were paid, and to one half if there were no children (Toby L. Ditz, *Property and Kinship: Inheritance in Early Connecticut, 1750–1820* [Princeton, N.J.: Princeton University Press, 1986], 49).

26. New Haven Colony Historical Society, *Ancient Town Records: New Haven Town Records, 1649–1769,* 3 vols. (New Haven, Conn., 1962), 3:788, 791.

27. The prosperity of eighteenth-century rural Connecticut is the subject for Bushman, *From Puritan to Yankee,* and Jackson Turner Main, *Society and Economy in Colonial Connecticut* (Princeton, N.J.: Princeton University Press, 1985).

28. Janet Moore Lindman, "The Body Baptist: Embodied Spirituality, Ritualization, and Church Discipline in Eighteenth-Century America," in *A Centre of Wonders: The Body in Early America,* ed. Janet Moore Lindman and Michele Lise Tarter (Ithaca, N.Y.: Cornell University Press, 2001).

29. Jonathan Heaton, 1799, Conn. Probate Records, New Haven District, Town of North Haven.

30. Hannah Heaton, 1794, ibid. Heaton's will, with her feeble signature, was written one and a half months before her death. She bequeathed to Calvin four acres in "Brandford" as well as three cows and seven sheep "that are presently let out"; this portion was valued by the court at £33. The remainder of her estate, which included household furniture, apparel, and money that "I have out at use, or not out," was valued at £44 and was divided equally between Jonathan and Calvin.

31. Conflict between father and son involved a "book debt," which was a legal procedure for suing to collect an amount of money, goods, or services long overdue. The written record, or book, was not necessary for settlement of the debt. All that was required was to sue within a fixed period after the debt had occurred. Debtors and creditors might be neighbors, relatives, or church members who resorted to this method in part because of the scarcity of currency. Calvin, with the aid of wife and friends, planned to sue his father for a piece of land in payment for farm work he had done after coming of age. The process was halted when a negotiator, Col. Street Hall, convinced Calvin he did not have a winning

case. Conflict between fathers and sons over land often involved, on the one hand, a desire for personal control and, on the other, a wish for autonomy (Bruce H. Mann, *Neighbors and Strangers: Law and Community in Early Connecticut* [Chapel Hill: University of North Carolina Press, 1987]; Philip J. Greven, Jr., *Four Generations: Population, Land, and Family in Colonial Andover, Massachusetts* [Ithaca, N.Y.: Cornell University Press, 1970]).

32. Thorpe, *North Haven Annals,* 261; Calvin Eaton, 1820, Conn. Probate Records, New Haven District, Town of North Haven.

33. Church-related associations of women are discussed by Nancy F. Cott, *The Bonds of Womanhood: "Woman's Sphere" in New England, 1780–1835* (New Haven: Yale University Press, 1977). For numerous examples of "contentious women" censured by the church, see Ruby Parke Anderson, comp., *The Parke Scrapbook* (Baltimore: Port City Press, 1965).

34. The sermon was printed: Samson Occom, *A Sermon, Preached at the Execution of Moses Paul, an Indian* (New London, Conn., [1772]).

35. Seeman, *Pious Persuasions,* xiii.

36. Concerning Jonathan Heaton, Thorpe declares that "there is no state record of his military service, but there is his mother's journal, and the muster roll of Captain Benjamin Trumbull." Jonathan was discharged May 26, 1777 (Thorpe, *North Haven Annals,* 236). Concerning Calvin, Brusic prints a list of Revolutionary War soldiers that includes his name (Brusic, *Amidst Cultivated and Pleasant Fields,* 279).

37. British raids on towns in the New Haven area are discussed in Albert E. Van Dusen, *Connecticut* (New York: Random House, 1961), 167–68. On the investment of current events with significance drawn from the Bible, see Mark A. Noll, "The Image of the United States as a Biblical Nation, 1776–1865," in *The Bible in America,* ed. Hatch and Noll.

38. The interweaving of religion and politics in sermons is discussed by Nathan O. Hatch, *The Sacred Cause of Liberty: Republican Thought and the Millennium in Revolutionary New England* (New Haven: Yale University Press, 1977). See also Stout, *The New England Soul;* James West Davidson, *The Logic of Millennial Thought: Eighteenth-Century New England* (New Haven: Yale University Press, 1977); Melvin B. Endy, Jr., "Just War, Holy War, and Millennialism in Revolutionary America," *William and Mary Quarterly,* 3d Ser., 42 (1985): 3–25.

39. The concept of "woman's sphere" as it began to evolve in the eighteenth century has received much attention. While viewed as restrictive to women by Barbara Welter and as conditionally enhancing to women by Nancy F. Cott, Linda Kerber argues that the concept of spheres has been exhausted and detracts from the more fundamental question of the social construction of a dependent female subject (Barbara Welter, "The Cult of True Womanhood: 1820–1860," *American Quarterly* 18 [1966]: 151–74; Cott, *The Bonds of Womanhood;* Linda K. Kerber et al., "Beyond Roles, Beyond Spheres: Thinking about Gender in

the Early Republic," *William and Mary Quarterly,* 3d ser., 46 [1989]: 565–85).

40. Elaine Forman Crane, "'I Have Suffer'd Much Today': The Defining Force of Pain in Early America," in *Through a Glass Darkly,* ed. Hoffman, Sobel, and Teute, 370–403.

41. According to Robert V. Wells, life expectancy at birth for eighteenth-century white women was probably about forty years. About 30 percent of these women could expect to reach age sixty-five; most of them would spend the majority of their lives engaged in childrearing (Robert V. Wells, "Demographic Change and the Life Cycle of American Families," in *The Family in History: Interdisciplinary Essays,* ed. Theodore K. Rabb and Robert I. Rotberg [New York: Harper and Row, 1971], 91).

42. Paula A. Scott, "'Tis Not the Spring of Life with Me': Aged Women in Their Diaries and Letters, 1790–1830," *Connecticut History* 36 (spring 1995): 12–30; Terri L. Premo, *Winter Friends: Women Growing Old in the New Republic, 1785–1835* (Urbana: University of Illinois Press, 1990); Reis, "The Devil, the Body," 19; Ulrich, *A Midwife's Tale;* Seeman, *Pious Persuasions.*

43. According to Mary Hewitt Mitchell, the Second Great Awakening was preceded by a wave of revivals in the last decades of the eighteenth century (Mary Hewitt Mitchell, *The Great Awakening and Other Revivals in the Religious Life of Connecticut* [New Haven: Yale University Press, 1934], 23). On the greater reception of women than men to these revivals, see David W. Kling, *A Field of Divine Wonders: The New Divinity and Village Revivals in Northwestern Connecticut, 1792–1822* (University Park: Pennsylvania State University Press, 1993); Harry S. Stout and Catherine A. Brekus, "Declension, Gender, and the 'New Religious History'," in *Belief and Behavior: Essays in the New Religious History,* ed. Philip R. Vandermeer and Robert P. Swierenga (New Brunswick, N.J.: Rutgers University Press, 1991).

44. Ps. 72:8; Zech. 9:10.

45. Following the diary, from p. 440 to p. 489, is a set of correspondence consisting of copied letters to and from family members, ministers, and friends, dated from the 1750s to the 1790s. Notable for their religious exhortations, biblical references, and emotional language, similar to that found in Heaton's diary, they suggest the existence of a network of friendships among lay evangelicals. Some of the letters are difficult to read compared to the diary, suggesting that Heaton's literacy was above average for her time.

46. Ruth H. Bloch, *Visionary Republic: Millennial Themes in American Thought, 1756–1800* (New York: Cambridge University Press, 1985).

47. Payne, *The Self and the Sacred.*

48. Linda K. Kerber and Jane De Hart Mathews, eds., *Women's America: Refocusing the Past* (New York: Oxford University Press, 1982). Kerber also has pointed to the changing social behavior of women in some churches beginning in the Revolutionary era (Kerber et al., "Beyond Roles, Beyond Spheres"). On

the other hand, Cornelia Hughes Dayton argues for the existence of a patriarchal household and a sexual double standard (*Women before the Bar: Gender, Law, and Society in Connecticut, 1639–1789* [Chapel Hill: University of North Carolina Press, 1995]).

1721–1740

1. Biblical citations throughout are from the Authorized King James Version. In some instances, Heaton herself supplies biblical references; in others, the editor has identified them.

2. This comment suggests the early pages of the diary were written in retrospect, following the death of her father in 1754 [p. 66].

3. John 13:34; Matt. 7:8.

4. The French and Indian War was that part of the Seven Years' War between Protestant England and Catholic France fought in America from 1754 to 1763.

5. The devil is a ubiquitous figure throughout the diary, casting doubts and fears into Heaton's mind. Conversations with Satan were ordinary ones but posed a threat to her mental stability by challenging her religious faith.

6. Northern lights, or aurora borealis, are luminous bands or streamers of light sometimes visible in the night skies of northern regions. They are a dramatic natural phenomena, part of the wondrous world described in Hall, *Worlds of Wonders.*

1741–1749

1. Three Presbyterian clergymen, Gilbert Tennent (1703–1764), William Tennent (1705–1777), and their father, the educator William Tennent (1673?–1745), took part in the mid-century religious revivals; which Tennent Heaton meant is not known. George Whitefield (1714–1770), British evangelist and organizer of the Calvinistic Methodists, was educated at Pembroke College, University of Oxford, and was an ordained Anglican minister. Because of his unconventional manner of preaching, many Church of England pulpits were closed to him; he therefore began to preach in the open air and attracted vast crowds by his eloquence. During a stay in America, together with Congregational minister Jonathan Edwards, Whitefield catalyzed the revival movement that later became known as the Great Awakening.

2. Election day was a day set by law for the election of public officials and considered by some as an occasion for celebration with a party or frolic.

3. John 14:26. The Comforter is the Holy Spirit.

4. Mal. 4:1.

5. Joseph Alleine, *Alarm to Unconverted Sinners* (1671; Boston, 1703). Heaton makes frequent reference to approximately two dozen books in addition to the Bible. Her reading remained exclusively religious throughout the century. On reading, see Hall, *Worlds of Wonder.*

6. Josh. 6. Jericho was an ancient city north of Jerusalem and near the Dead

Sea that was captured and destroyed by Joshua and the Israelites after its walls were miraculously leveled.

7. Dan. 5. Belshazzar, king of Babylon, was warned of his doom by Daniel's interpretation of mysterious handwriting on the palace wall.

8. Heaton narrates her conversion experience at the time of the Great Awakening. The literature on revival religion is vast; one overview is Bushman, *The Great Awakening.*

9. Matt. 7:7; 11:28.

10. Heaton writes verse throughout the diary, particularly in her later, more serene years. Some lines are of her own creation. Others are drawn from Isaac Watts, *The Psalms of David, Imitated;* the Bible; and possibly other sources. The insertion of poetry increases the artistic and religious appeal. Heaton's practice of beginning a couplet with a capital letter and her use of the colon have been retained.

11. Matt. 1:4.

12. Jonathan Parsons (1705–1776), educated at Princeton, was ordained and settled at the First Congregational Church of Old Lyme (1729–1745) and then was installed at the First Presbyterian Church of Newburyport (1746–1776). His text was John 14:6.

13. Possibly Reuben Judd (1716–1753), a graduate of Yale College, who was ordained and settled at Judae Parish in Washington, Conn.

14. Possibly Rev. David Youngs (b. 1719 in Southold).

15. Rev. 22:20.

16. James Davenport (1738–1757), ordained in Southold, Long Island, in 1738, met the charismatic Whitefield in Philadelphia in 1740, and spent the summer of 1741 preaching in Connecticut. He gained a considerable following, but his extremism aroused much controversy among the conservative Old Lights. Complaints against him led to his arrest on charges of disturbing the peace and order in Stratford, and he was ordered back to Long Island. Instead, he carried the revival to Boston.

17. 2 Cor. 6:14 ("Be ye not unequally yoked together with unbelievers."). Since Heaton married an unconverted man, she believed she must suffer the consequences.

18. Acts 9:16.

19. Song of Sol. 2:11–12.

20. Ps. 125:1.

21. Matt. 24:32.

22. Ps. 23:1.

23. Deut. 31:6; Heb. 13:5.

24. Jedediah Mills (1697–1776), educated at Yale, was the Congregational minister at Huntington (Ripton), Connecticut, 1723–1776.

25. Heb. 12:6. Heaton believed she was chastised with poor health for her weak faith, and thereafter the Bible became a "seald book" for her.

26. Luke 22:32.
27. 1 Kings 10:1, 7.
28. 1 Kings 21:17–29; 2 Sam. 22; Matt. 12:41.
29. James Mead, a teacher at Freetown, 1744, was ordained and settled in Middlebourough (New Light church).
30. Rom. 7:8.
31. Matt. 18:20.
32. Luke 15:11–32.
33. Luke 19:41.
34. Matt. 27:46; Mark 15:34.
35. Matt. 25:1–12.
36. Ps. 35:28.
37. 1 Cor. 10:10.
38. Ps. 42:1.
39. Ps. 114:4, 6; paraphrased.
40. Rev. 1:7; Acts 7:33; Deut. 3:27. Pisgah is the mountain ridge east of Jordan where Moses viewed the Promised Land.
41. Heb. 4:15; John 14:19.
42. A good example of the rhetorical cadences found throughout the diary.
43. Luke 7:38.
44. Ezek. 17:24.
45. John 5:25.
46. Zech. 2:8.
47. Job 5:19.
48. Deut. 7:3; 1 Kings 11:2.
49. Gen. 34:1–31.
50. Gen. 24:62.
51. Gen. 24:65. Heaton sees the veiled Rebekah as evidence of the difference between man and woman's "place," although she herself rarely expressed subjection to her husband.
52. The Hebrew captive who returned as leader of his people to rebuild Jerusalem and administer its affairs; the OT book of Nehemiah tells of his work.
53. 1 Kings 4:31. One of the sages whom Solomon excelled in wisdom.
54. John 17:24.
55. Jer. 20:7.
56. Gen. 17:7.
57. "My adversary" is a frequently used reference to her husband.

1750–1759

1. Gen. 22:2.
2. Prov. 30:4; Isa. 40:12.

3. Ps. 24:7.

4. 1 John 4:18.

5. Ps. 46:5.

6. Isaac Stiles (b. 1697), Congregational minister of the parish of North Haven, ran his church according to the principles of Solomon Stoddard. Anyone who lived a blameless moral life and professed belief in the orthodox doctrines of Christianity was entitled to full membership, including the privilege of the sacraments (Morgan, *The Gentle Puritan*).

7. Rev. 18:4.

8. Isa. 65:17.

9. Judg. 4, 5.

10. Ps. 25:3; Luke 1:45.

11. Heaton prayed that her hometown would be the center of a revival of religion. She later hoped that it would be the chief seat of Christ's kingdom on earth [p. 37].

12. 2 Tim. 4:16–17 (in part).

13. Mic. 7:8; Matt. 4:16.

14. Joel 2:21.

15. Here begins the controversy that would end in a trial. Since Heaton was no longer attending the regular church meeting, she was visited by neighbors and officials who tried to convince her to rejoin the fold. She countered that Stiles had "joined a confederacy with the world" [p. 63]. In 1758, she was charged formally with breaking the Sabbath, and her case was brought to court.

16. Deut. 33:25.

17. Joseph was the older son of Jacob and Rachel and the ancester of one of the tribes of Israel. Micah was a Hebrew prophet of the eighth century B.C.

18. Tirzah was a Canaanite town noted for its beauty.

19. Mic. 4:12.

20. Exod. 32:1–15.

21. Gen. 45:27.

22. Isa. 27:3.

23. She interprets her dream of a serpent as foreboding evil.

24. Ps. 116:5.

25. Isa. 30:20.

26. Isa. 59:1.

27. On the man and loaves, see Luke 11:5. Concerning the woman and daughter, see Matt. 15:22–28.

28. Exod. 12:37. Succoth was Israel's first camp on leaving Egypt.

29. 1 Kings 17:4.

30. Ps. 102:7.

31. 2 Chron. 20:12.

32. Matt. 10:30–31.

33. Heb. 11:27.

34. She expresses respect for her husband, but also concern about their differences of religious belief.

35. Samuel Bird (1724–1784), installed in the White Church of New Haven, served as a chaplain in the French and Indian War.

36. Ebenezer Frothingham (1717–1798) ministered to the Separate church in Middletown, Connecticut, 1747–1792, and was author of *The Articles of Faith and Practice . . .* (Newport, R.I., 1750), and *A Key to Unlock the Door . . .* (New Haven, 1767).

37. Acts 16:9.

38. "Wicked person" and "dear person" in this entry both refer to her husband.

39. Judg. 14:8–9.

40. Acts 16:24–25.

41. She wants to be the means of a revival, like Moses.

42. John 11:1–44.

43. Prov. 19:17. Throughout the diary, Heaton notes her charity to those in need.

44. John 4:9–29.

45. James Janeway, *Heaven upon Earth; or, The Best Friend in the Worst of Times* (1669; Boston, 1730).

46. 1 Sam. 13:2–3; 18:5; 19:1–7. Jonathan, the eldest son of King Saul, was his father's heir, which makes his loyalty to and affection for David, who succeeded Saul, the more remarkable.

47. 2 Sam. 17:27; 19:37. Barzillai was a faithful follower of David, and Chimham was a friend of David.

48. 2 Sam. 5:24.

49. Acts 6:15. Stephen, chief of the first seven deacons, is considered the first Christian martyr.

50. Acts 10:11.

51. John Bunyan, *Grace Abounding to the Chief of Sinners* (1666; Boston, 1717).

52. Heaton acknowledges the acute mental distress she underwent from time to time.

53. Not a Bunyan title. She may be referring to a work by William Saller, *The Two Covenants of Works and Grace* (London, 1682).

54. Luke 6:39.

55. Gen. 16:1–16, 21:9–21. Hagar, an Egyptian bond servant in Abraham's household, was Sarah's handmaid, Abraham's concubine, and mother of Ishmael.

56. For the account of Dinah, see Gen. 34.

57. 1 Kings 19:16–21. Elijah the prophet anointed Elisha as his successor.

58. Matt. 17:1–13.

59. Heaton, who had experience with four births, considered childbearing to be a spiritual trial.

60. Gen. 28:16.

61. Eleazer Wheelock (1711–1779) was Congregational pastor at Columbia, Connecticut, 1735–1770, and was the first president of Dartmouth College, 1769–1779.

62. She identifies with the Gaderene who was possessed by an "unclean spirit."

63. Heaton means Ps. 44:19.

64. This is not a title by John Bunyan. Heaton may be referring to Benjamin Keach, *The Travels of True Godliness* (London, 1683).

65. Ezek. 37:4.

66. Despite previous doubts, she now firmly believes she will go to heaven.

67. Rom. 16:20.

68. Rev. 20:3.

69. 1 Sam. 15. Saul spared the life of the Amalekite king Agag, who was killed by Samuel.

70. 1 Chron. ll:22–23; 2 Sam. 23:15–16.

71. She believes speaking rashly is one of her principal sins.

72. John Flavel, *Keeping the Heart* ([Boston], 1720).

73. Jer. 1:14, 6:1.

74. William Troop (1720–1756), who previously had been settled in Connecticut, was installed in the Congregational church of Southold, Long Island, in 1748 by the presbytery of Suffolk and other ministers. He died in 1756 at the age of 36.

75. Isa. 35:3.

76. Joel 2:13; Amos 5:6.

77. Ps. 120:5.

78. Isa. 62:5.

79. Philemon Robbins (1709–1781) graduated from Yale and was ordained and settled in Branford, Connecticut.

80. 2 Kings 21:1–18. Manasses, or King Manasseh, reigned at a time of religious retrogression.

81. Ps. 88:15.

82. Wolfgang Musculus, *Commonplaces of Christian Religion* (London, 1573).

83. Henry Ainsworth (1571–1622) was a English nonconforming divine who published extensively.

84. Eccles. 7:1.

85. Acts 12:9.

86. Isa. 30:18.

87. A reference to the French and Indian War.

88. 2 Sam. 17. Ahithophel was one of the conspirators with Absalom in a plot against King David.

89. Num. 14:6. Joshua was the son of Nun and Caleb the son of Jephunneh.
90. Acts 3:17.
91. 2 Cor. 4:9.
92. Ps. 97:8
93. Eccles. 7:2.
94. Isa. 63:9.
95. Isa. 54:10.
96. Lam. 3:32.
97. Job 14:22.
98. Heaton may be referring to Crown Point, on Lake Champlain, 90 miles northeast of Albany. In 1731 the French built Fort Frederic on the site and held it against repeated English and colonial expeditions. In 1759, during the French and Indian War, British commander Baron Jeffrey Amherst destroyed the fort and began construction of Fort Crown Point, which was never fully completed. On May 11, 1775, during the American Revolution, a company of Green Mountain Boys captured the British fort.
99. Jer. 17:5.
100. Cf. Isa. 35:1.
101. 2 Kings 7:2.
102. Cases involving breaches of the Sabbath were rare in eighteenth-century Connecticut. Such cases, with others that involved fines under forty shillings, came within the jurisdiction of the local justice of the peace, in this case Deodate Davenport (1706–1761). On women and litigation, see Dayton, *Women before the Bar.*
103. OT Book of Jonah.
104. Gen. 43:34.
105. Gen. 19:20–23. Zoar, ancient city of Canaan, was spared the fiery destruction that came upon Sodom.
106. Gen. 5:22.
107. Jonathan Edwards, *An Account of the Life of the Late Reverend Mr. David Brainerd* (Boston, 1749). Brainerd (1718–1747), missionary to the Indians, was ordained by the presbytery of New York.
108. Joshua Morse (1726–1795) was ordained in the Baptist church in Montville (part of New London) in 1751 and settled in New London, 1750–1776.
109. Joseph Marshall (1731–1813) was ordained in the Separatist church in the North Society of Canterbury, Connecticut. He served in North Canaan, Connecticut, for 16 years.
110. 1 Chron. 13:7–11. Uzza, a driver of the cart that bore the ark, was slain by the Lord for putting his hand to the ark when the oxen stumbled.
111. Daniel Miner ministered to the Separatist church in Lyme, 1757–1799.

1760–1769

1. Pss. 6:6, 42:3; conflated.

2. The Great Awakening increased divisive contention in communities and churches, causing some who were struck by the spirit of the revival and saddled with an unsympathetic minister to form their own congregations. By the 1770s, there were more than 120 Separatist congregations in New England. Voluntarism emerged as the dominant principle of the New England churches, and individuals grew increasingly confident in their ability to determine their own religious beliefs (McLoughlin, *New England Dissent*).

3. Gen. 42:36.

4. Jer. 16:16.

5. 1 Kings 17–21; Ps. 105:17–18.

6. Rev. 18:21.

7. Benjamin Trumbull (1735–1820), born in Hebron, Connecticut, graduated from Yale in 1759 and was ordained and settled in North Haven. He served as a chaplain during the American Revolution. His printed sermon is a source of information about early North Haven (Trumbull, *A Century Sermon*).

8. Isa. 10:5.

9. Isa. 34:13.

10. Exod. 8:17–32.

11. Luke 18:1–8.

12. Isa. 35:7.

13. Ps. 10:17.

14. Thomas Harrison, *Spiritual Pleadings and Expostulations with God in Prayer,* a reprint of *Topica Sacra: Spiritual Logick: Some Brief Hints and Helps to Faith, Meditation, and Prayer, Comfort and Holiness* (London, 1658).

15. 1 Tim. 1:15. Following biblical precedent, Heaton describes herself as one of the greatest of sinners.

16. 1 Sam. 30:10.

17. 2 Sam. 12:10.

18. Ps. 41:9; Matt. 10:36.

19. Zech. 8:4–5.

20. Benjamin Beech (1737–1812) was the Separate minister in Wallingford when Heaton joined the church. His brothers Nathaniel (b. 1741) and Elias (1741–1821) frequently preached at weekday meetings of the Separates.

21. Heaton may have read Foxe's *Book of Martyrs,* or she may be referring to popular works on martyrs printed in America, such as *Martyrology* (Boston, 1736); Thomas Mail, *The History of the Martyrs Epitomised* (Boston, 1747); [Ellis Hooke], *The Spirit of the Martyrs Revised* ([New London?], 1750).

22. Num. 17:8.

23. Ps. 42:3.

24. Luke 12:52. Three against one refers to her husband and two sons in opposition to her.

25. 1 Sam. 1:18. Hannah was the mother of Samuel.

26. Eph. 5:14.

27. Luke 6:38.

28. Isa. 33:24.

29. The image is drawn from *Pilgrim's Progress.*

30. Her "wicked action" seems to have been an outspoken expression of self-righteousness that she regretted.

31. 2 Cor. 4:17.

32. Matt. 18:3.

33. Ps. 38:2.

1770–1775

1. Hos. 14:4.

2. John 4:18.

3. In the manuscript, between pp. 189 and 190, there are eight unmarked pages, which the editor has designated 189 a through h.

4. In professing the pure church ideal, the Separates almost inevitably came to the conclusion that if churches were to be composed only of the regenerate, the rite of admission should be restricted to confessed believers. Furthermore, if the practice of infant baptism was not repudiated the unregenerate would not be excluded from church, and the quest for a true church in all its purity would be undermined (Goen, *Revivalism,* 208–58).

5. Isa. 60:8.

6. The impious behavior of her family had worn her out.

7. Isa. 42:3.

8. 2 Sam. 3:39.

9. 2 Sam. 6:10–12.

10. 1 Chron. 21:15–25.

11. Ps. 101:5; Lev. 19:18.

12. Mark 10:46.

13. Judg. 6:37–40.

14. Matt. 5:22–24; paraphrased.

15. Deut. 32:47.

16. Deut. 13:13; Judg. 19:22; 1 Sam. 2:12.

17. Isa. 26:20.

18. Jon. 4:6–7.

19. Possibly Nathan Cole, a layman. See Crawford, "The Spiritual Travels of Nathan Cole."

20. Ps. 5:12.

21. Samson Occom (1723–1792), a Mohegan convert to Christianity, was

widely known as a Presbyterian minister and missionary to other tribes. His execution sermon was delivered at the request of Moses Paul, an Indian convicted of murdering a white man. Execution sermons were meant to reinforce the prevailing attitudes of society; on this occasion, the sermon stressed the evils of alcohol, which had played a role in the murder.

22. Luke 16:19–31. Dives is the popular name for the rich man in the parable.

23. Rom. 6:23.

24. The established church was not always successful in collecting ecclesiastical taxes. The Separates refused to pay on grounds of conscience, contending that forced payment was an unwarrented invasion of their right to worship as they pleased; some went to prison rather than compromise.

25. Heaton gives biblical and personal reasons for keeping a diary.

26. John Bunyan, *Pilgrim's Progress* (Boston, 1681).

27. Isaac Watts, *The Psalms of David, Imitated . . .* (Philadelphia, 1729), 6.

28. Ps. 103:2.

29. Gen. 39:20.

30. Deut. 33:9.

31. 1 Sam. 7:12; a memorial stone.

32. Luke 8:2.

33. Ps. 27:8.

34. Gen. 28:12.

35. [Thomas Hooker], *The Soules Preparation for Christ* (London, 1643).

36. Ps. 37:5; Prov. 3:6; conflated.

37. John 14:26.

38. Increasing disorders in Boston, brought on by the Townshend Acts, caused the British government to instruct General Gage to send at least one regiment to Boston.

39. It seems clear that Heaton was a patriot, not a loyalist.

40. Liberty trees and poles, some with effigies of British officials, were first erected to celebrate repeal of the detested Stamp Act.

41. Allyn Mather, a graduate of Yale College, was ordained at the Fair Haven church in New Haven and settled in New Haven, 1772–1784.

42. Robert Campbell was an ordained elder in the Separate church at New Milford, preaching there for about twelve years.

43. Thomas Gage (1719?–1787) was British commander in chief in America when Boston became the focus of rebellion. Beginning in 1768, he gradually built up the Boston garrison. After Lexington and Concord, the Boston seige, and Bunker Hill, he was recalled and succeeded by Howe and Clinton.

44. Ps. 34:7.

45. Isa. 35:1.

46. George III (1738–1820) needed a major national issue in order to rally a substantial political following. Parliament's right to tax the colonies emerged as

the issue he needed, and he hastened to exploit it with the help of Lord North (1732–1792).

47. Rev. 5:4–5.

48. The Church of England was only nominally established in New York, and dissent was widespread (Bonomi, *Under the Cope of Heaven*).

49. *The Whole Works of the Reverend Mr. John Flavel* 2 vols. (London, 1701).

50. Ps. 72:8; Zech. 9:10. Heaton hopes the Rhode Island reformation will spread "from sea to sea." It is the kingdom of God, rather than the young republic, that she believes will flourish.

51. Gen. 35:8, 24:59; Deborah was the nurse of Rebekah.

52. John 8:56.

53. Job 29:3.

54. Phrase from Ps. 109:28. Sometimes Heaton recalls a few words or phrases from Scripture and joins them in her own composition.

55. 1 Sam. 18–19.

1776–1779

1. Job 38:39, 41; paraphrased.

2. R[ichard] S[ault], *Second Spira: Being a Fearful Example of an Athiest Who . . . Dyed in Despair at Westminster* (London, 1693).

3. Pleurisy, nervous fever, and camp distemper are eighteenth-century terms for inflammation and fevers associated particularly with typhus. For references to long fever and bilious fever, see pp. 295, 372.

4. Gen. 17:17–22. Isaac was the son of Abraham and Sarah.

5. Luke 24:38.

6. Isa. 44:28, 45:1–7.

7. Thomas Shepard, *The Sound Believer* (Boston, 1742).

8. Ps. 68:30.

9. Matt. 5:9.

10. Hos. 4:11.

11. Gen. 24; Rebekah was the wife of Isaac. Luke 1:36–80; Elisabeth was the wife of Zecharias.

12. Ps. 126:5.

13. 1 Chron. 4:9–10.

14. Joel 2:25.

15. Judg. 15:18.

16. Solomon Stoddard, *Safety of Appearing at the Day of Judgment* (Boston, 1687).

17. Arthur Hildersam (1563–1632) was an English Puritan divine.

18. John Burgoyne (1722–1792) was a British general whose invasion of New York in 1777 and surrender at Saratoga climaxed his military career.

19. Dan. 3.

20. Amos 3:2.

21. Matt. 6:33.

22. Eliphalet Wright (d. 1784) served as pastor in the Separate congregation of South Killingly, Connecticut, from 1765 to 1784.

23. John 8:24.

24. Matt. 5:4.

25. Luke 22:19.

26. Matt. 11:29.

27. Heb. 11:7.

28. 1 Cor. 9:7.

29. Heb. 13:14.

30. Deut. 3:27.

31. John Cornell is described by Ezra Stiles as one of the "spontaneous Separate Teachers" (*The Living Diary of Ezra Stiles* [New York: Charles Scribner's Sons, 1901]).

32. Isa. 48:9.

33. To punish Connecticut for attacking British shipping on Long Island Sound, and for supplying the rebel army, Sir Henry Clinton received approval from London to mount a large scale punitive expedition. New Haven was the objective of the first division, East Haven of the second division (July 5–6, 1779).

1780–1789

1. Prov. 6:6.

2. Rev. 3:21.

3. Ps. 34:20.

4. Ps. 34:3.

5. Acts 2:20.

6. Samuel Mather, *The Self-Justiciary Convicted and Condemned* (Boston, 1707).

7. Ps. 126:5.

8. Ps. 97:11.

9. Ps. 51:10.

10. Lev. 10.

11. Rev. 3:20.

12. Michael Wigglesworth, *The Day of Doom* (1662; Boston, 1701).

13. Isa. 24:16.

14. Maj. Gen. Charles Cornwallis, 1st Marquess Cornwallis (1738–1805), was the British general in the Yorktown, Virginia, campaign, whose surrender on October 19, 1781, virtually ended the American Revolution.

15. Ezek. 13:22; Ps. 74:11.

16. 2 Tim. 2:3.

17. 1 Pet. 5:6.

18. Luke 21:14.

19. 1 Thess. 5:19.

20. Ps. 40:5 or Job 37:16.

21. Edmund Bonner (1500?–1569), English prelate, figured prominently in the persecution of English protestants that marked the reign of Queen Mary, a Roman Catholic.

22. Ruth 4:11. Rachel, an Aramaean woman, was the wife of Jacob. She and her sister Leah were honored by later generations as those "who together built up the house of Israel."

23. Thomas Brooks, *Heaven on Earth* (London, 1654).

24. 2 Sam. 14:25. Third son of David, Absalom conspired against David's throne; defeated in battle, he died an ignominious death.

25. Heb. 11:16.

26. Ruth 1:4–15. Orpah was the daughter-in-law of Naomi.

27. Rev. 4:8.

28. Gen. 49:23–24.

29. Ps. 51:11.

30. 1 Tim. 6:12.

31. Eph. 6:12.

32. Gen. 50:20.

33. Possibly John Bunyan, *Defense of the Doctrine of Justification by Faith* (1672).

34. Either *The Spiritual Magazine; or, The Christian's Grand Treasure* (London, [1752]), or *The Royal Spiritual Magazine; or, The Christian's Grand Treasure* (Philadelphia, 1771).

35. Ps. 47:5.

36. Isa. 53:11.

37. Isa. 5:26.

38. Jer. 49:16; Song of Sol. 2:14.

39. Eph. 6:11

40. Jer. 31:3; the correct reference is to Isa. 54:5.

41. Exod. 5:7.

42. Possibly from Isa. 8:21.

43. John 20:15.

44. 1 Cor. 5:7.

45. Amos 4:12.

46. Elisha Paine (1693–1775), called "Moses" of the Separates by Isaac Backus, was trained as a lawyer. During the Great Awakening, he believed he was called by God to preach the gospel. Arrested for his preaching, he continued in this practice after release and carried on a running battle with the ministers of Windham County for many years.

47. Luke 6:38.

48. Hos. 11:9.
49. Phil. 3:21.
50. Reference to Isa. 63:3.
51. Matt. 5:47.
52. 1 Chron. 5:10, 19. In the time of Saul, the tribe of Reuben was united with Gad and Manasseh in an attack on the Hagarites, a nomad people.
53. Matt. 11:12; Lev. 25:38.
54. Lam. 3:23; Ps. 116:12; paraphrased.
55. Ps. 45:3–5.
56. Ps. 50:23.
57. Isa. 63:1, 3; Rev. 19:13. Edom was the name given to Esau, his country, and his people; the capital of Edom was Bozrah.
58. Phil. 3:12.
59. Matt. 26:36–56. Gethsemane, at the foot of Olivet, near Jerusalem, was the scene of Christ's agony and betrayal.
60. Ps. 50:10.
61. Matt. 25:34.
62. John 4:14.
63. Prov. 23:7.
64. John 20:17.
65. Heb. 7:25.
66. Isa. 54:13.
67. Matt. 8:14–15.
68. Isa. 53:4.
69. The doctor resorts to a practice of folk medicine, placing split fowls over Calvin's feet.
70. Dan. 10:12.
71. Eph. 2:10.
72. Luke 2:14.
73. John 3:16.
74. Ps. 119:71.
75. Judg. 14:14.
76. Isa. 66:14.
77. Isa. 65:10.
78. Acts 27.
79. Ps. 42:6–7; transposed.
80. Ps. 76:4.
81. Ps. 68:12.
82. 2 Sam. 11–12.
83. Ps. 8:4.
84. Ps. 60:8.
85. Gen. 37–48.

86. John 1:48–50.

87. 1 Cor. 2:9.

88. Ps. 62:8.

89. Mal. 4:2.

90. Ps. 116:7.

91. Gen. 8.

92. Josh. 10:12.

93. Should be Prov. 31:29.

94. Gen. 21:31–33. Beersheba was an ancient site in southern Palestine, so named by Abraham.

95. Exod. 15:11.

96. 1 Kings 21:1–16

97. John 10:3.

98. Gen. 49:19.

99. The squire is possibly Giles Pierpont, who in 1786 was the wealthiest man in town.

100. Matt. 10:23.

101. 2 Kings 4:12–17.

102. Judg. 16:4–20.

103. Benjamin Wildman (1737–1812), a Yale graduate, was ordained and settled in Southbury, Connecticut (1766–1812).

104. Ps. 32:10.

105. 1 Sam. 14:12.

106. Isa. 1:5.

107. Job 13:15.

108. Ps. 121:4.

109. 1 Chron. 16:21.

1790–1793

1. Rev. 5:12.

2. Matt. 15:24.

3. John 18:8, in part.

4. John 14:2.

5. Rom. 2:7.

6. Hab. 3:17.

7. 1 Kings 19:1–4.

8. Isa. 12:3.

9. Ps. 89:32–33.

10. John 10:14, 28–29, 15:9.

11. Ps. 103:1.

12. Heb. 6:12.

13. Gen. 27:45.

14. John 7:46.
15. 2 Cor. 10:4.
16. 2 Cor. 5:1.
17. Ps. 72:6.
18. Should be Ps. 91:3.

Letters, 1750–1792

1. A Richard Howell was party to several land transactions, according to Southold, Long Island, Town Records.
2. 1 Kings 22.
3. Rev. 12:7.
4. Luke 12:32.
5. Gal. 5:16.
6. Rom. 12:19.
7. Isa. 49:16.
8. Rom. 16:1–2.
9. The phrases are actually from 2 Sam. 22:47; Ps. 18:46.
10. 1 Kings 1:38.
11. Eccles. 11:6.
12. Gal. 6:8.
13. Mark 9:44.
14. Luke 16:31.
15. Rom. 8:38–39.
16. Song of Sol. 2:2; not a saying of Christ.
17. Rom. 8:28.
18. John 10:1.
19. Luke 21:19.
20. Jer. 51:7. Babylon, the capitol city of the Babylonian empires, reached the height of its splendor and strength under Nebuchadnezzar.
21. Phil. 4:6.
22. James 5:7–8.
23. Luke 19:13.
24. Acts 19:28.
25. Heaton concludes with a message of hope, also on p. 55.

Index

Note: Spelling for proper nouns has been modernized. Arabic numerals refer to the original diary page numbers.